T0332725

Artificial Intelligence of Things for Weather Forecasting and Climatic Behavioral Analysis

Rajeev Kumar Gupta
Pandit Deendayal Energy University, India

Arti Jain
Jaypee Institute of Information Technology, India

John Wang
Montclair State University, USA

Ved Prakash Singh
India Meteorological Department, Ministry of Earth Sciences, Government of India, India

Santosh Bharti
Pandit Deendayal Energy University, India

A volume in the Advances in Computational Intelligence and Robotics (ACIR) Book Series

Published in the United States of America by
IGI Global
Engineering Science Reference (an imprint of IGI Global)
701 E. Chocolate Avenue
Hershey PA, USA 17033
Tel: 717-533-8845
Fax: 717-533-8661
E-mail: cust@igi-global.com
Web site: http://www.igi-global.com

Library of Congress Cataloging-in-Publication Data

Names: Gupta, Rajeev Kumar, 1984- editor.
Title: Artificial Intelligence of things for weather forecasting and climatic behavioral analysis / Rajeev Gupta, Arti Jain, John Wang, Ved Prakash Singh, and Santosh Bharti, editors.
Description: Hershey, PA : Engineering Science Reference, [2022] | Includes bibliographical references and index. | Summary: "This book provides extensive research for weather forecasting and climatic behavioral prospects that contribute towards detection of weather hazards, mitigation and disaster management through technological enhancement including Machine Learning, Deep Learning, Internet of Things, Artificial Intelligence, Big Data Analytics and Geo-Spatial Visualization"-- Provided by publisher.
Identifiers: LCCN 2022001849 (print) | LCCN 2022001850 (ebook) | ISBN 9781668439814 (h/c) | ISBN 9781668439821 (s/c) | ISBN 9781668439838 (eISBN)
Subjects: LCSH: Meteorology--Data processing. | Artificial intelligence. | Weather broadcasting--Data processing.
Classification: LCC QC874.3 .A78 2022 (print) | LCC QC874.3 (ebook) | DDC 551.630285/63--dc23/eng20220420
LC record available at https://lccn.loc.gov/2022001849
LC ebook record available at https://lccn.loc.gov/2022001850

This book is published in the IGI Global book series Advances in Computational Intelligence and Robotics (ACIR) (ISSN: 2327-0411; eISSN: 2327-042X)

British Cataloguing in Publication Data
A Cataloguing in Publication record for this book is available from the British Library.

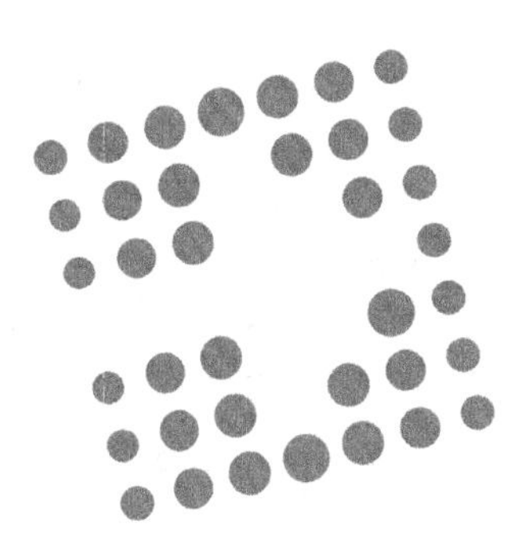

Advances in Computational Intelligence and Robotics (ACIR) Book Series

Ivan Giannoccaro
University of Salento, Italy

ISSN:2327-0411
EISSN:2327-042X

Mission

While intelligence is traditionally a term applied to humans and human cognition, technology has progressed in such a way to allow for the development of intelligent systems able to simulate many human traits. With this new era of simulated and artificial intelligence, much research is needed in order to continue to advance the field and also to evaluate the ethical and societal concerns of the existence of artificial life and machine learning.

The **Advances in Computational Intelligence and Robotics (ACIR) Book Series** encourages scholarly discourse on all topics pertaining to evolutionary computing, artificial life, computational intelligence, machine learning, and robotics. ACIR presents the latest research being conducted on diverse topics in intelligence technologies with the goal of advancing knowledge and applications in this rapidly evolving field.

Coverage

- Cognitive Informatics
- Adaptive and Complex Systems
- Neural Networks
- Artificial Intelligence
- Heuristics
- Computational Logic
- Cyborgs
- Robotics
- Brain Simulation
- Computational Intelligence

IGI Global is currently accepting manuscripts for publication within this series. To submit a proposal for a volume in this series, please contact our Acquisition Editors at Acquisitions@igi-global.com or visit: http://www.igi-global.com/publish/.

The Advances in Computational Intelligence and Robotics (ACIR) Book Series (ISSN 2327-0411) is published by IGI Global, 701 E. Chocolate Avenue, Hershey, PA 17033-1240, USA, www.igi-global.com. This series is composed of titles available for purchase individually; each title is edited to be contextually exclusive from any other title within the series. For pricing and ordering information please visit http://www.igi-global.com/book-series/advances-computational-intelligence-robotics/73674. Postmaster: Send all address changes to above address.

Titles in this Series

For a list of additional titles in this series, please visit: www.igi-global.com/book-series/advances-computational-intelligence-robotics/73674

Unmanned Aerial Vehicles and Multidisciplinary Applications Using AI Techniques
Bella Mary I. Thusnavis (Karunya Institute of Technology and Sciences, India) K. Martin Sagayam (Karunya Institute of Technology and Sciences, India) and Ahmed A. Elngar (Faculty of Computers & Artificial Intelligence Beni-Suef University, Beni Suef City, Egypt)
Engineering Science Reference • © 2022 • 300pp • H/C (ISBN: 9781799887638) • US $245.00

Principles and Applications of Socio-Cognitive and Affective Computing
S. Geetha (VIT University, Chennai Campus, India) Karthika Renuka (PSG College of Technology, India) Asnath Victy Phamila (VIT University, Chennai, India) and Karthikeyan N. (Syed Ammal Engineering College, India)
Engineering Science Reference • © 2022 • 330pp • H/C (ISBN: 9781668438435) • US $245.00

Artificial Intelligence Applications in Agriculture and Food Quality Improvement
Mohammad Ayoub Khan (University of Bisha, Saudi Arabia) Rijwan Khan (ABES Institute of Technology, India) and Pushkar Praveen (Govind Ballabh Pant Institute of Engineering & Technology, India)
Engineering Science Reference • © 2022 • 335pp • H/C (ISBN: 9781668451410) • US $215.00

Artificial Intelligence for Societal Development and Global Well-Being
Abhay Saxena (Dev Sanskriti Vishwavidyalaya Hardwar, Uttrakhand, India) Ashutosh Kumar Bhatt (Uttarakhand Open University, India) and Rajeev Kumar (Teerthanker Mahaveer University, India)
Engineering Science Reference • © 2022 • 315pp • H/C (ISBN: 9781668424438) • US $245.00

Applications of Computational Science in Artificial Intelligence
Anand Nayyar (Duy Tan University, Da Nang, Vietnam) Sandeep Kumar (CHRIST University (Deemed), Bangalore, India) and Akshat Agrawal (Amity University, Guragon, India)
Engineering Science Reference • © 2022 • 284pp • H/C (ISBN: 9781799890126) • US $245.00

Challenges and Applications for Hand Gesture Recognition
Lalit Kane (School of Computer Science, University of Petroleum and Energy Studies, India) Bhupesh Kumar Dewangan (School of Engineering, Department of Computer Science and Engineering, O.P. Jindal University, Raigarh, India) and Tanupriya Choudhury (School of Computer Science, University of Petroleum and Energy Studies, India)
Engineering Science Reference • © 2022 • 249pp • H/C (ISBN: 9781799894346) • US $195.00

701 East Chocolate Avenue, Hershey, PA 17033, USA
Tel: 717-533-8845 x100 • Fax: 717-533-8661
E-Mail: cust@igi-global.com • www.igi-global.com

Table of Contents

Detailed Table of Contents

Meteorology refers to the scientific study and analysis of weather conditions and atmospheric phenomenon on the earth, processes that cause these weather conditions, and their future predictions. Not just limiting to prediction, meteorologists note the physical conditions of the atmosphere above them, and they study the raw data from maps, satellite, and radar. Forecasts are highly needed for applications that are tailored to specific industries. Gas and electric utilities, for example, may need to have forecasts of temperature within one or two degrees a day ahead of time, or government may need to have data of rainfall for the areas having lower ground level to predict water overflow. The chapter provides a detailed introduction to weather forecasting, its requirements and types of forecasts, and its applications. Importance of data, weather prediction models and methods, machine learning approaches, and baseline measures in the weather prediction are also discussed. Thereby, it opens up an area of research and future developments in this direction.

Because of global warming, pollution, and many other factors, the environment is changing at an alarming rate. Accurate forecasting can assist people in making appropriate plans for activities such as harvesting, traveling, aviation, etc. Satellites and radar have been increasingly popular in weather forecasting over the previous few decades. The information collected by the satellite and radar can be used to monitor climate movement, track hurricanes, and give barometrical estimations that can be turned into mathematical climate expectation (NWP) models for exact forecasting. Currently, more than 160 meteorological satellites are located in orbit, which generates approximately 80 million observations every day. This

Weather forecasting is an important role in meteorology and has long been one of the world's most systematically difficult problems. This plan deals with the structure of a weather display system that may be built by electronics hobbyists utilizing low-cost components. Severe weather occurrences present a challenging forecasting problem with just a partial explanation. Developing communication methods makes it possible. Technology-enabled applications provide severe weather alerts and advisories. The airline industry is highly sensitive to the weather. Accurate weather forecasting is crucial. The IoT enables agriculture, notably arable farming, to become data-driven, resulting in more timely and cost-effective farm production and management while lowering environmental impact. Applying AIoT and deep learning to smart agriculture combines the best of both worlds.

Section 2

The changes in the weather play a significant role in people's planning. It has attracted the attention of several study communities due to the fact that it has an impact on human life all over the world. But weather forecasting is a challenging task because it is dependent on a variety of factors such as wind speed, wind direction, global warming, etc. Deep learning-based solutions have seen a lot of success in the geospatial domain over the last few years. In the past few years, a variety of deep learning-based weather forecasting models have been proposed. The forecasting techniques used traditionally are highly parametric and so are complex. In this chapter, deep learning techniques which are used for weather forecasting, such as Multilayer Perceptron, Jordan Recurrent Neural Network, Elman Recurrent Neural Network, etc., are discussed in detail. This chapter presents a comparative analysis of various deep learning-based weather forecasting models that are currently available.

According to recent findings, deep learning algorithm outperforms in many tasks like image classification, image segmentation, image recognition, etc. in the field of computer vision. With the help of deep learning, classification tasks on remote sensing image data can attain better performance compared to traditional approaches. This chapter primarily demonstrates how residual neural networks are used to classify satellite images of cyclones in the Bay of Bengal (BoB) and the Arabian Sea (Arab Sea). The authors further discovered the cyclones' locations and investigated using satellite images in the infrared and visible bands of electromagnetic spectrum. From the evaluation metrics, the neural network looks to be capable of correctly identifying the cyclonic storm utilising Gradient Class Activation Mapping (Grad-CAM). Satellite images of both cyclone storm and non-cyclone storm are analysed for cyclonic storm recognition and classification.

Weather affects air quality globally since different aspects of the weather like humidity, temperature, wind speed, and direction essentially affect the movement, creation, and concentration of various major air pollutants like surface ozone, PM 2.5, methane, carbon dioxide, etc. Air pollution is caused when an excessive amount of harmful substances like gases, particles, etc. are poured into our atmosphere which can severely affect the health of any living organisms. In this chapter, the most relevant weather affected urban air quality prediction papers are studied along with recent IoT systems developed for air pollution, and the authors observed that modern artificial intelligence algorithms are better than traditional statistical models. However, artificial intelligence-based algorithms cannot be directly compared effectively due to the hybrid nature of data sources used. Also, a need is identified to develop a powerful end-to-end model based on artificial intelligence algorithms and IoT systems.

Preface

The erratic weather patterns in an agrarian country such as India can cause livelihood and food availability to be disrupted, resulting in decreased agricultural productivity. Consequently, precise knowledge of the past, present, and future weather patterns over a region can play an important role in increasing agricultural production and implementing polices to ensure the most efficient use of available soil and water resources. Nowadays, Weather forecasting has gained significant attention by the researchers and the industries as it directly influences the human life. Weather updates may have diverse implications for farmers, industries, and the government. Accurate weather forecasting can assist farmers in making important planting and harvesting decisions, airline firms in scheduling flights, and the government in disaster risk management. Because of technological advancements, a massive amount of weather data is now available that may be acquired from variety of sources such as satellites, Radars, weather balloons, and sensors. Problem of handling such large heterogeneous datasets can be addressed by the Artificial Intelligence (AI) where AI/ ML based approaches are used for analysis of dynamic patterns distribution and weather forecasting. Due to the public availability of weather and agricultural data, AIoT is expected to emerge as a primary tool for the researchers and professionals to solve the weather and climate related problems. Thus, there is a dire need to integrate all available modern technologies, earth science and agricultural science for environmentally sustainable socio-economic development.

The topics in this book include weather forecasting, role of Radars and satellites in meteorology, impact of weather in agriculture, tools and techniques used for weather analysis, flood assessment using hydrodynamic modelling, soil-water management, drought Management, how AI can be useful for the on-ground aspects of environmental sciences and many more.

ORGANIZATION OF THE BOOK

The book is organized into 15 chapters. These chapters are categorized into three sections *i.e.*, Remote sensing data collections for meteorology, Deep Learning methods for analyzing remote sensing data, and weather forecasting and utility of AIoT for the environmental applications. A brief description about each section is given below:

Section 1 comprises of five chapters. These chapters cover the introduction of remote sensing and weather forecasting, as well as how data is gathered for meteorological applications.

Chapter 1 provides a detailed introduction to weather forecasting, its requirement, types of forecasts and its applications. Weather variability gets highly influenced spatially and temporally due to the gaseous constituents in the earth's atmosphere and can prove to be an important factor for analysis and prediction. Meteorologists note the physical conditions of the atmosphere above them, and study the data from in-situ observations, satellite, and Radar. The effects of various atmospheric constituents on the weather are also discussed in this chapter.

Chapter 2 discusses the climate change analysis using the satellite imageries. The changes in the pattern of the seasons have been observed in different regions of the world due to climate change and this issue has been taken on a serious note. Researchers are now taking data from satellites using remote sensing to analyze and prepare reports for predicting future disasters, which will be helpful to tackle calamities. This chapter discusses how remote sensing plays an important role in collecting data using satellites, how satellites monitor objects and which satellite data is used for the climate change analysis.

Chapter 3 focuses on the various tools that are used to collect data for weather forecasting and application of Internet of the Things (IoT) in weather forecasting.

Chapter 4 discusses the different types of satellites used for the meteorological observations. Satellites and Radars have been increasingly popular in weather forecasting over the previous few decades. Currently, more than 160 meteorological satellites are located in different orbits, which generate approximately 80 million observations every day. This chapter discusses several meteorological satellites which are primarily used to extract weather patterns.

Chapter 5 discusses the tools and technologies that are used for the weather forecasting. Currently, with rapidly growing technology, image datasets are also produced in large volume, it is important to classify the datasets. As a result, researchers are working on identification and classifying images utilizing machine learning. This chapter provides a wide overview of the present state of art. It looks at existing and future uses and assesses the advantages and disadvantages. Finally, it addresses some of the IoT based sensors, satellites, radar and other technologies, as well as possible future paths in arable agriculture and climate.

Section 2 comprises of seven chapters. These chapters discuss the deep learning-based methods and interdisciplinary applications of remote sensing and weather forecasting data analytics.

Chapter 6 explores the comparative analysis of various existing deep learning-based weather forecasting models. Weather forecasting is a challenging task as its dependent on a variety of factors such as dynamic nature of parameters, regional variability, and global warming, *etc*. Deep learning-based solutions have seen a lot of success in the geospatial domain over the last few years. In this chapter, Deep learning techniques which are used for spatio-temporal data analysis and forecasting such as Multilayer Perceptron, Jordan Recurrent Neural Network, Elman Recurrent Neural Network, *etc.*, are discussed in detail.

Chapter 7 proposes a deep learning-based models for the rainfall prediction. A Long Short-Term Memory Network (LSTM), Artificial Neural Network (ANN), and 1-dimensional Convolutional Neural Network LSTM (1D CNN-LSTM) models are explored for predicting the rainfall at multiple lead times. The daily weather parameters datasets in the period of last 15 years or above, are collected for a station in Maharashtra. Rainfall data is classified into three classes: no-rain, light rain, and moderate to heavy rain. Moreover, the Principal Component Analysis (PCA) helped to reduce the input feature dimensionality.

Chapter 8 describes a deep learning-based model for SAR data analysis. The potential of Synthetic Aperture Radar (SAR) to detect surface and subsurface characteristics of land, sea, and mountain/ ice using polarimetric information has long piqued the interest of scientists and researchers. This chapter presents a small case study on CNN for land use land cover classification using SAR data.

Chapter 9 discusses the different types of draught and how AI can be helpful in efficient drought management. Adverse impacts of drought are non-structural and typically, spread over a larger geographical region than are damages resulting from other natural hazards. However, there are various conventional ways to mitigate these adverse impacts, but modern technological advancement has shown a path to harness Artificial Intelligence methods towards facing drought challenges more efficiently with accurate prediction of its location-specific occurrences and duration estimation.

Chapter 10 attempts to explore various AI enabled irrigation and soil-water management techniques. Artificial water applications directly affect soil quality, water availability for the plants and ultimately, affect the growth and yield of the crops. If irrigation is well managed with accurate soil-water availability estimation, minimizing water requirement, and intelligent techniques of applying the water in sites of interest, then significant reduction in overall farming cost can be achieved with retaining soil nutrients and reusable irrigation facilities in the fields. This chapter included the introduction to the models that can be used for irrigation and soil-water management.

Chapter 11 proposes a flood assessment model by using hydrodynamic HEC-RAS modelling for Balaghat district of South-East Madhya Pradesh State at regional level. The present study used HEC-RAS version 5 to develop a flood model for the Chandan River, situated in the southern part of Balaghat district. The Digital Elevation Model (DEM) used for this analysis is 30m open source CartoDEM V-3 R1. The peak floods of 1990, 2002, and 2006 are taken into consideration. The river reach was divided into 48 cross sections and a one-dimensional steady flow analysis is performed on HEC-RAS to assess the flood. The depths observed in the floods of 1990, 2005, and 2006 are 5.99 m, 3.2 m, and 3.49 m, respectively. The Coefficient of Correlation (R^2) is obtained as 0.954 which shows the consistency and accuracy of the model. This study can help governing bodies to plan the city and attenuate the losses caused by localized riverine floods.

Chapter 12 presents the comprehensive review on time-series forecasting of the water quality parameters using classical statistical and artificial intelligence-based techniques. Some of the most important approaches for calculating the Water Quality Index are discussed briefly. In addition, a problem formulation for the modelling of water quality parameters, performance metrics suitable for evaluating time-series approaches, comparative analysis, and important research challenges of the Water Quality time-series modeling and forecasting are discussed in detail.

Section 3 comprises of three chapters. These chapters discuss the geographical applications of remote sensing data analytics, opening the door towards AI enabled solutions for agriculture, disaster mitigation and preparedness.

Chapter 13 explains the effects of climate abnormalities that cause crop damage and how agrometeorological advisory can be used to deal with this issue. Agrometeorological inputs to the agriculture can play a significant role, mainly in the countries like India, where agriculture and allied sectors are the key pillars of its economy. To facilitate substantial growth in the sector and to improve the socio-economic status of farmers, it becomes inevitable to advise the agro-user community on how best they can avail the advantages of the meteorological parameters and to minimize the damage to agriculture, livestock, caused due to hazardous weather elements.

Chapter 14 proposes a deep learning-based model for the classification and localization of satellite cyclonic images using deep neural networks over the Bay of Bengal and Arabian Sea. This chapter primarily demonstrates how Residual Neural Networks are used to classify satellite imageries of cyclones in the Bay of Bengal (BoB) and the Arabian Sea (Arab Sea). Authors further discovered the cyclones' locations and investigated using satellite images in the infrared and visible bands of electromagnetic spectrum. Satellite images of both cyclone storm and non-cyclone storms, are analyzed for Cyclonic storm recognition and classification

Chapter 15 presents the comparative analysis of different AI and IoT based solutions for monitoring and forecasting air quality. Weather impacts the quality of air in various ways like a windy day can remove smoke from our atmosphere whereas a dust storm can decrease the air quality. In this chapter, a detailed review of several recent AI based models is given while exploring the latest IoT technologies used for monitoring and forecasting the air quality indices.

Rajeev Kumar Gupta
Pandit Deendayal Energy University, India

Arti Jain
Jaypee Institute of Information Technology, India

John Wang
Montclair State University, USA

Ved Prakash Singh
India Meteorological Department, Ministry of Earth Sciences, Government of India, India

Santosh Bharti
Pandit Deendayal Energy University, India

Section 1

Chapter 1
Introduction to Meteorology and Weather Forecasting

Ketaki Anandkumar Pattani
https://orcid.org/0000-0002-7942-8637
Institute of Advanced Research, Gandhinagar, India

Sunil Gautam
Nirma University, Ahmedabad, India

ABSTRACT

Meteorology refers to the scientific study and analysis of weather conditions and atmospheric phenomenon on the earth, processes that cause these weather conditions, and their future predictions. Not just limiting to prediction, meteorologists note the physical conditions of the atmosphere above them, and they study the raw data from maps, satellite, and radar. Forecasts are highly needed for applications that are tailored to specific industries. Gas and electric utilities, for example, may need to have forecasts of temperature within one or two degrees a day ahead of time, or government may need to have data of rainfall for the areas having lower ground level to predict water overflow. The chapter provides a detailed introduction to weather forecasting, its requirements and types of forecasts, and its applications. Importance of data, weather prediction models and methods, machine learning approaches, and baseline measures in the weather prediction are also discussed. Thereby, it opens up an area of research and future developments in this direction.

INTRODUCTION

The analysis and prediction of weather with respect to environmental conditions and its likely developments is termed as weather forecasting which has emerged as one of the most challenging task as toiling up a cliff (van de Wal et al., 2005, p. 311). Weather in the complex areas can even affect half of the earth's land, living beings, property, nature, surface runoff and may cause menacing outcomes (Fitzroy et al., 1863). Groups may be affected diversely due to it involving constructive as well as destructive effects. Weather may be overwhelming in any facet even when we are least expecting any such changes.

DOI: 10.4018/978-1-6684-3981-4.ch001

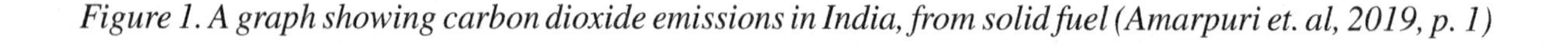

Figure 1. A graph showing carbon dioxide emissions in India, from solid fuel (Amarpuri et. al, 2019, p. 1)

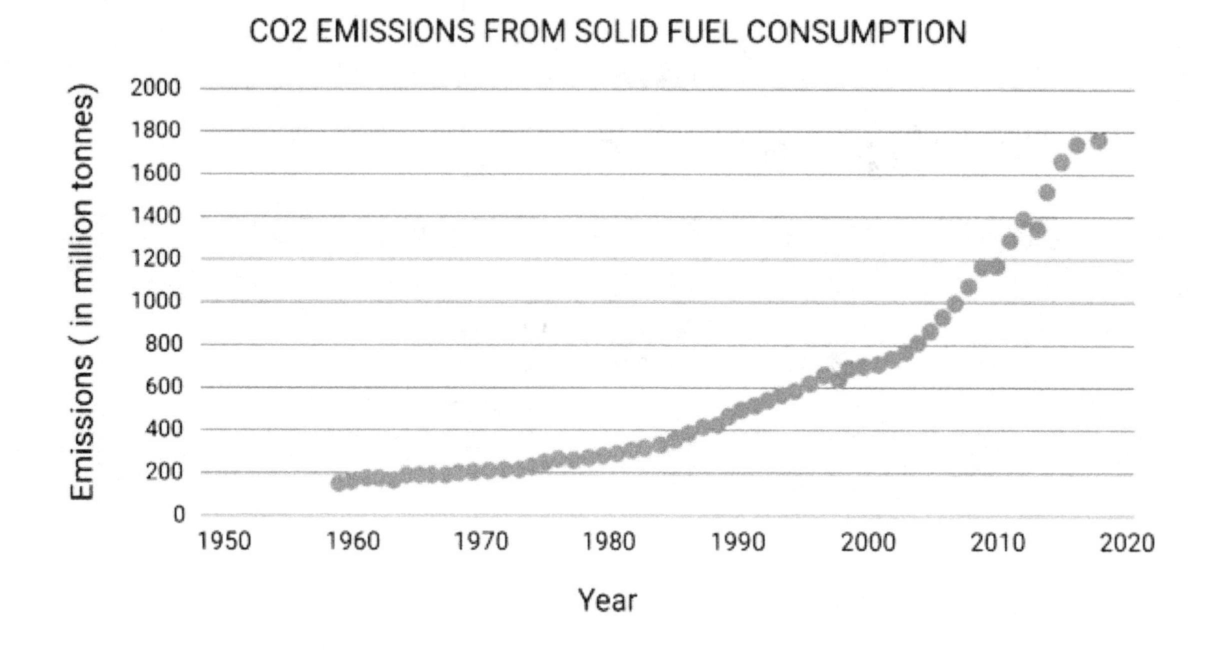

It can cause financial, social or organizational issues. Therefore, experts study these attributes and make predictions beforehand in order to prevent losses.

Air that covers and forms the earth atmosphere is elastic and mobile along with the properties of being compressible and expandable during extreme conditions. This forms a layer of fluids around the earth resulting into heat, cold, solar and lunar changes and other galactic interactions (Fitzroy et al., 1863). This demands for an understanding of the weather that surrounds every inch of earth. An analysis of these conditions can help us save from gigantic calamities (Fitzroy et al., 2011, p. 144). Climate gets highly infected due to the gases within it and can prove to be an important factor for analysis and prediction.

According to Scher & Messori (2018, p. 2830), Carbon Dioxide (CO_2) plays a hypercritical role in the balance of weather conditions. It ideally covers 1% of the total atmosphere and causes 81% of the total emissions thereby being a major greenhouse gas. Chine, United States of America, European Union and India have proven to be largest producers of carbon dioxide. Considering the weather adversaries, India has signed the Paris agreement to reduce the CO_2 levels to 30-35% based on predictive steps and models. Figures depict dangerous levels of CO_2 in India. Figure 1, 2 and 3 show a heavy rise in the levels of CO2 from 1950 to 2020 due to solid, liquid and gaseous fuels respectively (Amarpuri et. al, 2019, p. 1). This calls for an alarming state of requirement of study and better predictions in this direction. This chapter relates the need of weather cognizance to scientific developments and present standing in this regard. Section (ii) deals with the history of meteorology and forecasting. Section (iii) determines types of weather forecasting based on its range. Section (iv) entails existing approaches for weather forecasting and their impact showing the gap of requirement and availability. Finally, section (v) compares the approaches, leads to deductions and provides future scope of advancement.

Figure 2. A graph showing carbon dioxide emissions in India, from liquid fuel (Amarpuri et. al, 2019, p. 1)

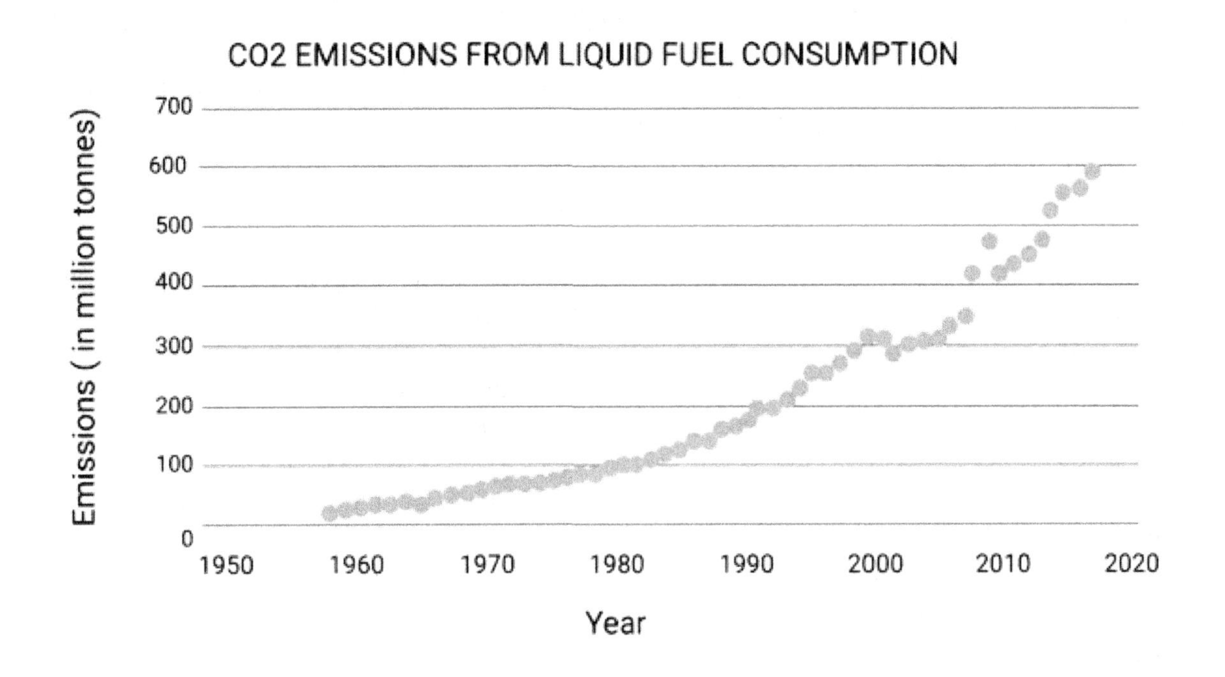

Figure 3. A graph showing carbon dioxide emissions in India, from gaseous fuel (Amarpuri et. al, 2019, p. 1)

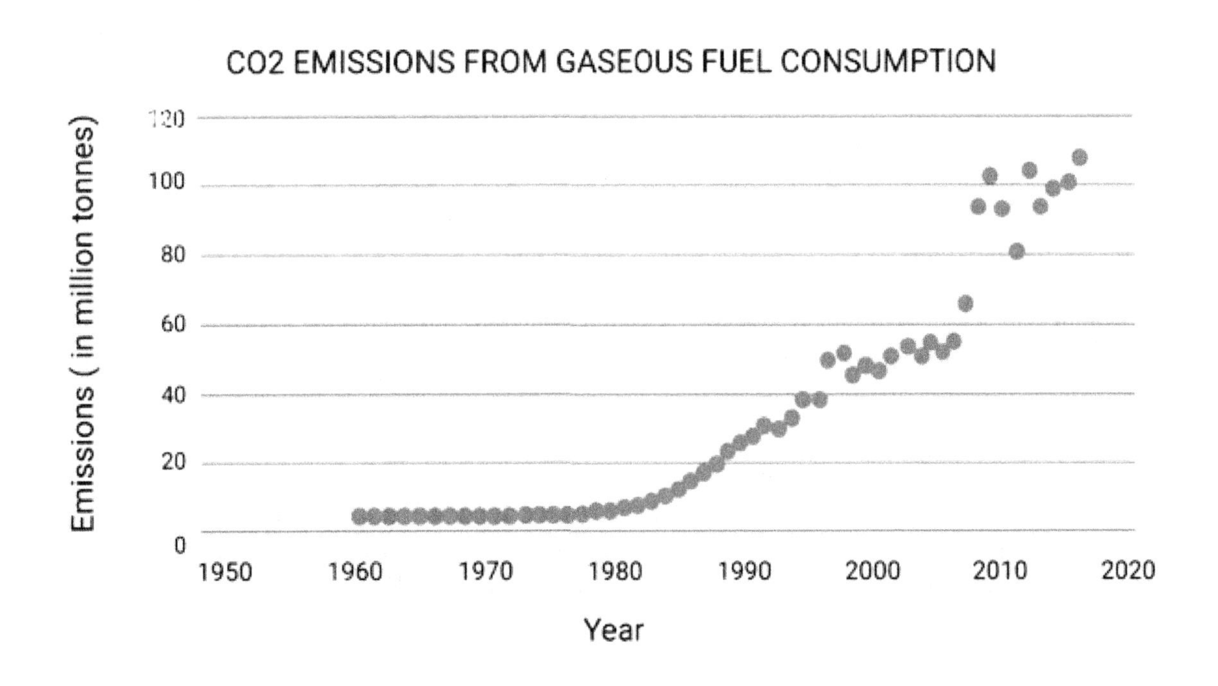

BACKGROUND & CLASSIFICATION

In ancient times, people used to analyze weather conditions through cloud patterns, weather patterns without knowledge of any physical processes (Glahn, 1985, p. 289). The term meteorology is derived from the ancient Greeks when in 340 B.C. Greek philosopher Aristotle defined weather and climate in his book named 'Meteorologica' the title he derived from 'meteoron' meaning 'high in the sky'. A meteorologist uses scientific principles and conditions to explain weather forecast and atmospheric phenomenon. National forecasts such as National Weather Service – US, military, research laboratories or private firms study the climate and give timely predictions. This may involve investigation regarding the changing climate, factors affecting weather conditions, temperature, region of the earth, etc. Meteorology is a part of atmospheric sciences. Mainly meteorology can be Boundary Layer Meteorology (BLM) or Dynamic Meteorology (DM). BLM deals with the atmospheric study just above the earth's surface dealing with heating, cooling, friction etc. Whereas DM focuses over the tiniest particles and their interactions dealing with fluid parcel in fluid dynamics of the earth. Also, meteorologists who study physical aspects like energy, radiation etc. are said to belong to physical meteorology. Operational meteorology is another sub-type dealing with meaningful interpretations of the atmospheric data and their analysis. During the early ages, the calculations were made manually and instruments such as barometer, thermometer, wind-gauge and similar others were used to measure and manipulate weather conditions. There were variety of scales creating a non-uniformity and difficulty in mapping. Units needed conversion for example Fahrenheit to Celsius leading to errors. Mathematical operations such as mean and median were used to publish the weather information. However, this could not be considered accurate. The issue was much sorted when Sir John Herschel proposed 'term days' for general purpose and chart for communication called Beaufort Notation as shown in figure 4 (Ahrens, 2018).

Another method was to scale the wind from 1 to 12 as increasing pressure and velocity. In the middle age, enthusiasts like Dampier took greater interest into the field. Now, there was a need of development into the field. Still, there were developments ongoing into the field. However, with advancing technology there rose a requirement to predict the weather well in advance accurately and automatically. Therefore, arose the need of computerized analysis. Some techniques are simple whereas some are complex. Different types of scatter diagrams and histograms fall into the first category whereas discriminant analysis and logical analysis fall into the second (Glahn, 1985, p. 289). Repackaging with malware: This technique develops he view of some popular app by disassembling the application and then adds its malware content to the app and again reassembles it and puts it on less monitored 3rd party market. For example: Amazon Application Store.

Weather forecasting can be performed in short range, medium range and long range analysis mode. The accuracy and ability to predict the weather depends upon the amount of available data, allotted time for analysis and most importantly the complexity of weather events and their inter-relation (Lee, 2014, p. 888). Instruments such as barometers measure air pressure in certain environment and can be helpful to determine pressure tendency through surface troughs, pressure systems and frontal boundaries. Radar systems measure the location and speed of clouds, position and intensity of snow, rain and other phenomenon. Thermometers, rain-gage, wind-gage, aneroid etc. are some of the instruments that measure the temperature component of the environment. Computer prediction models can be used to process the accumulated data and gain conclusions depending upon the situation (Kirkwood et. al, 2021).

Figure 4. Beaufort Chart (Ahrens, 2018)

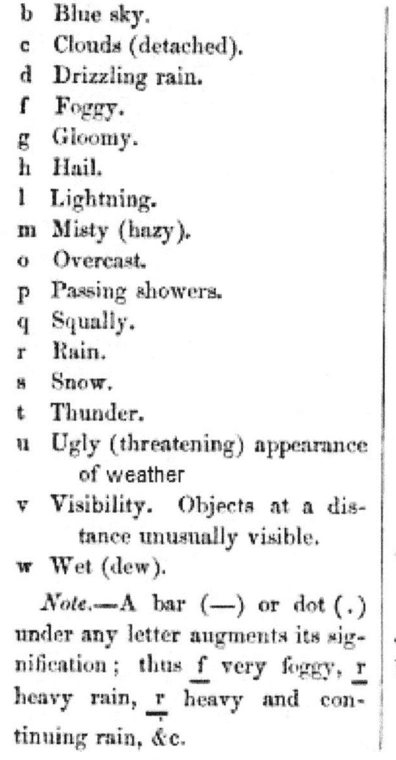

b Blue sky.
c Clouds (detached).
d Drizzling rain.
f Foggy.
g Gloomy.
h Hail.
l Lightning.
m Misty (hazy).
o Overcast.
p Passing showers.
q Squally.
r Rain.
s Snow.
t Thunder.
u Ugly (threatening) appearance of weather
v Visibility. Objects at a distance unusually visible.
w Wet (dew).

Note.—A bar (—) or dot (.) under any letter augments its signification; thus f very foggy, r heavy rain, r heavy and continuing rain, &c.

0 Calm.
1 Steerage way.
2 Clean-full from 1 to 2 knots.
3 " " 3 4
4 " " 5 6
5 With royals.
6 Top-gallant sails over single reefs.
7 Two reefs in topsails.
8 Three reefs in topsails.
9 Close-reefed topsails and courses.
10 Close-reefed main-topsail and reefed foresail.
11 Storm staysails.
12 Hurricane.

From 2 to 10 being supposed 'close-hauled.'

On land a gradually increasing estimate between 1 and 12 may be used as an approximation.

Figure 5. Types of weather forecasting

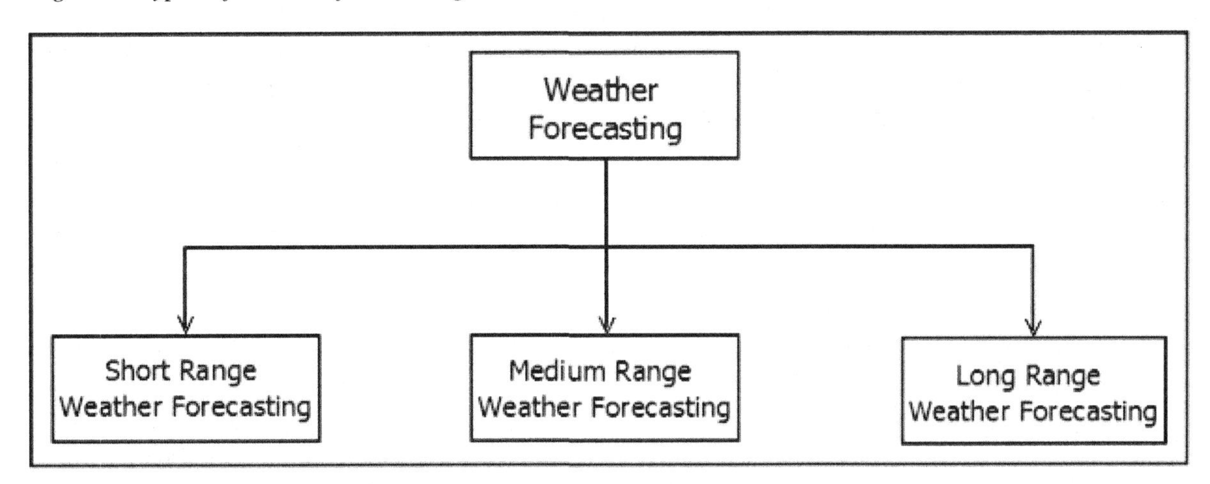

Data here, can be in forms like temporal data or time-series data. Temporal databases contain the data having time-stamp based information. Time-stamping can be done using a valid time, transaction time or both. One can apply time-stamping to tuples as well as attributes. (Mitsa, 2010) The databases that support time-stamping are mentioned below:

1. **Snapshot databases:** These are the databases that contain the latest or the most recent changes included data. Conventional database is one of its types.
2. **Rollback databases:** These are the databases that support the transaction time based processing.
3. **Historical databases:** These are the databases depending only upon the valid time.
4. **Temporal databases:** These are the databases that support both valid time as well as transaction time.

These databases may involve mining functionalities such as temporal prediction, temporal classification, temporal regression, temporal association, temporal clustering, temporal summarization and temporal outlier detection (Mahalakshmi et. al, 2016, p. 1).

Temporal data mining includes time series based study and prediction (Esling et. al, 2012, p. 1). This type of data is a collection of a large number of data values within a constant time period. Year, month, week, day, and so on may all be used to represent time. The time series is examined in order to forecast changes in the provided data as well as changes that will occur in the future. This includes short term, medium term and long term analysis.

Short range weather forecasting is done by 'now casting' wherein precisely radar and satellite observations of local weather conditions are analyzed and depicted rapidly by computers thereby forecasting weather well in advance. Whereas, long range weather forecasts provide expected weather conditions over a period of one to three months where the predictions are usually done by models like IFS (Integrated Forecasting System) coupled ocean-atmosphere model. Medium range forecasting falls in-between the short range and the long range. Also, forecasts are inherently uncertain and the parameters obtained from automatic weather stations play an important role in analysis and forecasting of a large variety of phenomena. Here, irregular values are abundant in meteorological data due to manifold issues in systems and hence affect the results (Ackerman et. al, 2011).

According to World Meteorological Organization (WMO), 'Long range weather forecast is defined as the forecast from 30 days' (i.e., one month) up to one season's description of average weather parameters of a particular area'.' (World Meteorological Organization, 2021) Long range weather forecast may include monthly weather forecast and seasonal weather forecast. Long-range weather predictions include information on future atmospheric composition and phenomena that are averaged over a period of one to three months. The IFS coupled ocean-atmosphere model generates long-range weather forecasts. Long-term models rely on characteristics of Earth system variability that occur over long time periods (months to years) and may be predicted to some extent. The ENSO (El Nino Southern Oscillation) cycle has been the most significant of them. Although ENSO is a linked ocean-atmosphere phenomena focused in the equatorial Pacific, its oscillations have an impact all over the globe (i.e., in India EL NINO phenomenon is related with the average rainfall in a year).

Monthly, long-range weather reports are provided, with a range of up to seven months. Every three months, the same process is used to make yearly weather forecasts, which can be extended up to 13 months in the future. The Indian meteorological department (IMD), which is part of the Ministry of Earth Sciences, is responsible for long-range weather reports in India. It contains data on yearly rainfall, and these projections have been found to be 60% accurate. Statistical regression model is the model employed by IMD for these projections.

The European Centre for Medium-Range Weather Predictions defines medium range forecasting as weather forecasts that contain an average description of the climate in a specific area for a period of 3 to 7 days. (World Meteorological Organization, 2021) Temperature, relative humidity, wind speed, wind direction, and cloud cover are all included in these forecasts. A 7-day time frame is typical for these things. These parameters are important in predicting the genesis of cyclones. Regional Climate Model produces medium-range weather predictions. The National Centre for Medium Range Weather Forecasting, which is a Centre of Excellence in Weather and Climate Modelling underneath the Ministry of Earth Sciences, is responsible for medium-range weather forecasting in India. These projections are generally correct to within 70-90 percent of the time (World Meteorological Organization, 2021).

Weather forecasts over the next 72 hours are known as short term weather forecasts. This prediction range is primarily concerned with the weather systems seen on the most recent weather graphs, as well as the formation of new systems throughout time. Because the prognosis is for a very short period of time, these forecasts are frequently 80 to 90% correct. Nested atmospheric model is the model that is utilized for short-range weather prediction. IMD is in charge of these forecasts in India. These predictions' data is accessible on the Indian Meteorological Department's internet page.

MACHINE LEARNING BASED APPROACHES

Weather prediction has progressed to heights of computational exploration from merely being laborious manual efforts in early ages. Techniques of Numerical Weather Prediction (NWP), Artificial Intelligence (AI), Statistical Weather Prediction (SWP), and Machine Learning (ML) algorithms and data science have led to advancement and refinement in the analysis. Predictive and descriptive data mining algorithms may be used to create these forecasting strategies. Predictive models may be used to forecast values from a variety of sample data of various forms, and they are classified into three types: classification, time-series, and regression. However, the descriptive model allows us to determine patterns in many forms of sample data, and it is divided into three types: clustering, association rules, and summarization (Nalluri et. al, 2019, p. 1977).

One of the early AI based approaches included Dynamic Integrated foreCast (DICast) which was built on algorithms to mimic the human-efforts in the development of weather prediction processes. It uses the past data records, creates analysis graphs for it, and uses intelligent models to predict the future outcomes. DICast shows an improvement of around 10-15% in the performance of the best of individual model of weather forecasting. Also, it is capable to perform predictions accurately with relatively small datasets. The Graphical Atmospheric Forecast System (GRAFS) is the gridded version of this system that can make predictions for data-sparse regions (Myers et. al, 2011). The applicability of prediction models determines the results and many have failed to identify abnormal patterns in the weather prediction systems (Dou et. al, 2011, p. 116) (Zang et. al, 2009, p. 313) (Li et. al, 2011, p. 1) (Zang & M., 2011, p. 761). Prediction systems such as RAMS (Pielke et. al, 1992, p. 69), used in Olympics of Atlanta, USA, use of back propagation algorithm (Krishna et. al, 2015), and integrated autoregressive moving average (Shariq et. al, 2022, p. 1) are some approaches have their own flaws and complications as these algorithms do not prove to be suitable for weather predictions all the time. Also, RAMS has failed to provide longer range prediction results and does not prove to be effective for medium-range models for weather analysis (Li et. al, 2010).

Simple k-means, Hierarchical clustering, Density based clustering, filtered clustering, and farthest first clustering are some of the clustering approaches available (Rai et. al, 2010). Clustering is the process of putting things together based on their commonalities. It's difficult to have a goal attribute that stays the same throughout the procedure when we're predicting the weather for the next day. Weather, on the other hand, tends to bring about more frequent changes in the environment. As a result, classification cannot be used for weather prediction. Structured date, Analysis, Precipitation type, Rainfall, Temperature, Humidity, Wind and its speed, cloud coverage and pressure, and daily summary are some of the features that may be found in the datasets. Pre-processing of data may also be required at times when data has inconsistencies.

Traditional forecasting systems included regression method, multiple regression method and exponential soothing types of methods. These were models using previous data to generate patterns used to predict the future. There are also stochastic forecasting methods that perform modelling based on different outcomes in different conditions using random variables and analysis. This may include algorithms like SVM, random-walk mean reversion etc.

METHODOLGY USED FOR FORECASTING

Weather observation has advanced significantly since the development of the very first weather instruments & methods in early ages. More advanced sensors and communication methods, as well as better-trained weather observers, have resulted in a more detailed, dependable, and representative record of weather and climate.

Data Assimilation

Data collecting has been classified into two groups in weather forecasting – earth surface layer weather analysis and upper layer dynamics namely Boundary Layer Meteorology (BLM) or Dynamic Meteorology (DM) respectively.

1. Boundary Layer Observations:
    a. The observations at the boundary layer can be easily taken from the earth's surface. However, it may vary at different locations. These analytics can be used for safety and future predictions depending upon the observations taken manually by humans or computerized analysis and markings.
    b. According to 'The Atmosphere' (Lutgens, 1989), a large network of weather stations is necessary to create a weather chart, which will cover enough ground to be helpful for short-term forecasts. The WMO, which is made up of over 130 countries, is in charge of gathering the necessary data and providing some broad prognostic charts on a worldwide basis. Surface weather measurements of air pressure, heat, wind patterns, moisture, and wetness are made by skilled observers or automatic weather stations in the earth's atmosphere. The WMO works to standardize instruments, determining techniques, and monitoring timing across the world. These atmospheric observations can be utilized by commercial aviation services also, for determination of suitable flying conditions. Apart from these, agriculture, industrial processes, research, marine and military applications also utilize these results.
2. Dynamic Meteorology:
    a. Radiosondes aboard weather balloons are used to find temperature readings, moisture, and pressure above the surface. A radiosonde is a device used in weather balloons that measures and transmits numerous atmospheric data to a fixed station. The radio frequency of radiosondes

can be set to 403MHz or 1680MHz, and both types can be changed slightly higher or lower as needed. Also, the weather analysis device communicates vertical profiles of air temperature, pressure, and relative humidity to the ground station up to a height of around 30 kilometers. In addition, winds at different atmospheric levels are calculated using a radio direction finding antenna to follow the balloons. Aircraft, dropwind sondes, radar, and satellites all collect upper-air weather data (Mohan et. al, 1991, p. 356).

b. According to Albert Arking (Mohan et. al, 1991, p. 356), clouds create a remarkable effect in weather forecasting and prediction. Cloud sensitivity, defined as the difference in responsiveness of top-of-atmosphere fluxes to variations in cloudiness characteristics, is an important aspect in cloud feedback. However, at this moment, observational data is insufficient to provide much more than preliminary estimations. There are significant differences between two estimates of cloud sensitivity. Future sensitivity research will need to distinguish between different types of clouds in order to be useful. Cloud sensitivity to clouds precipitation nuclei increases the possibility of a more direct role for clouds in climate change, where aerosols linked with S02 emissions can eventually lead to brighter clouds and less solar heating.

Analysis

The many functions of approximation and filtering data for use in numerical models, according to Ackerman and Knox (Arking et. al, 1991, p. 795), are collectively referred to as data assimilation. Analysis cycles are used to assimilate data. Observations of a system's present (and perhaps previous) state are merged with the output of a mathematical model (the prediction) in each analysis cycle to generate an analysis, which is regarded the best estimate of the system's current state. This is referred to as the analysis stage. Essentially, the analysis stage attempts to balance the facts and projected uncertainty. As the model is advances with time, the outcome will become the prediction.

Numerical Analysis for Weather Prediction

The term numerical is deceptive, as mentioned in 'The Atmosphere' (Lutgens, 2010), because all kinds of weather prediction are based on some numerical methods and hence fall under this category. The fact that the gases in the atmosphere obey a number of established scientific rules provides the basis for numerical weather prediction. In an ideal world, these physical rules may be utilized to forecast the state of the atmosphere in the future based on current conditions. This circumstance is analogous to projecting the moon's future location based on physical rules and current position knowledge. Nonetheless, the enormous number of factors that must be taken into account when considering the dynamic environment makes this a tough process.

According to (Linacre et. al, 1997, p. 321), Numerical Weather Prediction (NWP) is a reduced set of equations termed the basic equation that is used to calculate changes in circumstances. Meteorological data prediction is significantly used in modern forecasting.

Global Models for Forecasting

The following are some of the more well-known global numerical models (Rahman, 2017):

1. Global Forecast System (GFS) - The National Organization for the Atmosphere in America (NOAA) developed GFS. The output is open to the public.
2. Navy Operational Global Atmospheric Prediction System (NOGAPS) - A model developed by the US Navy to compare to the Global Forecast System (GFS).
3. Global Environmental Multi-scale Model (GEM) - Canada's Meteorological Service developed the GEM.
4. European Centre for Medium-Range Weather Forecasts (ECMWF) – It is a limited-access model run by Europeans.
5. United Kingdom Meteorological Office (UKMO) - Developed by the United Kingdom Meteorological Office. There is a limited supply, but it is hand-corrected by expert forecasters.
6. Globales Modell (GME) - a weather forecasting system created by the German Weather Service.
7. Action de Recherche Petite Echelle Grande Echelle (ARPEGE) - It was created by Meteo France, the French weather service.
8. The Intermediate General Circulation Model (IGCM) - It was created by members of the University of Reading's Department of Meteorology.

Regional Numerical Models

Following are some of the most well-known regional numerical models (Paul et. al, 2018, p. 269):

1. NCEP and the meteorological research community collaborated to build the Weather Research and Forecasting (WRF) Model. WRF comes in a variety of configurations, including:
2. The ERF Non-hydrostatic Mesoscale Model is the major short-term weather forecast model for the United States.
3. WRF – NMM: The WRF Non-hydrostatic Mesoscale Model is the primary short-term weather prediction model for the United States. WRF was largely developed at the National Center for Atmospheric Research in the United States (NCAR)
4. The Mesocale Model for North America (NAM)
5. Colorado State University, which conducts numerical models of atmospheric meteorology and other environmental phenomena on scales ranging from meters to hundreds of kilometers.
6. MMS (Mesoscale Modeling System) – This is the fifth generation of mesoscale models.
7. Advanced Region Prediction System (ARPS) - The University of Oklahoma created the ARPS. It's a multi-scale nonhydrostatic simulation system with a lot of features.
8. High Resolution Limited Area Model (HIRLAM)
9. Global Environmental Multi-scale Limited Area Model (GEM-LAM)
10. Aladin: A high-resolution limited-area hydrostatic and nonhydrostatic model created by Meteo-France and used by European and North African nations.
11. COSMO: The COSMO Model, formerly known as LM, aLMD, or LAMI, is a nonhydrostatic restricted area model created within the collaboration for small scale modelling.

Analysis Output

Before being presented as a prediction, the raw data is frequently altered. This can take the form of statistical approaches to eliminate known biases (a phrase used to express a predisposition or predilection towards a specific perspective, ideology, or result) from the model, or adjusting the model to take into consideration consensus among other numerical weather forecasts. Model Output Statistics, or MOS, is a technique for interpreting numerical model output and producing site-specific recommendations. This information is available in coded numerical form for virtually all National Weather Service reporting sites (Kotal et. al, 2022, p. 1) (Klein & Glahn, 1974, p. 1217).

APPLICATIONS OF METEOROLOGY & WEATHER FORECASTING

Accuracy in the weather forecasts and its need in current scenario are prominent and visible in every field. The analysis of weather forecasting is applicable in the following domains:

1. **Weather analytics:** Nationwide there exist advisories and associations that work in the direction of making analysis helpful in emergency. They warn and release alerts in expected hazardous conditions. They include news regarding winds, temperature, moisture, fog, cyclone and related parameters. Especially affecting farmers and common people, these conditions also affect certain industries. In rural areas, weather crisis may result in severe hazards. Therefore, precautionary measures are always taken through proper analysis and advisories (Prathap, 2017, p. 227).
2. **Aviation departments:** Because the aviation department has to deal with the weather conditions, thunderstorms, precipitation, heavy winds, large hail, lightening, turbulence and icing, forecasting becomes utmost necessary for safety. Volcanic ash can also lead to loss of engine power and even airplane crash. Taking advantage of headwinds, airplanes are made to fly according to suitable runways. This reduces the efforts in takeoff and makes the weather conditions favorable (Wolfson et. al, 2006, p. 31). Weather forecasts help in pre judgments of weather conditions and hence help making better decisions.
3. **Agriculture:** India is an agricultural country where most of the farming is done naturally through rainfall and depending upon weather suitability. Cotton, wheat, and corn crops may all be ruined by prolonged drought. Drought may destroy crops, but their dried leftovers, known as silage, can be utilized as a cow feed replacement. Both in the springtime and the autumn, frosts and freezes create disruption with crops (Putjaika et. al, 2016, 53). Therefore, precautionary measures taken with weather forecasts help gain better harvest.
4. **Marine:** Direction of the wind and pace, wave frequency and heights, tidal energy, and moisture can all limit commercial as well as recreational use of waterways. Each of these characteristics can have an impact on the safety of sea transportation. As a result, a number of codes have been developed to effectively send precise maritime weather forecasts to maritime pilots through radio, such as the MAFOR code (Marine forecast) (40. Sevastyanova, 2018).
5. **Armed forces:** Military, navy and air force are highly affected by the changes in the weather conditions. Weather forecasts help judging the probable scenarios or flight conditions or natural responses. As a result, precise and early forecasting of the same is required, as well as distribution to the public for human security and wellbeing (Miller, 1967).
6. **Industrial & Private Sector:** Industries such as oil industry, gas production, electricity boards are some of the huge sectors of the market that are affected by the weather conditions. Their demand and production is highly dependent upon weather conditions (Anđelković & Bajatović,

2020). Therefore, its supply and probable profits are also dependent upon the weather. Prediction of weather becomes the deciding factor in business decisions and reliable action.

FUTURE SCOPE OF ARTIFICIAL INTELLIGENCE & MACHINE L EARNING IN WEATHER FORECASTING

Weather forecasting is not just limited to prediction of weather but also impacts security in terms of climate and disaster management. Such is the case when not just the prediction, but its accuracy also matters to the world. According to a partnered research by the University of Washington and Microsoft Research, they demonstrate how artificial intelligence can examine previous weather patterns to anticipate future catastrophes far more quickly and accurately than current technologies (Anđelković & Bajatović, 2020) (Weyn et. al, 2020). The increasing interest in adapting successful deep learning (DL) approaches for image identification, natural language processing, automation, dynamic games, and other application areas to the field of meteorology has been prompted by the recent buzz around artificial intelligence. There is some notion that incorporating big data mining and neural networks into the weather prediction workflow can result in better weather forecasts. The subject of whether DL methods can totally replace present numerical weather models and data assimilation systems is accepted differently (Schultz et. al, 2021).

Traditional machine learning algorithms, according to Reichstein et al., may not be ideally adapted to solve certain data difficulties caused by Earth system data. 'To cope with complicated stats, numerous outcomes, varied noises, and high-dimensional spaces [of Earth system data], deep learning approaches are required,' they said. New system architectures are sorely in demand that take use of both local neighborhood (at various sizes) and long-range linkages (for example, for tele-connections), but the specific cause-and-effect correlations between parameters are unknown and must be determined.' (Reichstein et. al, 2019, p. 195)

Weather forecasting challenges can be effectively solved using DL ideas. Nonetheless, very few efforts to replace the full NWP workflow with a DL system have been confined to short-term forecasting (up to 24 hours less than that) or have only employed a small portion of the available meteorological data. There exist some of the obstacles that must be addressed before a comprehensive end-to-end DL weather forecasting system can give results that are equivalent to the existing NWP. While there have been some spectacular success stories from DL applications in other domains, and preliminary attempts to apply DL to meteorological data have been performed, this study is still in its early stages. DL solutions for several of these concerns are being developed, but there is presently no DL approach that can deal with all of these issues at the same time, as would be necessary in a full environmental forecast system (Schultz et. al, 2021). Therefore, this opens up a scope for better developments and efficient analysis in the field utilizing AI and ML approaches.

CONCLUSION

Weather forecasting is a complicated and difficult discipline that relies on the effective coordination of weather observation, data processing by meteorologists and computers, and speedy communication methods. Weather forecasters have attained a high level of proficiency in different aspects weather forecasting and is widely utilized in different applications. Denser surface and higher air observational networks, more exact numerical models of the atmosphere, larger and faster computers, and other advancements are predicted to make the analysis stronger. However, because the atmosphere is a continuous fluid that recognizes no political borders, continual international collaboration is needed. Also, automated methodologies and intelligent systems incorporating advanced machine learning algorithms will analyze past events and make future predictions more reliable.

REFERENCES

Van de Wal, R. S. W., Greuell, W., van den Broeke, M. R., Reijmer, C. H., & Oerlemans, J. (2005). Surface mass-balance observations and automatic weather station data along a transect near Kangerlussuaq, West Greenland. *Annals of Glaciology*, *42*, 311–316. doi:10.3189/172756405781812529

Fitzroy, R. (1863). *The weather book: A manual of practical meteorology* (Vol. 2). Longman, Green, Longman, Roberts, & Green.

Estévez, J., Gavilán, P., & Giráldez, J. V. (2011). Guidelines on validation procedures for meteorological data from automatic weather stations. *Journal of Hydrology (Amsterdam)*, *402*(1-2), 144–154.

Scher & Messori. (2018). Predicting weather forecast uncertainty with machine learning. *Quarterly Journal of the Royal Meteorological Society*, *144*(717), 2830–2841.

Amarpuri, L., Yadav, N., Kumar, G., & Agrawal, S. (2019, August). Prediction of CO 2 emissions using deep learning hybrid approach: A Case Study in Indian Context. In *2019 Twelfth International Conference on Contemporary Computing (IC3)* (pp. 1-6). IEEE.

Glahn, H. R. (1985). Statistical weather forecasting. Probability, statistics, and decision making in the atmospheric sciences, 289-335.

Ahrens, C. D., & Henson, R. (2021). *Meteorology today: An introduction to weather, climate, and the environment*. Cengage Learning.

Lee, M. K., Moon, S. H., Kim, Y. H., & Moon, B. R. (2014, October). Correcting abnormalities in meteorological data by machine learning. In *2014 IEEE International Conference on Systems, Man, and Cybernetics (SMC)* (pp. 888-893). IEEE.

Kirkwood, C., Economou, T., Odbert, H., & Pugeault, N. (2021). A framework for probabilistic weather forecast post-processing across models and lead times using machine learning. *Philosophical Transactions of the Royal Society A*, *379*(2194), 20200099.

Mitsa, T. (2010). *Temporal data mining*. Chapman and Hall/CRC.

Mahalakshmi, G., Sridevi, S., & Rajaram, S. (2016). A survey on forecasting of time series data. *International Conference on Computing Technologies and Intelligent Data Engineering (ICCTIDE'16)*, 1-8.

Esling, P., & Agon, C. (2012). Time-series data mining. *ACM Computing Surveys*, *45*(1), 1–34.

Ackerman, S., & Knox, J. (2011). *Meteorology*. Jones & Bartlett Publishers.

World Meteorological Organization. (2021). *Global producing centers for Long range forecasts*. Available: https://public.wmo.int/en/programmes/global-data-processing-and-forecasting-system/global-producing-centres-of-long-range-forecasts

World Meteorological Organization. (2021a). *Short range forecasts*. Available: https://library.wmo.int/index.php?lvl=categ_see&id=11189

World Meteorological Organization. (2021b). *Medium range forecasts*. Available: https://www.ncmrwf.gov.in/

Nalluri, S., Ramasubbareddy, S., & Kannayaram, G. (2019). Weather prediction using clustering strategies in machine learning. *Journal of Computational and Theoretical Nanoscience*, *16*(5-6), 1977–1981.

Myers, W., Wiener, G., Linden, S., & Haupt, S. E. (2011). *A consensus forecasting approach for improved turbine hub height wind speed predictions. Proc. WindPower.*

Dou, Y. W., Lu, L., Liu, X., & Zhang, D. (2011). Meteorological data storage and management system. *Computer Systems & Applications*, *20*(7), 116–120.

Zhang, C., Chen, W.-B., Chen, X., Tiwari, R., Yang, L., & Warner, G. (2009). A multimodal data mining framework for revealing common sources of spam images. *Journal of Multimedia*, *4*(5), 313–320.

Li, C., Zhang, M., Xing, C., & Hu, J. (2011). Survey and review on key technologies of column-oriented database systems. *Computer Science*, *37*(12), 1–8.

Zhang, M. (2011). Application of data mining technology in digital library. *Journal of Computers*, *6*(4), 761–768.

Pielke, R. A., Cotton, W. R., Walko, R. E. A., Tremback, C. J., Lyons, W. A., Grasso, L. D., ... Copeland, J. H. (1992). A comprehensive meteorological modeling system—RAMS. *Meteorology and Atmospheric Physics*, *49*(1), 69–91.

Krishna, G. V. (2015). An integrated approach for weather forecasting based on data mining and forecasting analysis. *International Journal of Computers and Applications*, *120*(11).

Shariq, M., Singh, K., Maurya, P. K., Ahmadian, A., & Taniar, D. (2022). AnonSURP: An anonymous and secure ultralightweight RFID protocol for deployment in internet of vehicles systems. *The Journal of Supercomputing*, 1–26.

Li, C., Zhang, M., Xing, C., & Hu, J. (2010). Survey and review on key technologies of column oriented database systems. *Computer Science*, *37*(12).

Rai, P., & Singh, S. (2010). A survey of clustering techniques. *International Journal of Computers and Applications*, *7*(12), 1–5.

Lutgens, F. K., & Tarbuck, E. J. (1979). *The atmosphere, an introduction to meteorology*. Prince-Hall.

Mohan, J. M., & Morgan, M. D. (1991). *Meteorology: The Atmosphere and Science of Weather* (4th ed.). Macmillan Ontario.

Arking, A. (1991). The radiative effects of clouds and their impact on climate. *Bulletin of the American Meteorological Society*, *72*(6), 795–814.

Ackerman, S. A., & Knox, J. A. (2003). *Meteorology: Understanding the Atmosphere*. Brooks/Cole USA.

Linacre, E., & Geerts, B. (2002). *Climates and weather explained*. Routledge.

Rahman, M. (2017). *Effect of Differently Interpolated Geographical Data on WRF-ARW Forecast* (Doctoral dissertation). Khulna University of Engineering & Technology (KUET), Khulna, Bangladesh.

Paul, S., Wang, C. C., Chien, F. C., & Lee, D. I. (2018). An evaluation of the WRF M ei-yu rainfall forecasts in T aiwan, 2008–2010: Differences in elevation and sub-regions. *Meteorological Applications*, *25*(2), 269–282.

Kotal, S. D., & Sharma, R. S. (2022). Development of a NWP based Integrated Block Level Forecast System (IBL-FS) using statistical post-processing technique for the state Jharkhand (India). *G eofizika*, *39*(1), 1–31.

Klein, W. H., & Glahn, H. R. (1974). Forecasting local weather by means of model output statistics. *Bulletin of the American Meteorological Society*, *55*(10), 1217–1227.

Prathap, G. (2017). Micro-level Agromet Advisory Services using block level weather forecast–A new concept based approach. *Current Science*, *112*(2), 227.

Wolfson, M. M., & Clark, D. A. (2006). Advanced aviation weather forecasts. *The Lincoln Laboratory Journal*, *16*(1), 31.

Putjaika, N., Phusae, S., Chen-Im, A., Phunchongharn, P., & Akkarajitsakul, K. (2016). A control system in an intelligent farming by using arduino technology. *Fifth ICT International Student Project Conference (ICT-ISPC)*, 53-56.

Sevastyanova, I., Motornaya, S., & Korepanov, A. (2018). Specifics of sea graduates' professional training in Russia. *SHS Web of Conferences*, 55.

Miller, R. C. (1967). *Notes on analysis and severe-storm forecasting procedures of the Military Weather Warning Center, Air Weather Service (MAC)* (Vol. 200). United States Air Force.

Anđelković, A. S., & Bajatović, D. (2020). Integration of weather forecast and artificial intelligence for a short-term city-scale natural gas consumption prediction. *Journal of Cleaner Production*, *266*, 122096.

Weyn, J. A., Durran, D. R., & Caruana, R. (2020). Improving data-driven global weather prediction using deep convolutional neural networks on a cubed sphere. *Journal of Advances in Modeling Earth Systems, 12*(9).

Schultz, M. G., Betancourt, C., Gong, B., Kleinert, F., Langguth, M., Leufen, L. H., ... Stadtler, S. (2021). Can deep learning beat numerical weather prediction? *Philosophical Transactions of the Royal Society A*, *379*(2194), 20200097.

Reichstein, M., Camps-Valls, G., Stevens, B., Jung, M., Denzler, J., & Carvalhais, N. (2019). Deep learning and process understanding for data-driven Earth system science. *Nature*, *566*(7743), 195–204.

Chapter 2
Role of Meteorological Satellites and Radar in Weather Forecasting

Divyang Dave
Pandit Deendayal Energy University, India

Rajeev Kumar Gupta
https://orcid.org/0000-0002-5317-9919
Pandit Deendayal Energy University, India

Santosh Kumar Bharti
https://orcid.org/0000-0002-0627-6433
Pandit Deendayal Energy University, India

Ved Prakash Singh
https://orcid.org/0000-0002-2281-5687
Ministry of Earth Sciences, Bhopal, India & Indian Institute of Technology, Patna, India

ABSTRACT

Because of global warming, pollution, and many other factors, the environment is changing at an alarming rate. Accurate forecasting can assist people in making appropriate plans for activities such as harvesting, traveling, aviation, etc. Satellites and radar have been increasingly popular in weather forecasting over the previous few decades. The information collected by the satellite and radar can be used to monitor climate movement, track hurricanes, and give barometrical estimations that can be turned into mathematical climate expectation (NWP) models for exact forecasting. Currently, more than 160 meteorological satellites are located in orbit, which generates approximately 80 million observations every day. This chapter discusses several meteorological satellites which are used to extract weather pattern. For the time being, the results of Observation System Simulation Studies (OSSE) utilising satellite information are presented in order to demonstrate the relationship between perceptions from satellite sensors and ground-based sensors.

DOI: 10.4018/978-1-6684-3981-4.ch002

INTRODUCTION

A weather satellite system is a kind of satellite which is used to track the temperature and environment of the Earth. Satellites can be polar circling (covering the entire Earth in a non-concurrent manner) or geostationary at times (floating over a comparable spot on the equator). Weather forecast is one of the most important issues in any technical or scientifically disruptive issue (Xie et al., 2022). Rigorous mathematical models were accelerated due to technological advancements in order to produce exact projections. The adoption of largely dependent machine learning models yields higher and better results. Not only are recent meteorological characteristics complex, but they are also numerical. The conditions of the earth's atmosphere change over time. There are seasons such as summer, winter, spring, autumn, Monsoon, and others. The weather conditions can change whenever. Around the world, this type of weather fluctuation is fairly regular and predictable. We can minimise or restrict losses in a range of industries, including agriculture, natural disasters, and many others, if we can effectively predict weather conditions (Rai et al., 2021). So, if we're talking about a particular agricultural region, these techniques can provide us a better view of what is exactly happening, but the technique and amount of data collected is too large for a farmer to absorb and make decisions on in real time. Artificial Intelligence (AI) enables computers to learn how to analyse data more effectively and autonomously, allowing for faster pattern recognition, categorization, and forecasting without the need for human intervention. Space information has turned into an unquestionable part in weather conditions checking and dynamic demonstrating because of late improvements in satellite innovation concerning high goal, multi-spectral band groups including infrared, visible and microwave domains. With the rising exactness of satellite recoveries, upgrades in models could be made, prompting better estimates, especially in the tropics. Since the dawn of human civilization, the importance of weather prediction has been recognised, and initial efforts to estimate the weather were depends upon human experience, intuition, and an understanding of the relationship between the weather and natural cycles. Scientists postulated around the turn of the twentieth century that the atmosphere must satisfy the fundamental principles of physics.

Predicting how the current state of the climate will change is what weather forecasting includes. Ground measurements, measurements from ships, observations from aeroplanes, radio noises, doppler radar, and satellites are all used to determine current weather conditions. This data is forwarded to meteorological centres, which collect, analyse, and present the data in a variety of charts, maps, and graphs. Thousands of observations are transferred onto surface and upper-air maps using modern high-speed computers.

The ability to anticipate the weather requires a sufficiently exact understanding of the on-going state of the climate as well as a sufficiently accurate understanding of the natural principles that regulate the formation of the weather. As a result, meteorological observation networks were seen as equally crucial to the evolution of weather prediction models. Because of the emphasis on atmospheric observations, a ground-based measurement network of surface and upper air weather parameters has been continuously developed. Despite the fact that the ground-based monitoring network is crucial for weather prediction and meteorology, it only delivers limited observations across difficult locations such as mountains, deserts, and extensive marine areas. When compared to observations from a traditional ground-based network, meteorological satellites give observations that are large in coverage, narrowly spaced, representative, and more frequent. As a result, it's no surprise that meteorological satellites have been at the frontier of earth observation. The Television InfraRed Observation Satellite (TIROS-1) was the first meteorological satellite, launched on April 1, 1960. Within the scope of the Global Observing System, the World Meteorological Organization (WMO) created an operational satellite surveillance network

Figure 1. Constellation of geostationary and low earth orbiting meteorological satellites (WMO, 2014)

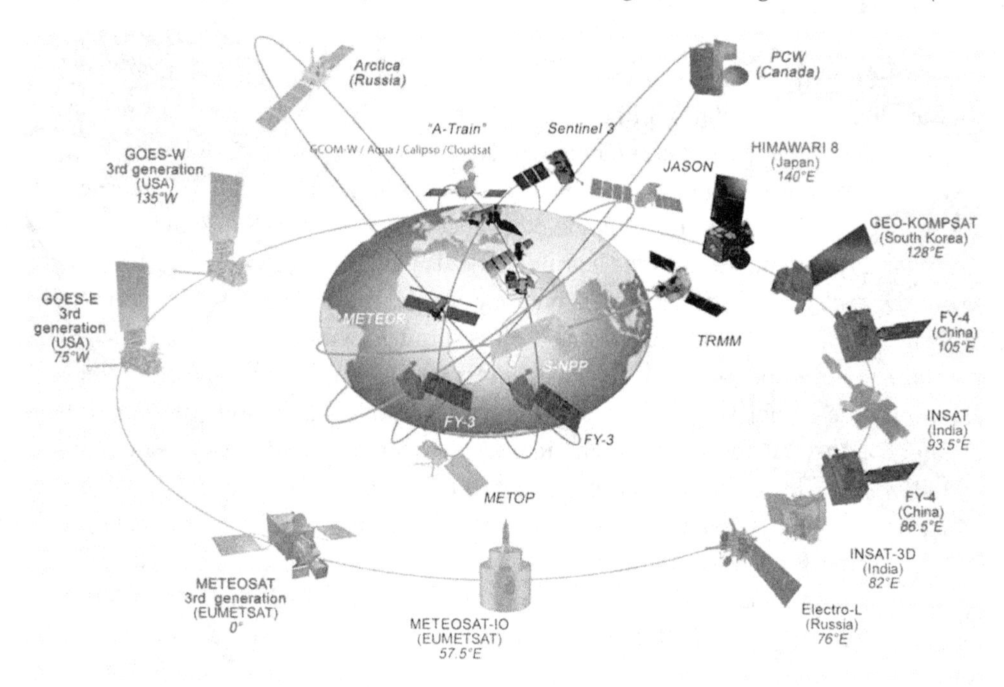

of geostationary and polar-orbiting meteorological satellites in 1963 (WMO, 2014). Figure 1 illustrates the Constellation of geostationary and low earth orbiting meteorological satellites.

TRADITIONAL WEATHER FORECASTING

To predict the weather condition at a certain location and time is called weather forecasting. Lots of quantitative data pertaining to the current atmospheric condition is gathered, and then scientific understanding of the climate system is applied in order to anticipate how the climate will evolve in the future (Maier et al., 2020). There is a wide range of people that utilize weather forecasts. Weather warnings such as earthquakes, heavy rain, floods, and other natural disasters can be extremely beneficial in protecting lives and property. In order to anticipate the weather, two types of satellites are used: polar-orbiting and geostationary satellites. However, according to the findings of the study, polar-orbiting satellites are more ideal for weather forecasting since they are closer to the planet whereas geostationary satellites are more suitable for solar monitoring (Perez et al., 2013). Several radiation ranging sensors from shorter to infrared wave are connected into all satellite to collect the useful information. The spectral bands of latest United States satellite (GOES 8–15) is shown is table 1.

Observing weather patterns was once the primary source of forecasting data. Over time, meteorological research has yielded a range of rainfall forecasting methods. A combination of computer models, interpretation, and knowledge of weather patterns is required to accurately predict rainfall in the modern world. The following method was used in existing weather forecasts. Z Mohamad et al. introduced a rainfall prediction system using satellite image (Mohamad et al., 2021). The data from the Multi-Functional Transport Satellite (MTSAT-1R) and the Terminal Doppler Weather Radar product were utilised

Table 1. Spectral bandwidth the series of GOES satellite

Satellite imager channel	Wavelength range (μm)	Ground resolution at nadir	Primary detection
Visible	0.55–0.75	1 km	Clouds, albedo, smoke
Shortwave IR	3.80–4.00	4 km	Clouds, smoke
Moisture IR	6.30–6.70	8 km	Clouds, water vapor
Surface Temperature IR	10.20–11.20	4 km	Clouds, water vapor, surface temperature

in this study to estimate rainfall in Malaysia. Figure 2 depicts the flow diagram of whole process used for rainfall prediction.

Figure 2. Flow diagram of an approach used for rainfall prediction (Z. Mohamad, 2021)

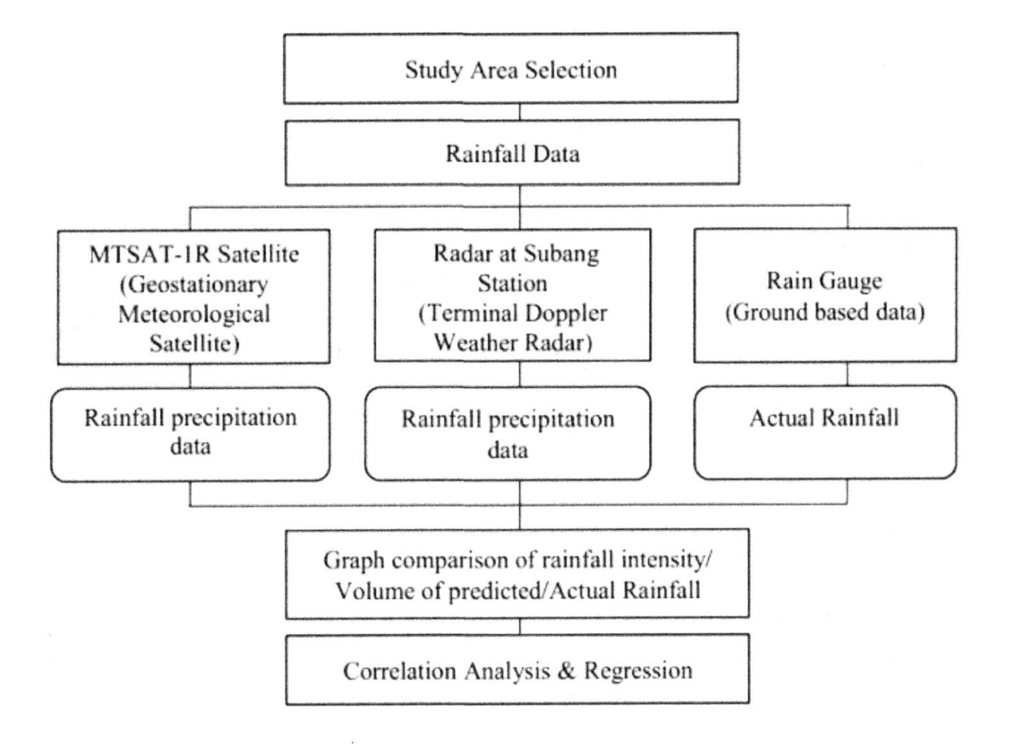

Use of A Barometer

Since the late 1800s, barometric pressure and pressure tendency measurements have been employed in weather estimation. The greater the shift in pressure, the greater the likelihood of a change in weather. If the pressure drops quickly, a low-pressure system is forming, which means rain is more likely.

Looking at The Sky

The condition of the sky, together with pressure tendency, is one of the most essential characteristics used to predict weather in mountainous terrain. The attack of a higher cloud deck or the thickening of overcast cover are the two indications of approaching precipitation. High thin mists can form coronas around the moon in the evening, giving the illusion of a warm front with the accompanying downpour. Morning haze foreshadows pleasant conditions, as blustery conditions are preceded by wind or mists, which prevent hazy arrangement.

Nowcasting

Nowcasting is a word used to describe the process of predicting the weather for the next six hours. In this time frame, smaller details such as individual showers and rainstorms, as well as other factors too small to be resolved by a computer model, can be calculated with reasonable accuracy. Given the most current radar, satellite, and observational data, a human would wish to do a more thorough analysis of the limited scale highlights offered, and thus develop a more precise conjecture for the next few hours. (Srinivasa et al. 2012)

Numerical Weather Prediction Model

The study of forecasting the weather using atmospheric models and computing tools is known as numerical weather prediction (NWP). To predict the weather, current meteorological conditions are fed into statistical model of the weather. This model typically offers surrounding points with a spatial resolution of a few kilometres around the wind farm. To generate a forecast, NWP relies on the computing capability of computers. A forecaster looks at how the computer's projected features will combine to create the weather for the day. The NWP technique is problematic because the models' equations for simulating the environment are not accurate.

A number of meteorology organisations have modelling centres where supercomputers are utilised to perform global NWP models. The National Center for Environmental Prediction in the United States, the European Centre for Medium-Range Weather Forecasts and the UK Meteorological Office are among them. A universal method for NWP is necessary, even if it is costly, especially for long-term forecasting (Gerrity et al., 2020). As a result, generating accurate forecasts necessitates a thorough investigation from which to build the model. This is accomplished by a computer-assisted process known as data assimilation, in which the newest meteorological measurements from all over the world are integrated with the model forecasts to produce a global study of present circumstances. This is the computer version of the human assessment cycle that forecasters perform on a regular basis, it becomes the beginning for the next cycle of the NWP model. Modern weather forecasting relies heavily on global models, and Met Service meteorologists utilise the NCEP, UKMO, and ECMWF models on a daily basis to help them produce predictions and weather warnings. These models give data about the way of behaving of atmospheric conditions for a huge scope without zeroing in on local subtleties.

Ensemble Forecasting

Meteorologists have shaped anticipating models that estimate the air by utilizing this ensemble method of predicting to characterize how climatic temperature, pressure, and dampness will change according to the long run to foresee weather estimation. The equations are encoded into a computer, which is then supplied data on the current atmospheric conditions. The conditions are addressed by the PC to decide how the different climatic factors will vacillate over the accompanying couple of moments (Nies et al., 2016). The PC then, at that point, rehashes the strategy, utilizing the outcome from one round as the hotspot for the coming step. The computer produces its computed information for a future date and time. The data is then analysed, and lines showing the projected positions of the various pressure systems are drawn. The prognostic chart is used by forecasters to anticipate the weather. The atmosphere is represented by a variety of atmospheric models, each of which interprets the atmosphere in a somewhat different way. Forecasts for the next 12 and 24 hours are usually accurate. Two- or three-day forecasts are frequently accurate. Forecast accuracy dramatically decreases after five days.

Remote sensing, particularly radar and satellites, can also provide weather data.

Radar

Radar is an abbreviation for Radio Detection and Ranging, and it is primarily used to detect any thing, such as a word, vehicle, building, aeroplane or any other object, using electromagnetic waves. It was originally developed for military purposes prior to World War II, but it is now also employed in other fields such as weather forecasting, flood detection and many more. Figure 3 shows the satellite.

A transmitter puts out radio waves in radar. The radio waves are reflected off the nearby object before returning to the receiver. Weather radar can detect a wide range of precipitation parameters, including its position, motion, severity, and the possibility of upcoming precipitation. The majorly common type of radar is Doppler radar, that can trace the rate at which precipitation occurs. Radar can map out a storm's structure and predict whether or not it will cause severe weather (Roshni, 2022). Figure 4 illustrate the basic principles of radar.

Weather Satellites

Since the first weather satellite was launched in 1952, satellites have become increasingly essential providers of meteorological data. Large-scale processes, such as storms, are best monitored by weather satellites. Satellites can also track the spread of volcanic ash, smoke from raging fires, and pollution. They can keep track of long-term progress. Figure shows one of the geostationary satellites that monitors conditions throughout the world. Weather satellites could see all energy from all wavelengths in the electromagnetic spectrum. The visible light and infrared (heat) frequencies are the most important (Maier et al., 2020 & Hait, n.d.). Figure 5 shows the weather satellite.

Weather Maps

Weather maps illustrate meteorological conditions in the environment in a simple and pictorial way. Weather maps can show only one or numerous aspects of the atmosphere. They can display data from

Figure 3. Radar system (Roshni, 2022)

computer models as well as human observations. Newspapers, television, and the Internet all include weather maps (Hait, n.d.).

In general, two approaches to weather forecasting are used: empirical and dynamical approaches (Gerrity et al., 2020). Analogue forecasting is a term used by meteorologists to describe an empirical approach that is based on the frequency of analogues. If there are a lot of recorded occurrences, this method is usually beneficial for predicting local-scale weather. The dynamical technique, often known as computer modelling, is based on equations and forward ns of the environment. This dynamical method is valuable

Figure 4. Basic Principals of Radar System (Roshni, 2022)

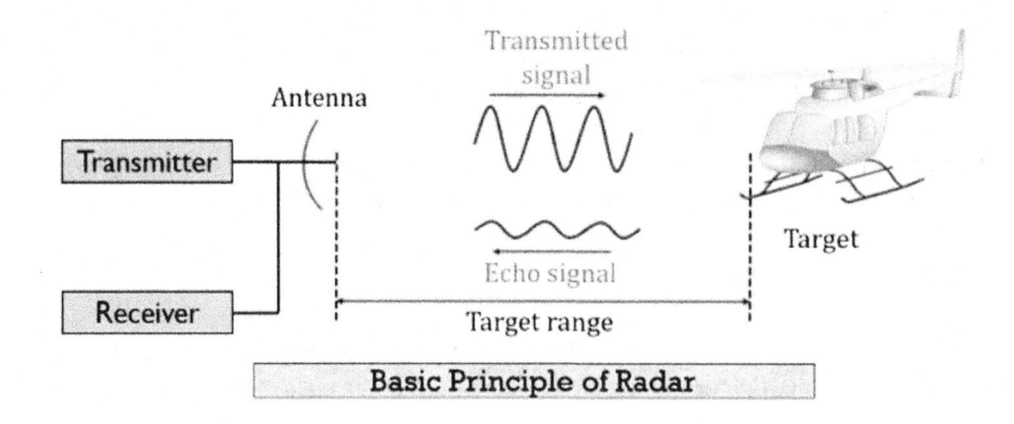

for displaying enormous scope climate scenario, yet it may not be proficient as far as foreseeing climate. Most of weather conditions estimating frameworks utilize both observational and dynamical techniques.

Figure 5. Weather Satellite (Maier et al., 2020 & Hait, n.d.)

TYPES OF WEATHER FORECASTING

Several scientists and meteorologists from all around the world contribute to the daily weather report. Weather satellites orbiting the earth capture images of clouds from space, and modern machines make forecasts more precise than ever. Forecasters create their predictions based on observations from the ground and space, as well as algorithms and principles based on historical experience. Meteorologists create their daily weather predictions using a combination of various distinct approaches. They are:

- **Persistence Forecasting:** Persistence forecasting is the most basic approach of weather prediction. It makes predictions about tomorrow's weather conditions based on the current weather circumstances. Whenever the weather conditions is steady, for example, throughout the mid-year season in the jungles, this can be a decent strategy to conjecture the climate. The event of a fixed weather condition is basic for this sort of determining. It tends to be utilized in both short-and long-haul projections. This presupposes that the weather will continue to behave as it does presently. Meteorologists perform weather observations to determine how the weather is behaving.

- **Synoptic Forecasting:** This technique makes advantage of the fundamental rules for forecasting. To construct a short-term forecast, meteorologists combine their observations with the laws they've studied.
- **Statistical Forecasting:** Meteorologists are perplexed as to what it does on a regular basis at this time of year. Using historical records of average temperatures, rainfall and snowfall, weather forecasters can gain an idea of what the weather is "supposed to be like" at a certain time of year at a given location.
- **Computer forecasting:** Forecasters use their observations to enter numbers into complex calculations. These many equations are executed on several ultra-high-speed computers to create computer "models" that provide a forecast for the following several days. Because different equations often produce different outcomes, meteorologists must always combine this strategy with other forecasting methods.

Forecasters utilize all of the above techniques to provide their "most realistic estimation" for what the weather conditions will resemble over the course of the following couple of days.

Weather conditions estimation today incorporates a different arrangement of functional things that are by and large separated into the Very short-range forecast, Short-range forecast, Medium-range forecast and Long-range forecast.

CONTRIBUTIONS OF SPACE OBSERVATIONS

TIROS-1, the world's first meteorological satellite, was launched into orbit in April 1960, ushering in a new era of space perceptions and providing the first glimpses of the Earth's dynamic cloud formations for the first time. From that point forward, innovation has progressed in parts of monitoring capacities as far as spatial, temporal and spectral resolution. With both geostationary and polar circling satellites, a worldwide arrangement of space perceptions has arisen (Bhatta and Priya, 2017).

Several variables contribute to the benefits of space observations, including:

- Finding the interrelationships of cycles with different spatial dimensions from a synoptic perspective on vast expanses.
- Regular perceptions from geostationary satellites give ceaseless checking whereas polar circling satellites perform two times day to day analysis; such information is applicable for investigation of climate framework elements.
- The inborn spatial averaging is more delegate than the point in-site perceptions and promptly usable for climate forecast models.
- As a result of the high degree of consistency in space perceptions, the issue of intercalibration, which is required for ground-based sensors, is eliminated.
- Filling of holes in surveillance; Space information covers huge sea regions and distant and remote land regions, subsequently giving worldwide inclusion.
- Information and surveillance boundaries that have never before existed, such as ocean surface temperature (skin temperature), ocean surface breeze pressure (ocean level), cloud fluid water content (radiation balance), and aerosol are just a few of the exceptional boundaries that can only be obtained through satellite technology.

- Concurrent perception of a few powerful boundaries given by various sensors in same stage works with investigation of inter-connections and information on processes (e.g., Ocean Surface temperature and profound convection).

METEOROLOGICAL SATELLITES AND THEIR WORKING / PAYLOADS

Right now, a few functional meteorological satellites are giving worldwide and local inspections. Six unique sorts of satellite frameworks at present being used are:

- Visible / Infrared,
- Microwave Imagers,
- Infrared Sounders,
- Scatterometers,
- Radar Altimeters,
- Microwave Sounders.

On geostationary and polar circling satellites, visible and infrared imagers are accessible. On polar circular frameworks, the last four are currently only achievable. In this portion of the world, the INSAT satellite system is the major satellite for weather observation (Joshi, Simon and Bhattacharyam, 2011). Meteorological and correspondence demands are supplied by a multipurpose geostationary satellite. It has a Very High-Resolution Radiometer (VHRR) payload that allows world to see pictures in apparent, infrared, and presently even water fume (Parker et al., 2021). Offering the accompanying types of assistance is planned:

- Nonstop observation of climate frameworks including serious climate occasions around the Indian zone.
- Overcast cover, ocean surface temperature, cloud top temperature, cloud movement vector, active long wave, snow cover radiation, and so on are operational boundaries for anticipating weather conditions.
- Information Collection Platforms collect and transmit meteorological, hydrological, and oceanic data from remote and out-of-reach locations.
- When imminent calamities such as typhoons are detected, they can be warned in advance using Cyclone Warning Dissemination Systems (CWDS).
- Meteorological data scattering, including handled climate framework images, through a Meteorological Data Spread Framework.

With the launch of the INSAT-1 series of satellites in the mid-1980s, the INSAT applications programme began. Following that, the INSAT-2 series was created in response to client feedback. The VHRR payload on INSAT-2A and 2B, launched in 92 and 93, had a 2 km apparent and 8 km warm band goal. Three modes were accessible: full edge, normal mode, and a 5-minute area mode for quick inclusion of extreme climate frameworks to picture modes.

INSAT-2E, sent off in 1999, was outfitted with a modern VHRR payload that could work in three channels: visible (2 km), warm (thermal), and water fume (8 km). In the centre lower atmosphere, the water fume channel can give water fume conveyance and stream designs. INSAT-2E also carried a CCD camera with three channels – visible, near infrared, and short-wave infrared – and a resolution of 1 km, which was used to map the vegetation cover in the region.

INSAT-3A payloads are quite identical to INSAT-2E, which was launched in April 2003. INSAT-3D will likewise have a climatic sounder for temperature and water fume profiles, as well as divided warm channels empowering precise ocean surface temperature recovery later on. A variety of quantitative products are being retrieved using data from INSAT satellites. INSAT symbolism is used widely for brief investigation and weather conditions determining (F. Olaiya et al., 2012). The products offered by INSAT, as well as their applications, are discussed in the following sections.

Visible/ Infrared/Water Vapour Imagers

The first satellite measurements were recorded utilising visible/infrared wavelengths, and these observations are still considered crucial for operational weather applications today. Almost all geostationary satellites, which orbit the earth at various points along the equatorial belt, use visible/infrared sensors to collect routine measurements of the earth-atmosphere system at set intervals. India's INSAT-3D and INSAT-3DR 6-channel imagers, for example, give full earth coverage every 30 minutes with a spatial resolution of 1 kilometre in visible and short-wave infrared (SWIR) channels.

Infrared Sounders

On geostationary and polar circling satellites, these frameworks are accessible. Direct osmosis of IR radiation in NWP models is the primary application. Infrared channels in IR sounders will be increased by two orders of magnitude. This will work on the first temperature and dampness fields and work on the upward goal of inferred temperature and dampness profiles in clear regions or more cloud high level.

Microwave Imagers

The proxy variables VIS and IR are used. In VIS and IR imaging, precise eye placement is feasible. Uninvolved microwave sensors measure radiation beneath the cirrus haze, providing data on barometrical WV, precipitation, power, cloud fluid water, and convective action zones. The primary satellite from the Defense Meteorological Satellite Program was sent off in June 1987, conveying a microwave radiometer named Special Sensor Microwave/Image. It collided into the European Research Satellite, which conveyed a scatterometer and was sent off in July 1991. Because the ERS-I scatterometer's width was just 500 km, it only provided partial coverage of tropical regions each day.

Several measurements may be limited in their value due to the low horizontal resolution of some modern radiometers, while the SSM/I 85 GHz channel's 15 km resolution gives meso-scale information. These channel displays imagery same as radar and can detect circulation centres. The benefit of centre-fixing in TCs utilising 85 GHz imagery over traditional VIS and IR images with cirrus cloud is described by Velden et al. (1989). NASA launched a special satellite with the goal of taking new meteorological measurements in the tropics. In the year 2000, the Tropical Rainfall Measuring Mission concluded three years of prosperous data collection. It will be having sufficient fuel to perform nonstop estimations until 2005 subsequent to changing its elevation from 350 km to 400 km. The TRMM Microwave Imager has an even goal of 5-7 km for the 85 GHz channel, which is 2-3 times more noteworthy than the SSM/I, and the higher goal TMI 37 GHz channel can infiltrate further into hurricanes to uncover more highlights.

Microwave Sounders

Microwave sounder fetch information from the NOAA-15-17/AMSU sensor, similar to microwave imager information, can give indispensable data beneath cloud high level. In a non-coming down cloudy climate, tropospheric warm readings can be gathered. The information from microwave sounders has demonstrated to be very important in laying out upper tropospheric warm inconsistencies. This is then used to decide the force and change in power of typhoons.

Scatterometers

The scatterometers are generally used to quantify Sea Surface Wind Vectors. The backscattered power from the unpleasantness components on the world's surface is estimated by a scatterometer, which sends microwave heartbeats to the surface. The level of the wind stress and also the direction of the wind relative to the radar beam affect the back scatter power. Due to a lack of validation, the relationship between backscatter signal and ocean surface winds is not well established under the severe wind and rainy circumstances of a cyclone. Following the collapse of NASA's NSCAT system, a second system dubbed Quikscat was quickly deployed, covering an area of 1800 km and providing unprecedented global ocean coverage. Most TC figure workplaces approach wind fields from Quikscat in close to continuous. The fundamental breeze item has a spatial goal of 25 kilometers. The information uncovered the hurricane's external breeze structure. It's additionally used to work out the span of 35 bunches, track down shut flows in creating frameworks, and give lower cutoff points to greatest supported breezes. Sarkar (2003) assessed procedures for estimating surface breeze over the world's seas from satellite stages. The microwave scatterometer has been the most efficient wind sensor.

Radar Altimeters

The function of subsurface thermal structure in tropical storm intensification has been demonstrated in numerous research. Satellite altimetry data can often be used to infer subsurface structure. Improved tropical cyclone intensity estimates could be achieved by better incorporating these data into statistical forecast methods and linked ocean-atmosphere models. Improved temporal and spatial sampling would make altimeter measurements more useful. Multi-beam altimeters have the ability to expand geographic sampling and fill data gaps considerably.

SIGNIFICANCE OF WEATHER FORECASTING

Weather estimation is employed in a variety of contexts, including severe weather alerts and advisories, cloud behaviour prediction for air transportation, waterway prediction in a ocean, farm development, and forest fire prevention (Parker et al. 2021 & Roshni, 2022).

Severe Weather Alerts and Advisories

Basic weather conditions cautions and warnings, gave by the National Weather Service ahead of serious or hazardous climate, are a significant part of present-day weather conditions estimating. This is done to ensure the safety of people and property. Winter weather, severe winds, floods, tropical cyclones, and fog are all examples of tornado and thunderstorm advisories Serious climate warnings and cautions are communicated through the media, by means of crisis frameworks, for example, the Emergency Alert System, which interrupts customary programming.

Predicting the Behavior of The Cloud for Air Transport

The aeronautics business is extremely climate delicate, so exact weather conditions forecasting is basic. Many planes are unable to land or take off due to fog or extremely low ceilings. Turbulence and ice are two more major in-flight dangers. Rainstorms make huge disturbance because of updrafts and outpouring limits, icing because of weighty precipitation, colossal hail, solid breezes, and lightning, all of which can make disastrous harm an airplane in flight. Volcanic debris is likewise a major issue for flying, as debris mists can make planes lose motor power.

Prediction of Waterways in Ocean

Wind direction, speed, wave periodicity, high tides, and precipitation can all interfere with commercial and recreational river use. Each of these factors can have an effect on the safety of marine travel. As a result, a variety of codes, such as marine forecast, have been created to efficiently communicate detailed maritime weather forecasts to vessel pilots through radio. Standard weather forecasts at sea can be obtained through radio fax (WMO, 2018)

Agricultural Development

Agricultural production is greatly influenced by the weather in many ways. Because of changes in nutrient mobilisation as a result of water stressors, it has a significant impact on crop growth and development, yield, insect and measles occurrence, water needs, and fertiliser requirements. It also has a significant impact on the timely and effective implementation of preventive and cultural crop activities. Weather abnormalities can cause agricultural damage and soil erosion, among other things. It is possible that the weather will have an affect on the quality of a crop produced when it is being transported from its original location in the field to storage and then to sale. Bad weather has the potential to degrade the quality of produce while it is being transported, as well as the viability and vigour of seeds and planting material while they are being stored.

Avoiding Forest Fire

Predictions of wind, precipitation, and humidity are essential for averting and managing wildfires. Various indexes, such as the Forest fire weather index and the Haines Index, have been developed to predict whether a given location is more likely to have a fire due to natural or human causes. It is also possible to utilise weather forecasting to predict the development of potentially dangerous insects.

Military Applications

Weather forecasters in the military provide weather updates to the warfighting community. Pilots receive pre-flight and in-flight weather briefings from military weather forecasters, as well as real-time resource protection services for military facilities. The waterways and ship weather forecasts are covered by naval forecasters. The Navy's Joint Typhoon Warning Center offers an interesting support to both themselves and the remainder of the central government by delivering hurricane gauges for the Pacific and Indian Oceans.

Air Force

The Air Force and the Army uses utilize the latest technology to forecast weather During both war and peacetime, Aviation Force forecasters assist the Army by providing coverage for air operations. Civilian and military weather forecasters collaborate on the analysis and development of weather forecast products.

Observations of Air Quality and Gases from Satellites

Due to an increase in air pollutants in the atmosphere, such as SO2, NO2, CO2, NO, CO, NOx, and O3, there has been a focus on air quality forecast in recent years. Existing numerical models include advanced strategies for predicting precise air quality. The accuracy of air quality models, like weather prediction models, is dependent on the correct initial conditions. Aside from ground observations, satellite-based sensors provide information on a variety of air pollution species, including aerosols. Today, advanced satellites such as the MODerate resolution Imaging Spectroradiometer (MODIS), Multi-angle imaging Spectro Radiometer (MISR), Atmospheric infrared Sounder (AIRS), Visible Infrared Imaging Radiometer Suite (VIIRS), Ozone Monitoring Instrument (OMI), Measurements of pollution in the troposphere (MOPITT), Greenhouse gases orbiting satellite (GOSAT), Orbiting carbon observatory (OCO), Tropos The data from the above-mentioned satellites' various multispectral and hyperspectral channels is valuable for determining global, regional, and local air quality over daily, seasonal, interannual, and decadal timescales.

IMPACT OF SATELLITE OBSERVATIONS ON WEATHER PREDICTION

It is impossible to overestimate the relevance of satellite observation in practical weather forecast. Scientists have been able to answer fundamental questions about weather and climate thanks to decades of satellite data. Satellite observations provide meteorologists with continuously updated information about the status of the atmosphere, allowing them to develop models that predict weather into the future with far greater accuracy than pre-satellite forecasts. As a result, the accuracy of 7-day forecasts has more than doubled in the last three decades. These advancements are saving many lives and have a huge economic impact. Weather observations, which are used to make forecasts, are one of the most widely utilised satellite products. No hurricane or typhoon has gone unreported since satellite photos became widely available, giving impacted coastal areas ample warning and essential time to prepare. Over the last few decades, a variety of new satellite sensors have been developed that provide new types of observations that could improve the quality and accuracy of weather forecasting. The majority of operational weather prediction today is based on NWP models, which use data assimilation to incorporate a large volume of satellite observations. The impact of data from various satellite sensors on the accuracy of weather prediction is an important subject that aids in the development of new satellite sensors and the development of strategies for obtaining valuable observations for enhanced weather forecast.

SUMMARY

At this time, satellite measurements are without a doubt the most essential source of observations for numerical weather prediction. Satellites use various sensors and orbital configurations to offer valuable data of temperature and humidity profiles, surface temperature, and winds, among other things. In operational synoptic scale analysis and prediction, as well as nowcasting of major weather phenomena, satellite observations are critical. The capacity of satellite observations to see huge maritime regions and sections of the southern hemisphere where traditional data are sparse is one of the key reasons that satellite observations have a large impact on weather forecasting at a worldwide scale. On a regional scale, however, the impact of conventional radiosonde observations is still greater than that of satellite observations, particularly in the Indian monsoon region, because in a cloudy environment, these observations resolve the vertical dynamic and thermodynamic structure of the atmosphere better than satellite observations. However, observations of atmospheric temperature and humidity profiles from satellite microwave sounders, vertically integrated water vapour from microwave radiometers, ocean surface winds from microwave scatterometers, and atmospheric winds from geostationary satellites all have a significant impact on NWP models' monsoon forecasting. Satellite observations are improving in terms of data volume, geographical and spectral resolution, coverage, temporal frequency, and other factors as sensor technology advances.

REFERENCES

Bhatta, N. P., & Priya, M. G. (2017). *RADAR and its applications*. https://www.researchgate.net/publication/316696944_RADAR_and_its_applications

Gerrity, J. P. (2020). Weather forecasting and prediction. *Earth Science*. doi:10.1036/1097-8542.742600

Hait, S. (n.d.). *SEC4T (Weather Forecasting), Topic: Basics of Weather Forecasting (part-I)*. Narajole Raj College. Retrieved from https://www.narajolerajcollege.ac.in/document/sub_page/20210607_094906.pdf

Joshi, P. C., Simon, B., & Bhattacharyam, B. K. (2011). Advanced INSAT Data Utilization for Meteorological Forecasting and Agrometeorological Applications. *Challenges and Opportunities in Agrometeorology*, 273–285. doi:10.1007/978-3-642-19360-6_21

Maier, M. W., Gallagher, F. W. III, St. Germain, K., Anthes, R., Zuffada, C., Menzies, R., ... Adams, E. (2021). Architecting the future of weather satellites. *Bulletin of the American Meteorological Society*, *102*(3), E589–E610.

Mohamad, Z., Bakar, M. Z. A., & Norman, M. (2021). Evaluation of Satellite Based Rainfall Estimation. *Proc. of IOP Conf. Series: Earth and Environmental Science*. 10.1088/1755-1315/620/1/012011

Nies, H., Behner, F., Reuter, S., Meckel, S., & Loffeld, O. (2016, June). Radar imaging and tracking using geostationary communication satellite systems-A project description. In *Proceedings of EUSAR 2016: 11th European Conference on Synthetic Aperture Radar* (pp. 1-4). VDE.

Parker, A. L., Castellazzi, P., Fuhrmann, T., Garthwaite, M. C., & Featherstone, W. E. (2021). Applications of Satellite Radar Imagery for Hazard Monitoring: Insights from Australia. *Remote Sensing*, *13*(8), 1422.

Perez, R., Cebecauer, T., & Šúri, M. (2013). Semi-empirical satellite models. *Solar energy forecasting and resource assessment*, 21-48.

Rai, P. V., Singh, P., & Mishra, V. N. (2021). *Recent Technologies for Disaster Management and Risk Reduction*. Earth and Environmental Sciences Library. doi:10.1007/978-3-030-76116-5

Roshni, Y. (2021, March 1). *What is radar system? definition, basic principle, block diagram and applications of Radar.* Electronics Desk. Retrieved from https://electronicsdesk.com/radar-system.html

Srinivasa, K. G., Harsha, R., Sunil, K. N., Arhatha, B., Abhishek, S. C., Harish, R. C., & Anil, K. M. (2012). Weather Nowcasting Using Environmental Sensors Integrated to the Mobile. In Mobile Computing Techniques in Emerging Markets: Systems, Applications and Services (pp. 183-203). IGI Global.

World Meteorological Organization (WMO). (2018). *Guide to Marine Meteorological Services*. https://library.wmo.int/doc_num.php?explnum_id=5445

World Metrological Organisation (WMO). (2014). *Preparing the Use of New Generation Geostationary Meteorological Satellites*. https://public.wmo.int/en/resources/bulletin/preparing-use-of-new-generation-geostationary-meteorological-satellites

Xie, Y., Xie, B., Wang, Z., Gupta, R. K., Baz, M., AlZain, M. A., & Masud, M. (2022). Geological Resource Planning and Environmental Impact Assessments Based on GIS. *Sustainability*, *14*(2), 906. doi:10.3390u14020906

Chapter 3
Enhancement of Meteorological Observational Systems Using the Internet of Things

Ramesh Chandra Goswami
Indus University, India

Hiren Joshi
Gujarat University, India

Sunil Gautam
Institute of Technology, Nirma University, India

ABSTRACT

Weather forecasting has an important role in meteorology, and it has long been one of the world's most systematically difficult problems. Severe weather events pose a threat to a complex method of weather forecasting with only a partial explanation. Weather forecasting accuracy has a significant impact on various sectors of the economy, necessitating the development of a system that allows for greater accuracy in real-time monitoring and future weather prediction. Farmers are exposed to weather threats since the agricultural process, such as soil preparation, sowing, irrigation, harvesting, and storage of crops, is directly dependent on weather conditions. The internet of things is currently growing at an exponential rate. The internet of things (IoT) is becoming increasingly important in a variety of fields. In this chapter, the authors look at how the IoT can help with weather forecasting.

INTRODUCTION

Weather is one of the most useful environmental limitations in our lives at every level. Weather forecasts play a crucial role in our day-to-day lives. A fantastic weather forecasting system aids in the recovery of development as well as any required planning in the event of inclement weather. IOT has been employed in fields such as healthcare, farming, and home automation, and has grown into a smart city contract, as

DOI: 10.4018/978-1-6684-3981-4.ch003

well as weather forecasting (Nikesh et al., 2016; Asghar et al., 2015; Das et al., 2021; Ojha et al., 2015). Weather forecasting is particularly valuable in a variety of businesses, such as the power industry and the agricultural transportation, and thus forecasting is an important aspect of economic development.

The application of science and technology to forecast atmospheric conditions for a specific location and time period is known as weather forecasting. People have been attempting to forecast the weather informally for generations, and formally from 19th century. Weather forecasting, which used to be done by hand and focused mostly on variations in barometric pressure, existing weather patterns, and sky status or cloud cover, is now done using computer-based models that take a variety of atmospheric variables into account (Sampathkumar et al., 2020 & Banara et al., 2022).

Weather forecasts are made by obtaining objective data on the current state of the atmosphere in a specific region and predicting how the weather will behave in the future using meteorology.

Weather forecasting is an important part of meteorology, and it has long been one of the world's most difficult problems. Severe weather occurrences present a challenging forecasting problem with just a partial explanation. There are various parameters to weather phenomena that are impossible to detail and compute (Y. Zhou et al., 2012). As communication channels improve, weather forecasting specialist systems might integrate and exchange assets, resulting in hybrid systems. Even with various advancements in climate prediction expert systems can't be completely dependable as long as weather forecasting is a major issue.

Weather monitoring is getting a lot of attention these days. People desire to know the current weather conditions of any location, such as industrial sites, offices, and visitor locations, at that time. Weather stations of various types are put in various locations to provide weather information (Rao et al., 2016 & Keshavkumarsingh et al., 2013).

There is various type of sensors such as environmental sensors are utilized for measurements at any given location and are reported on the cloud in real time by weather forecasting. Temperature, wind and humidity all had a role in forecasting (Yerpude et al., 2017). Various sensors were employed in weather forecasting to continuously sense weather parameters and broadcast them to an internet server through a Wi-Fi. The cloud is used to store weather data, which then provides weather forecasting in real time Smart cities have weather forecasting systems in place to forecast the weather. With the advent of high-speed Internet, an increasing number of individuals all over the world are becoming connected. The Internet of Things expands on this by connecting not only people but also technological objects that can communicate with one another (Pandey et al., 2022 & Wankhede et al., 2014).

With the decreasing cost of Wi-Fi-enabled gadgets, this trend will only gain traction. The IoT is based on the idea of connecting several electronic devices and then retrieving data from them, which can be dispersed in any way, and uploading it on cloud where it can be analyzed and processed. These data can be used in the cloud service to warn people in many ways, as an example activating ringing a bell, sending them an SMS, or sending them an email.IoT allows for not just human to human connection, but also human to device and interaction between devices. This advancement comes in the form of new interaction channels will have an impact on almost major businesses, including healthcare, logistics, energy and transportation and so on (Satyanarayana et al., 2013).

Wearable technologies like smart watches, fitness bands, connected healthcare is one of the most promising areas of IoT. However, this new technology platform brings with it a new series of challenges and roadblocks, such as deciding what to do with the vast volumes of data gathered.

LITERATURE REVIEW

A literature review is a comprehensive summary of previous research on a certain topic. It looks at scholarly publications, books, and other sources that are relevant to a certain research topic. Information mining was used, which is a tool that forecasts current and future patterns, allowing businesses to make proactive decisions. offers a survey of techniques for weather prediction based on data mining, with a focus on its benefits (Chauhan and Thakur, 2018). In (Radhika et al., 2019) author uses Support Vector Machines (SVMs) are used to forecast the weather. Temperatures at their most extreme of each day at a location was evaluated to anticipate the temperature of the following day at that location. The SVM was prepared for this application using a non-straight relapse technique.

In (Talegaon et al., 2022 & Z. U. Khan et al., 2014) author focus for determining climate predictions, researchers used a variety of information-gathering methods, including K-Nearest Neighbor and Decision Trees. In comparison to other calculations that have attained acceptable precision, the decision tree has produced promising results among the order algorithms.

Author using machine learning constructed a system for prediction of weather on the Indian subcontinent. They used linear regression techniques (Majumdar et al., 2021 & Shivang et al., 2018).

SIGNIFICANCE OF WEATHER FORECASTING

Weather forecasting can be used for a variety of purposes in everyday life, like deciding whether or not to bring an umbrella to work or what to wear. Here are a few examples of places where weather forecasting is important (Baste et al., 2017; Mat et al., 2016; Lachure et al., 2015):

1. Farming of fruits, pulses and vegetables are based on weather and so weather forecasting play a significant role in this direction. Sometimes due to bad weather farmers are suffering from huge amount losses. If they have some idea about whether they can take precaution to avoid losses.
2. It assists in the implementation of livestock-protection programmers.
3. It facilitates the transportation and storage of food grains.

Types of Weather Forecasting

Long range

Forecasting longer than four weeks is known as long range forecasting. Long-term forecasting of significant impending strategic decisions to be made inside and for an organization, with a focus on how to best utilize resources. The disadvantage of such forecasts is that they can only be vague. Prediction planners criticize forecasts when things turn out completely differently than planned and forecasting as a result receives criticism from all those who are affected.

Medium range

This forecasting range from 3-4 days to two weeks. Medium-term forecasts are made for small strategic resolutions in correlation with the nature of the business. They are particularly essential in the domain of corporate planning and development, and firm budgets are determined based on this forecast. Inaccurate forecasting can have major consequences for the rest of the company; the company will be obliged to keep unsold goods and will have to spend more money on production.

Short range

The forecasting of 1-2 days comes in this category. Forecasting is crucial in the planning of existing and future operations. As a result, there are a number of other factors that could have a significant impact on the forecasting results. Accurate forecasting, on the other hand, is critical. Forecasting is a useful tool for a variety of investigations.

INTERNET OF THINGS

The Internet of Things refers to a collection of physical objects or sensors that are connected to a network and have unique identities. It allows things to be sensed and controlled remotely using existing network infrastructure, allowing for a closer connection between the real world and computer-based systems (Lachure et al., 2015). It is a future technology that will link the entire globe. Sensors or items are connected in IoT network and data is transferred across a network cloud. According to a poll of technological experts, by 2022, 45-55 billion things will be linked together using IoT technology. IoT technology allowed for a wide range of sensor connectivity with numerous protocols and application features to achieve total interactivity.

An IoT system is extremely complicated, and it should have the layers listed below:

- **Device Layer** – Makes up the system's physical aspect, or the "things," that can automatically recognize, actuate, detect, and connect to the internet.Thus this layer contains physical objects that use an internet connection to collect, measure, and communicate data.
- **Network Layer** – By using communication protocols, the data is relayed to the network or cloud. Thus this layer contains devices that wirelessly transmit data to the cloud, such as a platform or application, where it can be stored and delivered to actuators. Mobile devices have begun to play a role in the Internet of Things, and they are very adaptive to technical needs. They can act as a data gateway, store data, and perform administrative activities such as updating other critical systems.
- **Application Layer** – The processed data is saved and made available to end users.So in this layer the analysis of data and based on that analysis decision is taken.

Uses Of IOT-Driven Data

Forecasting with IoT will utilize several analyzed data streams and can predict concrete events such as frost forecasts. So, it is reasonable to say that inputting IoT data can more accurately assess weather conditions on a farm and field basis (Kiani et al., 2015). With a bit of work, these systems will evolve.

Using the data collected by IoT sensors, the weather models could include the finer grid points relating to fields and farms. Meteorologists would not have to assume, estimate, or project different scenarios to accommodate unknown conditions because microclimate data would be readily available. In terms of arable farming, IoT data could help with:

1. Analyzing historical temperatures and potential frost dates to determine trends and predict accurate planting timeframes.
2. Providing information on which fields consistently produce higher yields for particular crop production and predicting future production.
3. Simulating data, similar to how the American and European models generate forecasts, to assist in determining the best months to sow fields.
4. The pinnacle of IoT systems utilizes "machine learning" to automate data analysis, interpret patterns and make decisions with very little human intervention. However, IoT systems are not infallible, and utilizing the data will require that farmers use their expertise to analyze the information before making final decisions.

Impact of IOT On Weather Forecasting

In the past, weather data was collected and analyzed by humans. However, remote sensing has opened the door to real-time weather data analysis and has altered how data is collected and analyzed for building a database for weather prediction.Today, smartphones have revolutionized our lives as we can access a massive number of useful applications such as tickets booking, global time, and directions to specific locations with a tap(Gangopadhyay and Mondal, 2016; Bulusu et al., 2001). Have you considered using an application that notifies you about storms or unpleasant weather that may come your way?

There are several weather forecasting apps on your smartphone. However, learning does not always occur after committing a mistake, sometimes we can benefit from learning from others. Various vehicles moving on the road will wirelessly communicate weather and road condition data, which includes air temperature, barometric pressure, light levels, motion, and other data required. It allows for more accurate forecasting and flexible real-time monitoring over multiple periods. Cars come with sensors on the windshield, wiper, and tires. With the help of IoT, these sensors can collect weather data, which can be analyzed in the cloud (Riquelme et al., 2009).

Meteorologists gathered data from weather sensors at airports and ships in the early days of computerized weather forecasting. The aviation and shipping industries, which may store infrastructure in enormous spaces, require the greatest weather data (Kumar et al., 2010 & 28. Cerpa and Estrin, 2004). To connect amongst these devices and transport data through the Internet, microwave waves are used. According to scientists, weather factors have an impact on these waves. Your phone network's Internet reception may be sluggish during inclement weather.

Mobile networks and Internet service providers supply this wave of activity data to weather firms. To obtain accurate weather forecasts, data is entered into these companies' systems. In short, IoT data is collected from mobile network towers, personal weather stations, traffic cameras, and vehicles. By accumulating all this data, hyper-local weather forecasts can be made more accurate.

METHODOLOGY

Wireless sensor networks and the Internet of Things are widely employed to address important difficulties in a variety of weather-sensitive areas. The numerous sensors communicate with various small computers, such as the Raspberry Pi, and data is collected using the Internet of Things. The data collected about various characteristics will be transferred to the cloud, where it will be examined and processed against

Figure 1. Generalized architecture of weather prediction

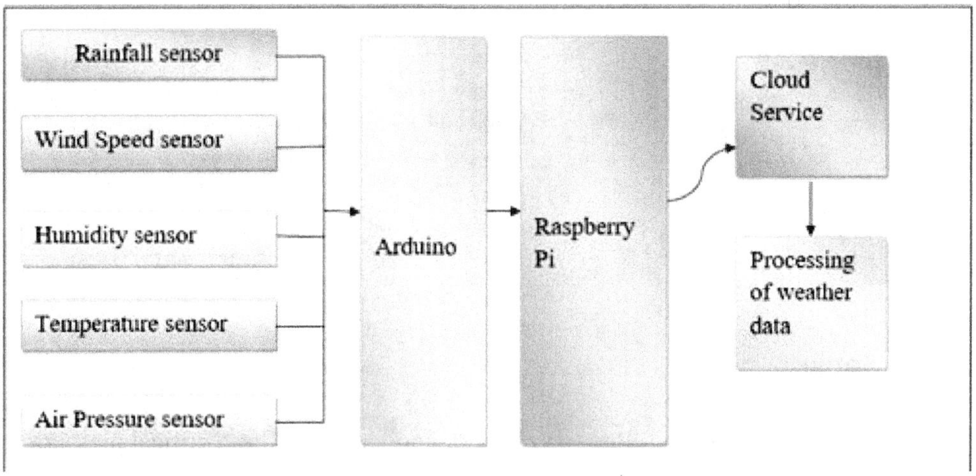

the appropriate climatic conditions. The data will then be used for weather prediction utilizing various application programming interfaces and Machine Learning techniques. The generalized architecture of weather prediction system is shown in given figure.

Major sensors which are used in weather forecasting are as follows.

Rain Sensor

Water is a vital need in everyone's existence, yet water conservation and good maintenance are critical. So here's a rain sensor that detects rain in the farm field and sends out an alert every time it rains, allowing us to take appropriate action.

Wind Speed Sensor

A wind speed sensor is a physical instrument that measures the speed of the wind. The top three wind cups rotate due to the wind generated by the airflow, while the internal sensing element is driven by the central axis to provide an output signal that may beused to figure out how fast the wind is blowing.

Humidity Sensor

The relative humidity of air is sensed, measured, and reported by the humidity sensor, which also detects the amount of water vapour present in a gas mixture (air) or pure gas. The process of water adsorption and desorption is linked to humidity sensing.

Temperature Sensor

A temperature sensor is a device that measures an object's temperature. The temperature of the air, the temperature of a liquid, or the temperature of a solid can all be considered. Temperature sensors come in a range of shapes and sizes, and they all use different technologies and concepts to detect temperature.

Figure 2. Weather forecast system data flow

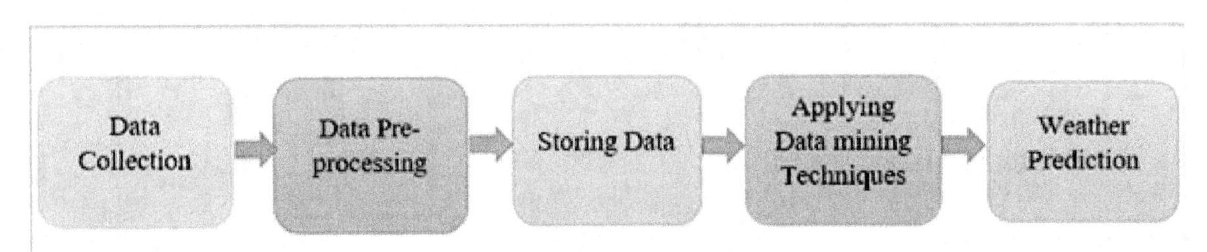

Air Pressure Sensor

In engineering, pressurised air is utilised for a variety of applications. These critical pressure levels can be measured and controlled using air pressure sensors. The measurement of atmospheric air pressure, often known as barometric pressure, is a separate issue that is described in another section.

The flow of the process is as follows:

Data from on-the-road vehicles is collected by IoT-enabled weather systems, which will wirelessly relay weather data such as air temperature, visibility or light, as well as any relevant information. This data aids in the construction of more precise forecasts and monitoring over a wide range of time frames. On the roads, sensors have been installed and also in the car. IoT-enabled sensors aid in the collection of weather data. IoT technology is advantageous to transportation system, farmersetc. Farmers can use to improve agricultural productivity and cost savings by cultivating vital actions to expand weather threats. Because it is so simple to go around weather forecasts, they will be more efficient and pose a lower risk of natural disasters. Thus data collection is the first step in the whole process.

After that preprocessing of the data take place because we require only valid data. The accuracy of the output depends on the input. This data should be in a format that allows the data mining process to work properly. After preprocessing the data is stored in database and data mining techniques are applied for predicting weather condition. In majority of the cases Chi Square and Naïve Bayes Algorithm are used. The model learns from the training set and this knowledge is used as data for test for weather prediction.

APPLICATION OF AN INTERNET-OF-THINGS (IOT) WEATHER FORECAST SYSTEM

1. Farmers might benefit from the IoT weather prediction system as well. In the subject of agriculture, weather forecasting is extremely significant.
2. The IoT weather prediction project has been shown to be quite useful in monitoring weather in locations such as volcanoes and rain forests. It is quite difficult for a human to dwell in such regions for an extended period of time, or even regions where there is a risk of radioactive leakage.
3. Logistics- For route planning, logistics will have more detailed weather information at their disposal.
4. Construction - Construction firms can predict when and where concrete should be poured, avoiding washouts and cost overruns.

5. Firefighters and energy companies - Firefighters and energy companies may organize their efforts based on weather forecasts that include micro variables like wind. This allows them to anticipate the paths that wildfires will take.
6. Major airlines - Airlines can plan routes, schedule events like deicing at airports, and eliminate dangers if they can obtain a more precise look at weather forecasts for the areas where they fly.

BENEFITS OF IOT WEATHER FORECASTING TECHNOLOGY

The accuracy of weather forecasting has a significant impact on different sectors of country which is a great deal, thus raising the necessity for a such type of system that facilitates more accurate monitoring and forecasting in real time (Lachure et al., 2015; Kiani et al., 2015; Gangopadhyay et al., 2016). Here are some industries that can benefit from IoT weather forecasting technology:

1. **Agriculture Sector –** Farmers are vulnerable since the activities of preparing the crops require soil, seeding, irrigation, harvesting, and storage are all directly influenced by the weather conditions. With the enhanced IoT technology, farmers will be able to increase crop fertility and cost, as well as diversifying weather hazards. As a result, farmers may use IoT technology to supply critical weather information via weather forecasting. Meteorological forecasts that are timely and accurate will increase productivity while lowering weather hazards.
2. **Transportation –** Uncertainty of unpleasant weather and associated risk factors are well known. When remote sensors are installed on a vehicle moving on the road, the system allows the reporting of real-time weather conditions to cover even minute details, such as road conditions, fog lightening, flooding, stormy and other conditions that increase the report's reliability and accuracy.

Therefore, the promotion of Smart and intelligent driving technologies will help improve driver safety and security by reducing road accidents caused by weather risks in a predictable manner. Various sectors of the economy rely on transportation, and if this is affected, the GDP will fall and the economy will slow. If an adverse economic impact occurs, it will take time to recover

3. **Other Industries –** Airline, energy, and construction industries are all influenced by adverse weather and are vulnerable to calamities.

CONCLUSION

Temperature, humidity, air pressure, wind direction, and rainfall are examples of climatic conditions in a given place at a given time interval. Weather monitoring and forecasting are playing an important role in the growth of an economy and it is a the major factors in the process of agriculture. The weather change in a zone is quick, so it needs to forecast. Weather forecasting requires the use of sensor devices in the environment for data collection and processing. Thus, the forecasting can be enhanced by using IoT and several other advanced technologies. The predicted weather conditions were used to alert the various person involve in different sector of economy affected by weather so that they can take proper step to avoid the loss of living and nonliving things. The accuracy of prediction is depending on how much earlier we are predicting. A short term prediction has more accuracy than long term prediction. Apart from this if two or more models may be employed for the same dataset, the accuracy of each model must be determined, as well as which model is best for predicting weather parameters. The number of parameters can be taken into account for increasing accuracy of prediction. Thus weather forecasting playing a significant role in human life.

REFERENCES

Asghar, M. H., Negi, A., & Mohammadzadeh, N. (2015, May). Principle application and vision in Internet of Things (IoT). In *International Conference on Computing, Communication & Automation* (pp. 427-431). IEEE.

Banara, S., Singh, T., & Chauhan, A. (2022, January). IoT Based Weather Monitoring System for Smart Cities: A Comprehensive Review. In *2022 International Conference for Advancement in Technology (ICONAT)* (pp. 1-6). IEEE.

Baste, P., & Dighe, D. (2017). Low Cost Weather Monitoring Station Using Raspberry PI. *International Research Journal of Engineering and Technology, 4*(5).

Bulusu, N., Estrin, D., & Girod, L. (2001). *Scalable coordination for wireless sensor networks: self-configuring localization systems.* In *International Symposium on Communication Theory and Applications (ISCTA 2001)*, Ambleside, UK.

Cerpa, A., & Estrin, D. (2004). ASCENT: Adaptive self-configuring sensor networks topologies. *IEEE Transactions on Mobile Computing, 3*(3), 272–285.

Chauhan, D., & Thakur, J. (2014). Data mining techniques for weather prediction: A review. *International Journal on Recent and Innovation Trends in Computing and Communication, 2*(8), 2184–2189.

Das, L., Kumar, A., Singh, S., Ashar, A. R., & Jangu, R. (2021, January). IoT Based Weather Monitoring System Using Arduino-UNO. In *2021 2nd International Conference on Computation, Automation and Knowledge Management (ICCAKM)* (pp. 260-264). IEEE.

Gangopadhyay, S., & Mondal, M. K. (2016, January). A wireless framework for environmental monitoring and instant response alert. In *2016 international conference on microelectronics, computing and communications (MicroCom)* (pp. 1-6). IEEE.

Keshavkumarsingh, S. (2013). *Design of Wireless Weather Monitoring System.* Department of Electronics & Communication Engineering, National Institute of Technology Rourkela.

Khan, Z. U., & Hayat, M. (2014). Hourly based climate prediction using data mining techniques by comprising entity demean algorithm. *Middle East Journal of Scientific Research, 21*(8), 1295–1300.

Kiani, F., Amiri, E., Zamani, M., Khodadadi, T., & Abdul Manaf, A. (2015). Efficient intelligent energy routing protocol in wireless sensor networks. *International Journal of Distributed Sensor Networks, 11*(3), 618072.

Kumar, D., Aseri, T. C., & Patel, R. B. (2010). EECHDA: Energy Efficient Clustering Hierarchy and Data Accumulation For Sensor Networks. *BIJIT*, *2*(1), 150–157.

Lachure, S., Bhagat, A., & Lachure, J. (2015). Review on precision agriculture using wireless sensor network. *International Journal of Applied Engineering Research*, *10*(20), 16560–16565.

Majumdar, P., Mitra, S., & Bhattacharya, D. (2021). IoT for Promoting Agriculture 4.0: A Review from the Perspective of Weather Monitoring, Yield Prediction, Security of WSN Protocols, and Hardware Cost Analysis. *Journal of Biosystems Engineering*, 1–22.

Mat, I., Kassim, M. R. M., Harun, A. N., & Yusoff, I. M. (2016, October). IoT in precision agriculture applications using wireless moisture sensor network. In *2016 IEEE Conference on Open Systems (ICOS)* (pp. 24-29). IEEE.

Nikesh, G., & Kawitkar, R. S. (2016). Smart agriculture using IoT and WSN based modern technologies. *International Journal of Innovative Research in Computer and Communication Engineering*, *4*(6), 12070–12076.

Ojha, T., Misra, S., & Raghuwanshi, N. S. (2015). Wireless sensor networks for agriculture: The state-of-the-art in practice and future challenges. *Computers and Electronics in Agriculture*, *118*, 66–84.

Pandey, A. K., & Mukherjee, A. (2022). A Review on Advances in IoT-Based Technologies for Smart Agricultural System. *Internet of Things and Analytics for Agriculture*, *3*, 29–44.

Radhika, Y., & Shashi, M. (2009). Atmospheric temperature prediction using support vector machines. *International Journal of Computer Theory and Engineering*, *1*(1), 55.

Rao, B. S., Rao, K. S., & Ome, N. (2016). Internet of Things (IoT) based weather monitoring system. *International Journal of Advanced Research in Computer and Communication Engineering*, *5*(9), 312-319.

Riquelme, J. L., Soto, F., Suardíaz, J., Sánchez, P., Iborra, A., & Vera, J. A. (2009). Wireless sensor networks for precision horticulture in Southern Spain. *Computers and Electronics in Agriculture*, *68*(1), 25–35.

Sampathkumar, A., Murugan, S., Elngar, A. A., Garg, L., Kanmani, R., & Malar, A. (2020). A novel scheme for an IoT-based weather monitoring system using a wireless sensor network. In *Integration of WSN and IoT for smart cities* (pp. 181–191). Springer.

Satyanarayana, G. V., & Mazaruddin, S. D. (2013, April). Wireless sensor based remote monitoring system for agriculture using ZigBee and GPS. In *Conference on advances in communication and control systems* (*Vol. 3*, pp. 237-241). Academic Press.

Shivang, J., & Sridhar, S. S. (2018). Weather prediction for indian location using Machine learning. *International Journal of Pure and Applied Mathematics*, *118*(22), 1945–1949.

Talegaon, N. S., Deshpande, G. R., Naveen, B., Channavar, M., & Santhosh, T. C. (2022). Performance Comparison of Weather Monitoring System by Using IoT Techniques and Tools. In *Intelligent Data Communication Technologies and Internet of Things* (pp. 837–853). Springer.

Wankhede, P., Sharma, R., & Pote, C. (2014). A review on weather forecasting systems using different techniques and web alerts. *International Journal of Advanced Research in Computer Science and Software Engineering*, *4*(2), 357–359.

Yerpude, S., & Singhal, T. K. (2017). Impact of internet of things (IoT) data on demand forecasting. *Indian Journal of Science and Technology*, *10*(15), 1–5.

Zhou, Y., Zhou, Q., Kong, Q., & Cai, W. (2012, April). Wireless temperature & humidity monitor and control system. In *2012 2nd International Conference on Consumer Electronics, Communications and Networks (CECNet)* (pp. 2246-2250). IEEE.

Chapter 4
Satellite Remote Sensing and Climate Behavioral Analysis

Mink Virparia
Pandit Deendayal Energy University, India

Rajeev Kumar Gupta
https://orcid.org/0000-0002-5317-9919
Pandit Deendayal Energy University, India

Ved Prakash Singh
https://orcid.org/0000-0002-2281-5687
Ministry of Earth Sciences, Bhopal, India & Indian Institute of Technology, Patna, India

ABSTRACT

In recent years, the most debatable topic across the world is climate change. The changes in the pattern of the season have been observed in different regions of the world due to weather change, and this issue has been taken on a serious note. Researchers are now taking data from satellites using remote sensing to analyze and prepare reports for predicting future disasters, which will be helpful to tackle calamities. Satellites like INSAT, IRS, Landsat 8, SARAL, Oceans 2, RISAT 2B, and many others have been launched for monitoring the soil conditions, land used, sea temperature, rainfall measurement, groundwater level, vegetation growth. Particularly, IRS satellites provide us the observation of land, water, crop growth, grassland, and INSAT satellites provide us the information of ocean temperature, outgoing long-wave radiation, cloud motion vectors, etc. This chapter discusses how remote sensing plays an important role in collecting data using satellites, how satellites monitor objects, and which satellite data is used for the climate change analysis.

INTRODUCTION

Climate change has emerged as the most contentious issue on the global stage in recent years. Changes in the seasonal pattern have been seen in several parts of the world as a result of climate change, and this

DOI: 10.4018/978-1-6684-3981-4.ch004

subject has been raised as a matter of considerable concern. The launch of satellites such as the INSAT, IRS, Landsat 8, SARAL and many more has allowed scientists and researchers to monitor soil conditions, land use, sea temperature, rainfall measurement, groundwater level, and vegetation development, among other things (Wirth, 2021). Then a database is made using the above data and it is used for the climate change in the specified region. Some studies that are made using this data include measuring the glaciers level in himalayan mountains, predicting the early flood, drought that can happen, Biodiversity mapping and regular weather reports. Remote sensing which has played a main role in detecting and monitoring the objects, land, water and grassland without direct contact with it. Therefore, it is called a replacement for onsite surveillance. It is used in different areas like water management, controlling environment issues, aerology, climatology, geoscience as well as military purposes, war situations, human applications. Remote sensing is generally satellite based; it can be used to discover the target bodies of the Earth using electromagnetic rays. There are two types of techniques: Active and Passive, which are based on satellite and aircraft. Active technique refers to a signal that is released by the satellite or by an aircraft, which is later returned by an object and it is detected by the sensor. In passive technique, sensors have to detect the radiation which is released by the object or nearby areas (Remote Sensing, 2021). Therefore, in current time, remote sensing has been numerously used in climate change analysis and it permits us to examine the surface of earth, ground water level, mountain peaks, etc.

There are various examples where remote sensing has achieved better responses are water management, land preservation, advancing in agriculture, tackling fire problems in the forest using real-time decision-making systems, observing land usage after tragedy events, assessment of carbon and related important elements, simulation of climatic change, glaciers monitoring, evaluation of ocean level (Satellite Data, 2021). Geographic information systems using satellite technology are used to build before time alarms which can lower the climate related calamities (for example developing a greater prediction system to tackle hurricanes, fire events), also it can be useful to develop predeveloped actions. Satellite image from remote sensing used for holocaust destruction detection by using relative analysis of damage areas on and before the calamity. Now we will see the details of satellite data and its properties which will be used in climate change analysis and later we will see areas in which satellite data from remote sensing is actually used in real life.

ROLE OF SATELLITE DATA

Nowadays data from the satellite is in huge demand in the research area. So, it is very necessary to launch satellites in orbit to generate a large amount of data of the land, water, forest and glaciers. We all know that to collect data of various kinds we have to put many satellites in different types of orbits around the Earth. Therefore, there are specifically three classes of the orbit which includes low-Earth orbit (160 to 2000 km above the earth), medium-Earth orbit (2000 to 35,550 km above the earth) and high-Earth orbit (above 35,500 km from orbit). Satellites which are orbiting at a distance 35,786 km from the earth at altitude level and match the orbital speed of the earth, are known to be in geosynchronous orbit (GSO). Moreover, satellites which are directly over the equator will be orbiting in geostationary orbit. Roger Saunders (Saunders, 2021) discussed the use of satellite data in weather forecasting. Geostationary orbit gives benefits to satellites to continue its position directly over the same place at earth's surface. Now we explore three classes of the orbit of earth.

Figure 1. Low Earth orbit satellite rotation (Source: European Space Agency)

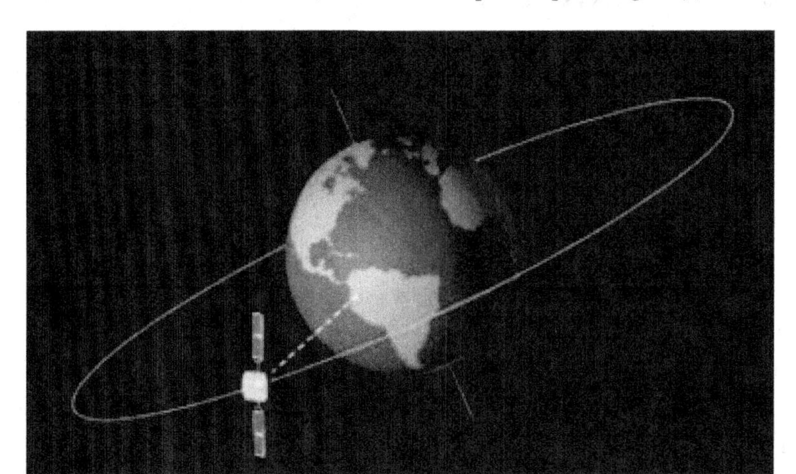

Generally Low Earth orbit is used for satellites which want to follow the specific orbital track around the earth (Thies and Bendix, 2011). Polar orbiting satellites are the example of this class because they are nearly inclined 90 degrees to the equatorial plane and can travel pole to pole as earth revolves. This gives satellites to collect data from the earth for the different regions.

Satellites which are in medium earth orbit take roughly 12 hours to complete the orbit. Therefore, we can say that a satellite meets two spots on the equator every day. Satellites which are used for telecommunication and GPS tracking are generally used in this orbit because medium earth orbits are consistent and foreseeable.

As geosynchronous and geostationary orbits are found at 35,768 above the planet, generally called as high earth orbit, so satellites which revolve around that geosynchronous orbit are found below or above the equator. Satellites in geostationary orbits are found to be in the same plane as the equator. Satellites take similar images of the earth from rotating in their respective orbits and provide us constant coverage of the particular area.

Figure 2. A satellite revolving in high Earth orbit (Source: Earth Data by NASA)

RESOLUTIONS OF THE SATELLITE

We know that to capture an image from a satellite, resolution has a vital role in how data from the sensor can be used for further applications. Resolution of the satellites depends on two things, one is how far the satellite is from the earth and second is the design of the sensor. Therefore, based on the resolution of the satellite, there are four types of resolution when we review a dataset: Spectral, Radiometric, Temporal and Spatial.

- Spectral resolution has a power to discern the wavelengths which have narrower bands. Therefore, based on the number of bands, there are two types of spectral resolution. First is multispectral which has 3 to 10 bands and Second is hyperspectral which has 100 or even 1000 of bands (Satellite Data, 2021). So we can say that narrower the wavelength of a band, finer the spectral resolution.
- Below given cube defines detail inside the data. In this detail, Distinguishing can be done between rock and foil types, herbage types and other elements. In the below image, the right corner, which has smaller parts of slightly higher response, has a red portion of clear spectrum (nearly 700 nm). It happened because of the presence of the 1 cm long red brine shrimp at the melting pond.

Figure 3. Image of spectral resolution of Earth (Source: Earth data by NASA)

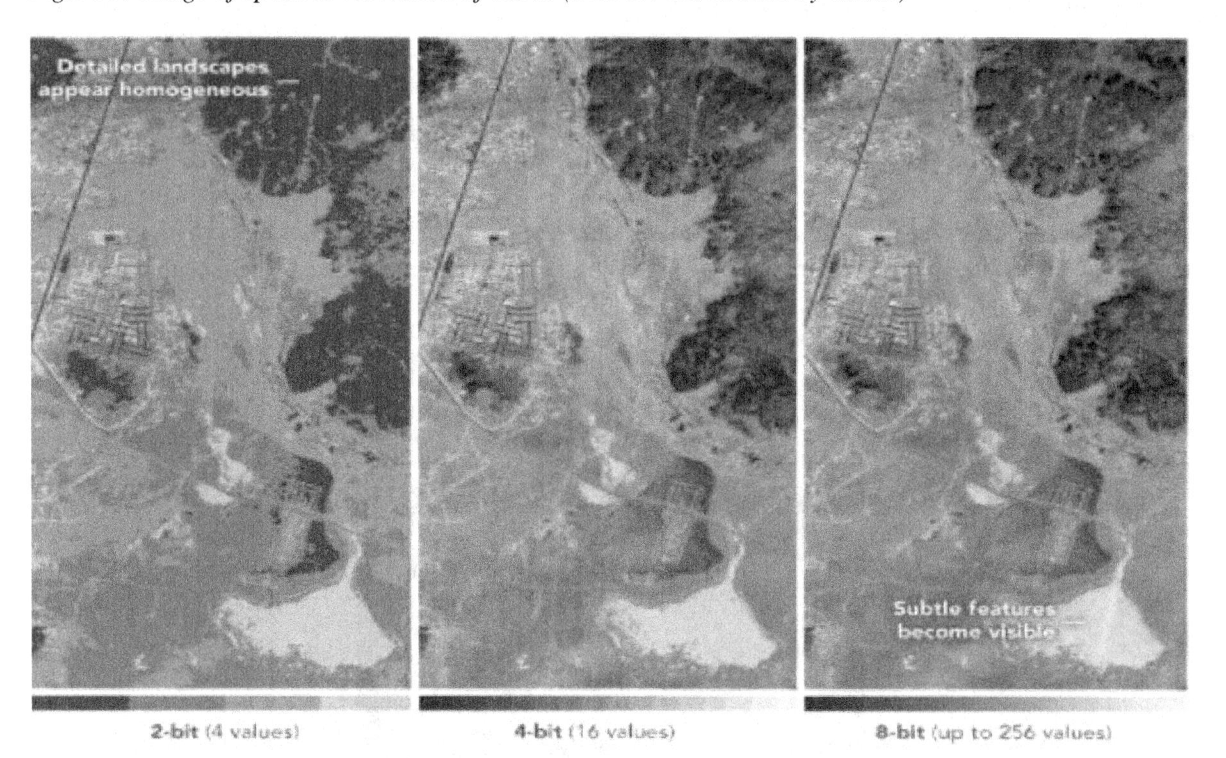

- Radiometric resolution represents the quantity of the data in each pixel, which is the number of bits that constitute for energy recorded. Each bit in this record is an exponent of the power 2. Suppose a 8 bit resolution is 28, that specifies that the sensor contains 256 digital values to store information (Remote Sensing, 2021). Therefore, it can dictate that higher the resolution, many values will be available to preserve the information by giving a good distinction between slightest differences in energy.

Figure 4. Image of radiometric resolution (Source: Earth data by NASA)

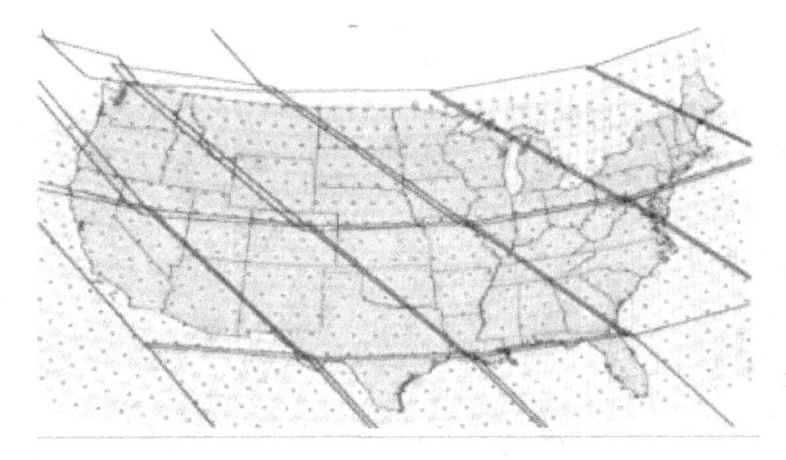

- Temporal resolution of a satellite is the time taken to revolve an orbit and revisit that same surveillance area. Such resolution depends on three things, sensor's features, orbit and swath width. We know that a geostationary satellite takes the same time as earth takes for completing one rotation, therefore, in such cases temporal resolution will be finer (Satellite Data, 2021). But polar orbiting satellites have a 1 to 16 days temporal resolution.

Figure 5. Image of temporal resolution (Source: Earth data by NASA)

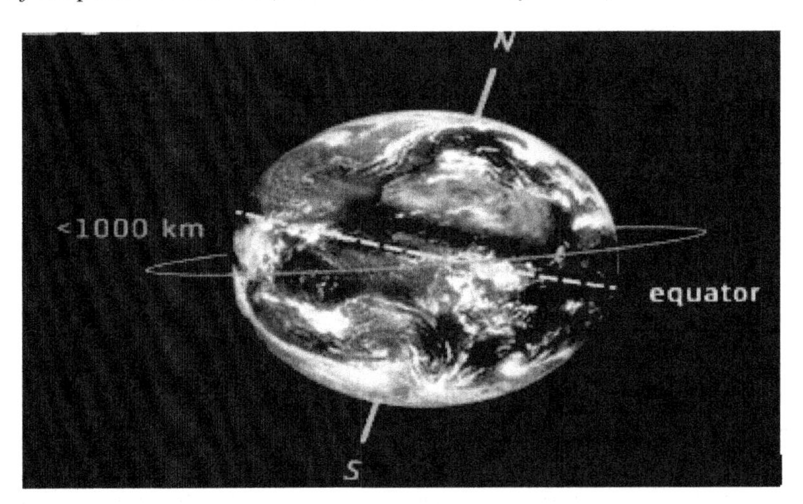

- This resolution is described from the size of the pixel which is in the digital image and area on earth surface represented by that pixel. Moderate Resolution Imaging Spectroradiometer (MODIS) examines the band and gives spatial resolution of 1 km with every pixel representing 1 km x 1 km area on the ground. It can also cover bands with a spatial resolution of 250 m or 500 m (Remote Sensing, 2021). Therefore, we can say that the more clear the image, the more detailed information will be seen. Below photo is an example of the aforementioned theory, it has a 30 m/pixel image (Left side image), 100 m/pixel image (centre image) and 300 m/pixel image (right side image).

Figure 6. Image of spatial resolution (Source: Earth data by NASA)

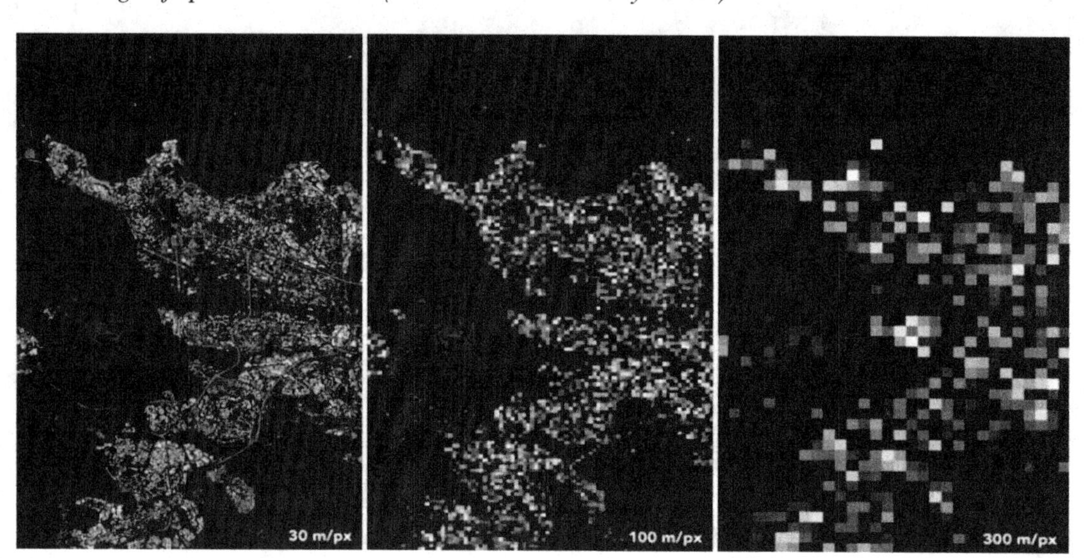

INTERPRETATION AND ANALYSIS OF SATELLITE DATA

Satellite data from remote sensing should be required to be pre-processed and analyzed before that data is given to researchers for the study purpose. Generally raw data from satellites are to be processed, For example if we have taken NASA Earth observation data then it is processed by NASA's Science Investigator Processing system. Data is processed to at least level 1, but most are associated at level 2 (geophysical variables) and level 3 (it is assigned to space-time grid scales) product. Also, there is a lot of data which are of level 4. This satellite data from remote sensing are stored in distributed Storage Centers in proper manner and it is available openly without any curtailment for research purposes.

Hierarchical Data format and Network Common data form format are other formats in which data can be stored in centres. Also, there are many tools which are used to convert, subgroup, visualize and export to any other file format. When data is ready for use then it can be applied in various fields like water management, cultivation, health sector and air quality checker. We all know that one sensor cannot be used for research, therefore a user needs many sensors and huge data that can answer their question with different resolutions which are discussed above.

MACHINE LEARNING AND DEEP LEARNING USING SATELLITE DATA

Bogdan Bochenek et al. (Bochenek, 2022) discuss the study of approx. 500 articles in the domain of geospatial data. According to this research, satellite data can be extremely beneficial in the development of machine learning and deep learning-based solutions to deal with a variety of real-world problems such as weather forecasting, earthquake prediction, flood prediction, and other similar issues. Waytehad Rose Moskolaï (Moskolaï, 2021) perform the study of satellite data in Deep learning-based solutions. According to this study, satellite data can be used for a variety of purposes other than weather forecasting, including monitoring and anomaly identification. Specifically, this study divides the deep learning approach into three categories: the feed forward model, the recurrent neural network (RNN), and the hybrid approach. In light of the fact that satellites provide sequential data, RNN produces superior outcomes when compared to a machine learning-based technique. A rain prediction model at the given location by using satellite image is introduced in (Wirth, 2021)

AVAILABILITY OF THE DATASET

The main thing which is required for studying climate change analysis is the high quality dataset of the climate, also the biggest challenges we are facing today are best realized through satellite data. Remote sensing analysis needs the data with the designated resolutions, locations, sensors and also it should be free of cost. There are many free sources present on the internet for the study and analysis purpose of climate change. Some of the sources are NASA Earth Observation (NEO), USGS Earth Explorer, ESA's Sentinel Data, NASA Earth Data, NOAA Class, NOAA Digital Coast, IPPMUS Terra, Lance, Vito Vision and Bhuvan India Geo Platform (Dataset, 2017). In these different sites, we get different types of authoritative datasets on land, ocean, energy, elevation, atmosphere, biosphere and cryosphere in daily, weekly, monthly and yearly frequency. These datasets are available in the form of JPEG, PNG, Google Earth and GeoTIFF. Datasets have a broad range of satellite, Aerial and LiDAR images, also they have a vast range search feature which lets you find your desired dataset from the site. Some of the platform provides services like Hyperion's hyperspectral data, Disperse radar data and MODIS land surface reflectance (Dataset Sources, 2021). In every platform, it is compulsory to make registration to download the dataset due to privacy issues. Therefore, there are various sources which are available for the dataset from which can get their required dataset for the research or analysis purpose and these datasets are preferred for large-scale applications that don't require finer details.

USE CASES OF REMOTE SENSING DATA FROM SATELLITE IN CLIMATIC CHANGE

The remote sensing data from the past and present is used for the examination of the spatiotemporal patterns by the environmental components and effects of living beings over the years. Remote sensing data from satellite is used for environmental data accretion for local and compact level (Eniolorunda, 2014). Therefore, in this section we will explore some of the areas where remote sensing data from satellites are used for the climatic behavior analysis.

- Fallback of many glaciers from the worldwide in recent years are observed and it is an obvious sign of the climatic change due to global warming. Quick stagnation of the glaciers, permafrost and ice gaps have been observed in regions like Mt kilimanjaro, US, Himalayan ranges, North America, Europe, Russia with accelerated environmental outcome. Some problems with disastrous release of water, menace of instability of mountain ranges, damage in the resident buildings and roads in the snowy areas are observed. Therefore, Satellite images of the glacier are very useful for scientists to monitor the growth and stagnation of the snow (Yang et. al, 2013). Because snowy regions where glaciers are found is also a prime essential of the Earth and it should be taken care of.
- Some studies states that using Global Land Ice Measurements from Space (GLIMS), we can build a database which contains image of glaciers like Ice sheets, ice caps, nearby ranges and then we can examine that image to produce meaningful information by various methods such as pre-programmed algorithms and manual simplification (Manikiam and Kamsali, 2015).
- Many studies have been proven that glaciers can be monitored by GLIMS (remote sensing) and field investigations. In some areas, glaciers have been pulled off naturally and were detected by satellite and field investigations. Therefore, from this we can say that remote sensing data is also used in this area and it is also easy to get information about the glaciers from the satellite.
- Hence, remote sensing data from satellites has been used to measure massive glacier movements, snow depth measurement, snow sliding in the glaciers, predicting calamities that can happen from glaciers.
- Vegetation is the one of the important parts of the earth where every living and non-living thing are dependent on it. Because it provides meals and shade for everyone. It is increasingly endangered mostly due to climate influences and anthropogenic. Forests throughout the world are increasingly appearing finite, vulnerable, dangerously finished and possibly already are subject to irreparable damage. Therefore, ecologists and biogeographers have an intellectual challenge to understand vegetation spatio-temporal patterns. Hence, remote sensing has been used to give vegetation changes that happen due to climatic change. One research was made in northeast china between 1988 to 2001, In that, using Landsat satellite data they have checked vegetation between grass and land surface. Different pre-processing and analyzing has been made and the result shows that land has been damaged due to irregular water supply and at last result was found to be water storage by climate change.
- For human survival, it is important to develop such strategies which can aim to prevent vegetation degradation because we all know that carbon is the main element on the earth and food, fuels and other things where carbon is used are part of the carbon cycle. If we don't maintain carbon then it leads to food insecurity, difficulty in living. So remote sensing is appropriately used for characterising vegetation phenology.
- Food insecurity will arise if climate change affects main biophysical components like quality of the soil, water accessibility and uneven climate behavior. Therefore, remote sensing data was found to be useful in managing irrigation, characterizing the soil type, crop condition assessment, finding well suited land for crop and agro climatic assessment, etc.
- Crop conditions can be assessed using satellite data in real time. During the 2000-2001 dry season of kaithal of haryana, india, researchers have used satellite images from Landsat7 to observe some data like properties of soil type, depth of the water, quality of the water, To build irrigation man-

agement methods. Results conclude that because of lesser water in that area, wheat crops are not growing properly so it should be taken care of in a concurrent manner (Yang et. al, 2018).

- Increase in the rise of flooding for the people around the world is due to climate change and it started a spark inside the researchers to forecast flooding events therefore it can save everyone in future. In one of the research papers, a theory claims that remote sensing data from satellites is very useful for the prediction of the flood happening zones, damage from flood and build strategies to tackle this scenario which can destroy things like roads, buildings, areas near sea,etc. Therefore, using satellite data, it can be used for safety management and decision making for the future.
- Researchers have examined a region called piemonte in italy during 1994. In which, they have used SAR (Synthetic aperture radar) images and also some local region map information to search the areas which can be affected by flood in the future. After some simplification and analyzing the images and area, they have found that 96.8% of the region can be wiped out if flood arrives. Therefore, from this research, we can say that satellite images are very useful to control this type of circumstances.
- Some of the scientific evidence has proven that climatic change has an impact on health and later it can cause various diseases from events like extreme temperature, flooding, desertification (Yang et. al, 2013). Disease like food poisoning, skin cancer, psychological stress, insect diseases due to increases in files are happening due to climate and temperature change. Therefore, remote sensing data plays an important role to overcome these situations by monitoring ecological components that have been detected remotely.
- Once a survey was made on malaria disease that has affected a region named Kisumu in kenya. They have used six similar variables that are similar to malaria and can be monitored using satellite data. So, they have used a technique called principal component analysis (PCA) on the data of those variables to find out existing factors of malaria in those areas. Result shows that using satellite data we can control the spread of the disease by taking proper precautions to tackle such situations and maintaining a proper environment.
- Another similar study was carried out on malaria disease in dry regions of the Sahara desert specifically on hilly areas of Mali. Results reflect that malaria disease remained steady in the villages where there was a ubiquity in the hotter dryer part than cold part. Also disease was found to be incomparable when there was a switch in the seasonality in that area.

CONCLUSION

In this chapter, we have demonstrated how data from satellite make crucial contributions to understand climate change behaviour. Also, satellite data from remote sensing provides us with accurate facts and figures of the specific region on a regular basis and on a larger scale. This knowledge of remote sensing data from satellite enhances the understanding in other areas like Energy monitoring, Sea traffic, Coastal Security, Land Use, Planning and Preservation of Forest, etc. In the upcoming time, remote sensing will be used on a quotidian basis for the monitoring of climate change and that includes numerical weather prediction, observation of functional missions of climate change. On the other hand, it can also be applied on building systems that have an assurance of making long term climatic dataset.

REFERENCES

Bochenek, B., & Ustrnul, Z. (2022). Machine Learning in Weather Prediction and Climate Analyses—Applications and Perspectives. *Atmosphere*, *13*(2), 180. doi:10.3390/atmos13020180

Dataset Sources. (2021). https://gisgeography.com/free-satellite-imagery-data-list/

Datasets. (2017). https://geoawesomeness.com/list-of-top-10-sources-of-free-remote-sensing-data/

Earth data by NASA. (2021). https://earthdata.nasa.gov/learn/backgrounders/remote-sensing

Eniolorunda, N. (2014). Climate change analysis and adaptation: The role of remote sensing (Rs) and geographical information system (Gis). *International Journal of Computational Engineering Research*, *4*(1), 41–51.

European Space Agency. (2020). https://www.esa.int/ESA_Multimedia/Images/2020/03/Low_Earth_orbit

Manikiam, B., & Kamsali, N. (2015). Climate Change Analysis Using Satellite Data. *Mapana Journal of Sciences*, *14*(1), 25–39. Advance online publication. doi:10.12723/mjs.32.4

Moskolaï, W. R., Abdou, W., Dipanda, A., & Kolyang. (2021). Application of Deep Learning Architectures for Satellite Image Time Series Prediction: A Review. *Remote Sensing*, *13*(23), 4822. doi:10.3390/rs13234822

Remote Sensing. (2021). https://earthdata.nasa.gov/learn/backgrounders/remote-sensing

Satellite Data. (2021). https://www.iceye.com/satellite-data

Saunders, R. (2021). The use of satellite data in numerical weather prediction. *Weather*, *76*(3), 95–97. doi:10.1002/wea.3913

Thies, B., & Bendix, J. (2011). Satellite based remote sensing of weather and climate: Recent achievements and future perspectives. *Meteorological Applications*, *18*(3), 262–295. doi:10.1002/met.288

Wirth, P. (2021, November 19). *Predicting rain from satellite images.* Medium. Retrieved from https://towardsdatascience.com/predicting-rain-from-satellite-images-c9fec24c3dd1

Yang, J., Gong, P., Fu, R., Zhang, M., Chen, J., Liang, S., Xu, B., Shi, J., & Dickinson, R. (2013). The role of satellite remote sensing in climate change studies. *Nature Climate Change*, *3*(10), 875–883. doi:10.1038/nclimate1908

Yang, W., John, V. O., Zhao, X., Lu, H., & Knapp, K. R. (2016). Satellite climate data records: Development, applications, and societal benefits. *Remote Sensing*, *8*(4), 331. doi:10.3390/rs8040331

Chapter 5
Tools and Techniques to Implement AIoT in Meteorological Applications

Jayashree M. Kudari
https://orcid.org/0000-0003-2720-8250
Jain University (Deemed), India

M. N. Nachappa
Jain University (Deemed), India

Bhavana Gowda
Jain University (Deemed), India

Adlin Jebakumari S.
Jain University (Deemed), India

Smita Girish
https://orcid.org/0000-0002-0337-3625
Jain University (Deemed), India

Sushma B. S.
Jain University (Deemed), India

ABSTRACT

Weather forecasting is an important role in meteorology and has long been one of the world's most systematically difficult problems. This plan deals with the structure of a weather display system that may be built by electronics hobbyists utilizing low-cost components. Severe weather occurrences present a challenging forecasting problem with just a partial explanation. Developing communication methods makes it possible. Technology-enabled applications provide severe weather alerts and advisories. The airline industry is highly sensitive to the weather. Accurate weather forecasting is crucial. The IoT enables agriculture, notably arable farming, to become data-driven, resulting in more timely and cost-effective farm production and management while lowering environmental impact. Applying AIoT and deep learning to smart agriculture combines the best of both worlds.

INTRODUCTION

Artificial intelligence (AI) technology is combining with the framework of IoT to improve AIoT, human-machine interactions, managing the data, and analyzing the data for IoT operations. AI might be used to transform IoT data into actionable information for enhanced decision-making, opening the path for

DOI: 10.4018/978-1-6684-3981-4.ch005

future technology like IoT Data as a Service (IoT DaaS). Through machine learning capabilities, Artificial Intelligence provides value to AI, and through connectivity, signaling, and sharing the data, IoT adds value to AI. Artificial IoT is transformative and both technology benefits from each other. The amount of unstructured data generated by humans and machines will grow as IoT networks spread across various industries. The Internet of Things (IoT) is a precise response to current operational difficulties including the high cost of good human capital.

THE ADVANTAGES OF AI-POWERED IOT

Businesses and consumers gain greatly from artificial intelligence in the Internet of Things. Some of the significant commercial benefits of merging two disruptive technologies are given below.

1. **Improving Operational Effectiveness:** IoT Enhances Operational Effectiveness Continuous streams of data are analyzed by AI, which uncovers patterns that basic gauges overlook. Machine learning combined with AI can also predict operation conditions and point out factors that need to be tweaked for the best results. Intelligent IoT will eventually indicate which procedures are abandoned and inefficient, as well as which duties may be fine-tuned to increase efficiency. Google, for example, uses artificial intelligence and the Internet of Things to cut data center cooling expenses.
2. **A better risk management system:** By merging AI with IoT, Businesses will be able to better comprehend and forecast the future, a wide range of threats, in addition to actions that are automated. As a result, they're more equipped to handle financial losses, personnel safety, and cyber risks. Fujitsu, for example, uses AI to analyze data from connected wearable devices to ensure worker safety.
3. **Starting the process of developing new and improved products and services:** People's ability to communicate with machines is improving thanks to NLP (Natural Language Processing). Combining IoT and AI may, without a doubt, assist businesses in developing new products or improving existing ones by allowing them to manage and analyze data more quickly. For instance, In the development of IoT-enabled airplane engine maintenance for Rolls-Royce plans. This method will aid in the detection of trends and the discovery of operational insights.
4. **Increase the scalability of the Internet of Things:** IoT devices include mobile phones, high-end computers, and low-cost sensors. Low-cost sensors generate huge amounts of data in the most common IoT ecosystem, on the other hand. An AI-powered IoT ecosystem analyses and summarizes data before passing it from one device to another. As a result, it compresses massive amounts of data into digestible chunks and enables the connection of a large number of IoT devices. The scalability of devices is the term for this. This is referred to as the scalability of devices.
5. **Saves money by reducing the expense of unplanned downtime:** Equipment failure can result in costly unscheduled downtime in a range of industries, including offshore oil and gas and industrial production. Predictive maintenance, which utilizes AI-enabled IoT, allows you to anticipate equipment failure and schedule routine maintenance procedures in advance. You'll be able to escape the negative impacts of downtime as a result. According to Deloitte, AI and IoT produce the following results: 20 percent to 50 percent reductions in the amount of time they spend planning maintenance. Reductions in the amount of time they spend planning maintenance of 20 percent to 50 percent.

The availability and uptime of equipment will increase by 10% to 20%. The cost of maintenance will be lowered by 5% to 10%.

BACKGROUND WORK

According to the authors Bauer, P., Thorpe, A., and Brunet G., (2015) weather (and climate) predictions are currently based on purely physical computer models in which the governing equations of the atmosphere and ocean, or our best approximation thereof, are solved on a discrete numerical grid (2015). In general, this strategy has shown to be rather effective. According to the researchers Vogel, P., Knippertz, P., Fink, A. H., Schlueter, A., and Gneiting, T., (2018) numerical weather prediction (NWP) model yet to have problems in various critical applications, such as forecasting mesoscale convective storms over Africa (2018). Furthermore, as illustrated on the website physics.ao-ph, massive quantities of computer power are required, particularly when producing approximate forecasts typically limited to fifty group members or less (2020). As a result of these. Because of these characteristics, as well as the reputation of machine learning (ML), more people are interested in employing data-driven methods to improve and gear up NWP.

Machine learning can be applied in a variety of methods to forecast the weather. The prediction of variables not directly generated by the somatic model – and the correction of statistical biases in somatic model output are two important applications of machine learning. (Gagne et al., 2014; Taillardat et al., 2016; Rasp and Lerch, 2018; Lagerquist et al., 2017; Rasp and Lerch, 2018; McGovern et al., 2017)

Although a somatic model still foresees the broad history of the atmosphere, these methods usually concentrate on very specific variables or regions. (Shi et al., 2015; Grönquist et al., 2021) illustrate short-range of 6 hours precipitation assumption using direct extrapolation of radar readings without the usage of physical model (Shi et al., 2015, 2017). Hybrid modelling, which blends a physical model with data-driven methods such shifting experience clouds or energy parameterizations, is another field of ML research. (Chevallier, et al., 1998).

The main idea behind these approaches is to only use machine learning emulation to replace uncertain (e.g. clouds) or computationally expensive model components, while leaving other model components (e.g. large-scale dynamics) alone. On the other end, hybrid vehicles have a variety of disadvantages. It's probable that the link between somatic and machine learning modules, as well as the drawbacks of uncertainty and biases, are underappreciated. Another issue is that they are difficult to implement since machine learning components must be integrated with compound climate model snippets written in Fortran. (Brenowitz and Bretherton, 2019).

The forecast range plays a significant role from a social perspective since it facilitates essential information for calamity preparedness, such as flooding, cold and hot times, and destructive winds (Lazo, Morss, and Demuth, 2009). It is required to figure out difficult atmospheric forecasts at work and the interplay between many variables over a range of scales in order to provide a solid medium-range forecast. This distinguishes it from post-processing and statistical forecasting, which use a physical model to forecast large-scale dynamics, and now casting, which uses a univariate, short-term evolution. To put it another way, this benchmark closely mimics the task that physical NWP models execute. There are a number of strong reasons to consider implementing a data-driven strategy.

The forecast range is vital from a societal standpoint, according to Lazo, Morss, and Demuth, since it gives crucial information for disaster planning, such as flooding, cold and hot times, and destructive winds (2009). To offer a credible medium-range forecast, it is necessary to understand sophisticated atmospheric dynamics and the interplay between several variables over a range of scales. This distinguishes it from post-processing and statistical forecasting, both of which employ a physical model to forecast large-scale dynamics, and now casting, which employs a single-variable, short-term evolution. This benchmark, to put it another way, closely resembles the task that physical NWP models perform. There are a slew of compelling reasons to pursue a data-driven strategy.

Despite the fact that the performance is complex and nonlinear, (Hamill and Whitaker, 2006), claim that it does exhibit some consistent patterns on small scales. (Ronneberger, Fischer, and Brox, 2015). Finally, the proposed benchmark provides useful trials for deep learning algorithms, such as comparing various architectures (Ronneberger et al., 2015; Huang et al., 2016). Many researchers have established data-driven, universal, medium-range weather prediction in recent years.

Researchers approve this technology has some promise, but they all emphasize the importance of more investigation. Right now, No shared benchmark challenge to help to accelerate the progress. Standard datasets have a huge influence because they compare various approaches empirically and encourage healthy competition, which is especially significant in a new field of study. MNIST (LeCun et al., 1998) and ImageNet (LeCun et al., 1998) are two well-known computer vision datasets (Russakovsky et al., 2015). In addition, well-edited standard datasets make it simple for experts from numerous domains to cooperate on an issue (Ebert-Uphoff et al., 2017).

AUTHORIZING AI TO TAKE CARE OF WEATHER FORECASTING

Weather forecasting is traditionally focused on developing composite numerical models in order to make more exact estimation. Because of weather uncertainty and model flaws like coordinate compatibility, this technique may fall short of meeting the goals of many use cases. To reduce the gap, artificial intelligence (AI) and data-driven solutions have been used.

In 1980 neural networks were implemented, artificial intelligence (AI) was introduced in weather forecasting. AI prototypes have been achieved in various industries in recent era, meteorologists are now using the technology into various domains such as satellite data processing, casting, Storm weather forecasting, and other profitable and eco-friendly analytics.

As per the journal Earth and Space Science, AI technologies are life-threatening for sinking the workload of human forecasters while supplying more accurate and timely predictions. The view of the US National Oceanic and Atmospheric Administration (NOAA) conveys that using AI and machine learning to forecast extreme weather like thunderstorms and hurricanes enhances exactly.

In this industry, Google is a prominent indicator. In December of 2019, the tech giant discovered new investigation on the construction of deep learning models for precipitation forecasting. The team used the widely used UNET convolutional neural network to address anticipating as an image-to-image translation problem. During an encrypting phase, layers iteratively reduce the resolution of images passing through them, and during the succeeding decoding phase, the low-dimensional depictions of the image formed during the encoding phase are enlarged back to higher resolutions.

Large firms are also banding together to profit from the current trend. IBM has purchaed the Weather Company in the year 2015. The Atmospheric Forecasting System was created by combining the two firms' technology and expertise in meteorological data, and it provides consumers all over the world with individualized, actionable knowledge. The system employs machine learning techniques to deliver a variety of prediction facilities, including 72-hour forecasts for weather-related power disruptions. To accommodate the better resolution and more frequent updates, the method is stated to be the first-ever working global weather model to run on GPU-accelerated servers.

The climatological agency in Shenzhen has been inspecting numerous methods to progress weather forecasting in Guangdong's challenging coastal region, which is prone to powerful convective weather. The agency collaborated with Huawei to develop a climatological cloud platform that combines 5G and AI to reduce forecast model development, training, and placement time from 2-3 days to just a few hours.

Entrepreneurs are gradually offering themselves as game-changers in their respective diligences. ClimaCell's MicroWeather engine advances weather forecasting accuracy by using machine learning to historical gridded weather data. The company just made available a historical weather data archive acquired from a global network of wireless signals, connected automobiles, planes, street cameras, drones, and other internet of things (IoT) devices for AI model training.

Despite the fact that artificial intelligence (AI) will continue to play a significant role in weather forecasting, attracting AI talent to the sector has proven difficult. IT firms specializing in showy fields such as computer vision and self-driving cars pay far too much for meteorological services to compete. Instead, internet behemoths are creating alliances with local meteorological organizations or taking over the work themselves. On a worldwide scale, we may see more of these types of agreements in the future.

AIoT APPLICATIONS

Applications of AIoT are now focused on repairing cognitive computing in consumer products, and many are retail goods centered. Smart home technology, for example, because smart equipment learns from human contact and response, it would be designated as AIoT. AIoT technology balances machine learning with IoT networks and systems to generate data "learning machines" in terms of data analytics. This is applied to enterprise and industrial data on use cases at the network's edge, such as automating tasks in a connected workplace. All AIoT use cases and solutions require real-time data. For HR professionals, AIoT systems, for example, might be combined with social media and HR-related platforms to establish an AI Decision as a Service function. Project financing is increasing for AI-focused IoT start-ups. The number of IoT start-ups focusing on AI is growing. IoT stage providers include Amazon, GE, IBM, Microsoft, Oracle, PTC, and Salesforce. Industries are working on incorporating AI abilities. Large organizations in a variety of industries are even now utilizing or exploring the power of AI in conjunction with IoT to deliver new contributions and drive more proficiently. AI will be used in 80 percent of IoT projects by 2022 said by Gartner.

THE AI KEY TO UNLOCK IoT POTENTIAL

In Internet of Things (IoT) applications and deployments AI plays a significant role. Both reserves and endeavors in organizations that combine AI and IoT have increased in the last two years. As outcome, primary IoT platform software suppliers include AI competences such as machine learning-based analytics in their contributions. The ability of artificial intelligence (AI) to rapidly extract perceptions from data, as well as its application in this perspective. Artificial intelligence (AI) technique which allows smart sensors and devices to identify patterns and perceive the changes in factors such as temperature, pressure, humidity, air quality, vibration, and sound. Traditional business intelligence tools, which often seek for numeric inceptions to be passed, can generate working forecasts up to 20 times faster and with more accuracy than machine learning systems. Rest of AI technologies, such as speech recognition and computer vision, can assist in data mining tasks that previously required human review. Businesses may use AI applications for IoT to reduce inadvertent downtime, improve operational efficiency, develop new goods and services, and better manage threat.

AVOIDING EXTREMELY EXPENSIVE UNPLANNED DOWNTIME

In some industries, such as industrial manufacturing and offshore oil & gas, an unintentional interruption caused by equipment failure can be costly. Predictive preservation, which employs analytics to predict equipment breakdown in advance so that maintenance procedures can be scheduled in a timely manner, can help to reduce the expenses of unplanned downtime. Machine learning allows for the identification of numerous patterns in data supplied by current technologies in order to predict equipment failure. According to Deloitte, predictive preservation can cut maintenance schedule time by 20–50%, improve equipment uptime by 10–20%, and lower range of upkeep outlays by 5–10% in business.

ENABLING NEW AND IMPROVED PRODUCTS AND SERVICES

IoT and AI can lead to the development of new products and services. NLP performance has increased. It also enables machines to converse with one another without the intervention of a human. Without the need for human intervention, drones and robots controlled by AI and IoT technology can be employed.

Artificial intelligence has altered the way marketable vehicles are examined at sea. Artificial intelligence has changed the way commercial vehicles are monitored at sea. Every quantifiable data point in a fleet of planes, trains, trucks, or cars in real time to advance routing and planning and prevent unexpected downtime which is monitored by AI.

NEW AND IMPROVED PRODUCTS AND SERVICES ARE AVAILABLE

AI-assisted the Internet of Things can lead to the creation of the latest products and facilities. Natural language processing is persistently educating, permitting humans to communicate with machines without the need for human assistance. Artificial intelligence-controlled drones and robots can go where humans can't, offering up previously unimagined possibilities for observing and examination.

AI is changing navy management for commercial vehicles because it can monitor every quantifiable data point in a fleet of planes, trains, trucks, or cars to identify more effective routing and scheduling, as well as prevent unscheduled downtime. Currently Cloudera entitlements that their navy management AI has condensed navy vehicle downtime by up to 40% using Navistar devices.

TOOLS USED BY METEOROLOGISTS TO FORECAST THE WEATHER

1. Doppler Radar
2. Satellite data
3. Radiosondes

Doppler Radar

J. Christian Doppler, an Austrian physicist, was the first to explain why a train's whistle sounds louder as it approaches than as it moves away, and this radar was named after him. (almanac.com,2021)

The Doppler Effect

The Doppler radar makes use of the Doppler effect. When precipitation approaches the radar, the frequency of the radar signal rises, and when precipitation retreats, the frequency of the radar signal falls. This is useful for predicting tornadoes and wind gusts. (ibm.com, By The Weather company, 2020)

Working of a Doppler Radar

The instrument Doppler radar (also known as Weather radar) transmits electromagnetic energy in the form of pulses into the atmosphere to detect precipitation, track its motion, and assess its intensity. It also distinguishes between different kinds of precipitation, such as rain, hail, and snow.

When an electromagnetic pulse collides with an object, such as a snowflake or a drop of rain, the wave returns to the radar and transmits data to the Doppler radar (as shown in the Example- 1 Doppler Radar). The quantity of energy returned to the radar is proportional to the size of the object. The precipitation intensity is estimated using this information.

The time it takes for a pulse to strike an item and return to the radar can be used to identify its location. Once an object's location has been determined, it can be plotted on a map. The status or severity of the weather is determined by the intensity of an object. This information is used to pinpoint places where extreme weather conditions are likely to occur. (ibm.com, By The Weather company, 2020).

Figure 1. Doppler radar
[Source: https://www.weareiowa.com/article/weather/weather-lab/weather-lab-how-does-doppler-radar-work/524-7484ba0c-d014-4dd6-b7f5-77b33bfcbf07]

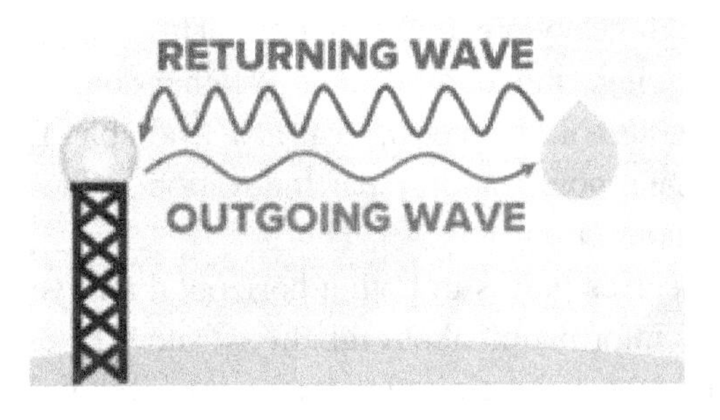

Doppler radar depicts rainfall or snowfall conditions using different colors. Colors of weather radar or doppler radar: The information obtained by the Doppler radar is sent to a computer, which provides output in the green, yellow, and red colors seen in weather forecasts. Light rain is indicated by green, moderate rain is shown by yellow, and heavy rain is indicated by red.

Classification of Doppler Radar

The wavelength of the Doppler radar can be used to classify it. Doppler radars are classified into three categories:

1. S-Band radars
2. C-Band radars, and
3. X-Band radars

S-Band Radars

The wavelength considered for S-band radars is 8 – 15 cm. Frequency range is 2 to 4 GHz. But Mostly, S-band Doppler radars operate within the 3 to 3.8 GHz frequency range. The radar's wavelength range permits the pulse to pass through layers of precipitation. By analyzing the wave (the wave that returns to the radar) that returns from wider distances, meteorologists generate the alarm in advance. This radar is the most expensive of the three radar bands.

C-Band Radars

The wavelength for C-band radar is between 4 and 8 cm. The acceptable frequency range is 4–8 GHz. Typically operations are handled at frequency range 5.3–5.6 GHz.

Short-range weather readings and medium to long-range precipitation analysis is done using C-band radars. The precipitation rates are incorrectly interpreted due to the attenuation of the radar beam. These radars are less expensive than radars used in the S-band.

X-Band Radars

For X-band radars, the wavelength ranges from 2.5 to 4 cm. The frequency range that is being examined is between 8 and 12 GHz. Because of their shorter wavelength, X-band radars are more sensitive to lighter particles. The shorter wavelength makes the wave attenuate to a greater amount, preventing the radar beam from passing through heavy precipitation areas. This can prevent radar from detecting the type of weather that will be arriving. (ibm.com, By The Weather company, 2020)

These days, an improved version of Doppler radar is also available. Let's discuss it.

Advanced Doppler radar- Dual polarization radar Innovation in the field of Doppler radar is dual polarization (as shown in Example 2).

This radar transmits and receives pulses in both a horizontal and vertical orientation. This means that the returning echoes carry information about raindrops, snowflakes, sleet, and hailstones in both directions. Improved discrimination of precipitation type is possible with more information regarding the apearence, volume, and composition of the targets. There are even products that can be created using dual-polarization radar that can help to locate tornado debris in real time and enhance rainfall estimation. (weareiowa.com, Weather Lab, 2021)

Figure 2. Dual polarization radar
[Sourec: https://www.weareiowa.com/article/weather/weather-lab/weather-lab-how-does-doppler-radar-work/524-7484ba0c-d014-4dd6-b7f5-77b33bfcbf07]

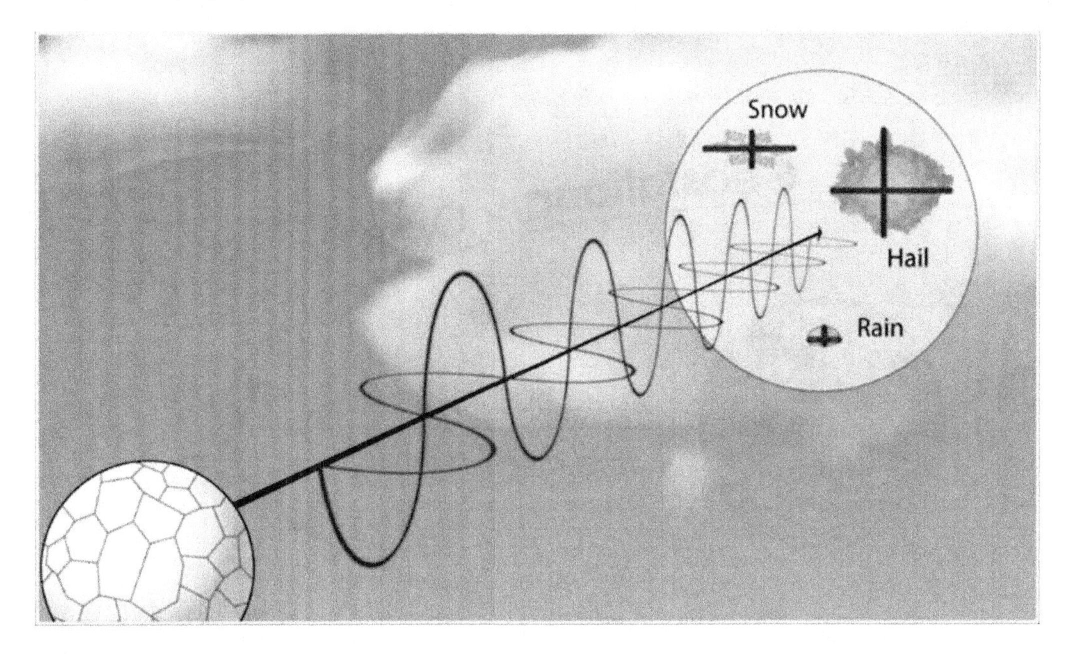

Benefits of Doppler radar:

- Predicts upcoming weather situations.
- Reliable and quality of data is high. (weareiowa.com, Weather Lab, 2021)

Satellite Data

One of the most useful sources of global weather data is weather satellites. These little satellites orbit larger objects and revolve around them. The satellite's speed and momentum protect it from dropping to the ground, and gravity stops it from going into space.

By flying high above the Earth's surface, weather satellites acquire photographs of cloud patterns and storm systems in order to track their movements. Such visuals are regularly seen on television weather reports. Weather satellites record breeze, temperature, rain, water vapor, tidal range, moisture levels in soil, and snow cover.

Polar Orbiting Environmental Satellites and GOES - Geostationary Orbiting Environmental Satellites are the two categories of weather satellites and they are categorized based on their orbits around the earth.

GOES: Geostationary Orbiting Environmental Satellites

GOES keeps taking photos of the same spot-on Earth, for the entire day. At an altitude of nearly 36,000 kilometers, GOES hover over a single location over the Earth's equator (22,300 miles). GOES satellites capture images which are seen on the weather news. These images are believed to be an excellent source for weather monitoring.

Figure 3. GOES
[Source: http://cimss.ssec.wisc.edu/SCALE/grade5/satellites.html]

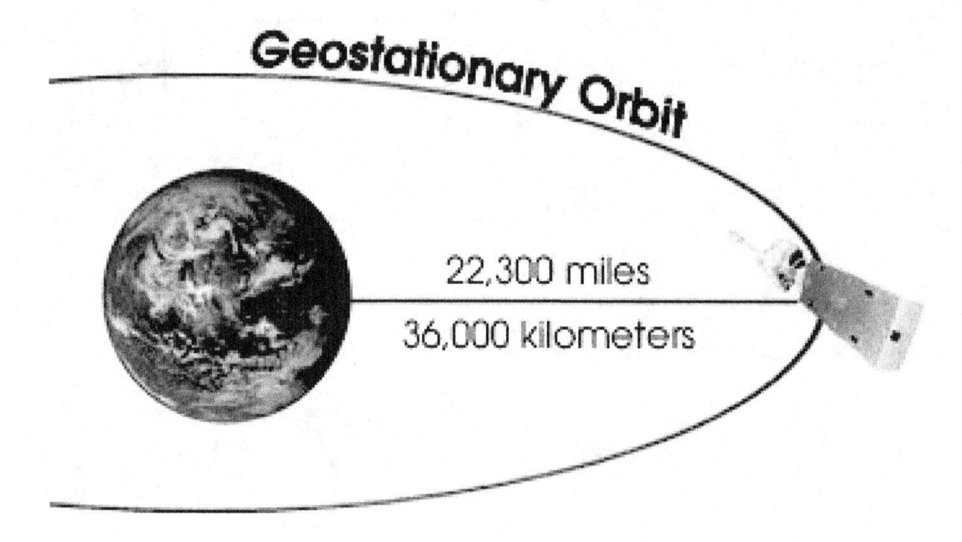

GOES are divided into three categories: visual, infrared, and water vapor.

Storms, clouds, fires, and smog belong to the domain of visual light images. The clouds, water and surface temperatures of the Earth, and properties of the ocean, such as marine currents. The water vapor imagery observes and tests the moisture content in the upper atmosphere.

POES: Polar Orbiting Environmental Satellites

Polar orbiting satellites travel a complete distance of one pole to another in a circular orbit. It orbits the Earth at a considerably lower distance than GOES. The Earth rotates on its axis then satellites collect data from the same path. These satellites collect data from the path as the earth rotates on its axis. In a 24-hour period, a polar orbiting satellite may see the entire earth twice. (weatherstreet.com, 2020)

Advantage of Satellite Data

Satellite imagery has been an incredible advantage to climate forecasters. With it, they can examine space dangers which include low clouds, screen thunderstorms, and track the dust plumes.

Figure 4. POES
[Source: http://cimss.ssec.wisc.edu/SCALE/grade5/satellites.html]

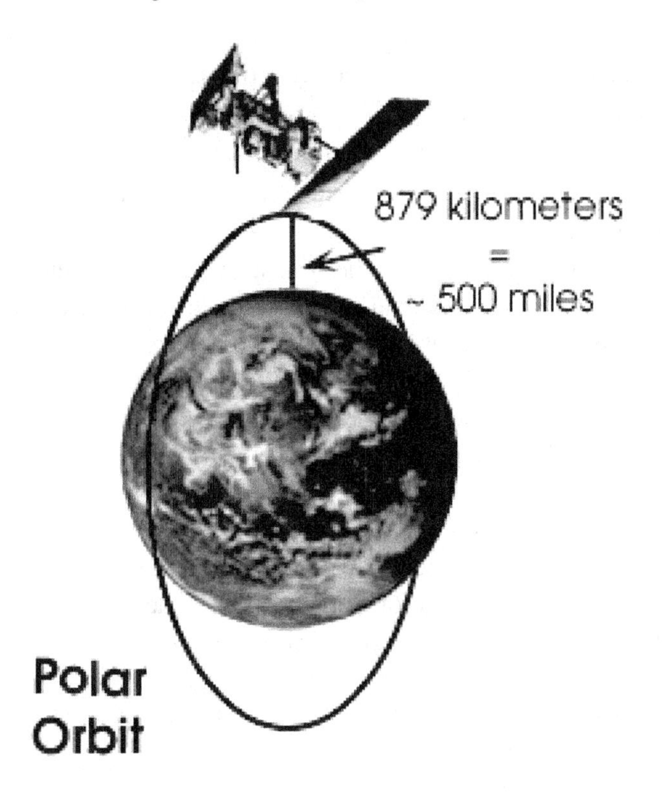

Radiosondes

Radiosondes are telemetry tool packages that run on batteries and are typically deployed into the ecosystem using weather balloons. At high elevation, it measures pressure, altitude, temperature, humidity levels, and breezes (both velocity and direction), and cosmic radiation. Rawinsonde is an abbreviation for radar wind sonde, a type of radiosonde whose function is monitored as it climbs in the atmosphere to verify the wind velocity and direction. Another type of radiosonde that may be dropped from planes and fall instead of being carried by weather balloons. Dropsondes are the name for these radiosondes. Radiosondes are used in a variety of operational atmospheric data integration applications. (Fletcher, in Data Assimilation for the Geosciences, 2017)

Radiosonde data are used in different types of application including:

- computer-based weather prediction models take Radiosonde data as inputs.
- Forecasts for severe storms, aviation, marine traffic, and fire weather in a specific or local area.
- Weather forecasting in winter and analyzing precipitation types.
- Forecasting of temperature.
- Analyzing air pollution based on data.
- Radiosonde data is used in research of weather and climate change.

AUTOMATED SURFACE OBSERVING SYSTEMS (ASOS)

The collaboration between the NWS (National Weather Service), the DOD (Department of Defense), and the FAA (Federal Aviation Administration) ITSELF IS THE ASOS. It is a critical surface weather monitoring network, intended to aid in weather forecasting and aerospace operations, as well as the needs of research communities such as hydrological, meteorological, and climatological organisations. ASOS units are automated sensors that give information to an air charter operator before and during their flight about the local weather.

Frequently it updates the observations about the atmosphere (every minute and every day of the year) from different locations to improve the weather forecasting and warnings and data are deposited in the Global Surface Hourly database. The accuracy and timeliness of forecasting will be improved using the recorded information which is the overriding aim of NWS Modernization. The basic strength of ASOS is to provide safety which is the main concern of the Aviation community.

A weather map will be used to record weather data such as temperature, covering of clouds, wind, humidity, air pressure, rain or snow, and the direction a weather system will continue or anticipated to move for a certain location at a given time. It also observes the important changes through the network and it automatically transmits the observations and a special report also when it encounters the abnormal condition i.e. if the threshold value defined for elements is exceeded in weather.

Example: As per the weather.gov, ASOS reports basic elements of weather:

1. The condition of the sky like whether the sky is clear/ scattered, visibility.
2. Basic current weather information like rain intensity, snow etc.
3. Any obstructions are there or not. Ex: Fog, haze
4. Speed and direction of the wind.
5. Ambient and dew point temperature.

SUPERCOMPUTERS

As per the website thoughtco.com, supercomputers are school-bus sized computers which are highly powerful. As mentioned, supercomputer size is very large and this is due to the hundreds of thousands of processor cores. The outcome of this mass computing capacity makes the supercomputers powerful and very large in size.

The Role of Supercomputers in Meteorology

Supercomputers are used to collect and store the periodic or recurrent event of weather observations. These observations are recorded every hour of every day by weather satellites, weather balloons, radars etc around the world.

By processing and analyzing stored data, supercomputers develop weather forecast models. The forecast model's output will then be utilized as a guide to develop their own forecasts. The output data provides information on all layers of the atmosphere as well as forecasts for the next few days. To make their forecast, forecasters will use this information, as well as their knowledge of meteorological systems, familiarity with regional weather trends, and personal experience.

According to the website weather.gov, the NWS (National Weather Service) uses supercomputers close to the clock, so that accurately it produces the data related to forecasts, watches, and warnings for the public. Supercomputers make use of these observed data that NWS collects and help to predict all kinds of weather hazards including extreme heat, hurricanes, tornadoes. Computer's data is also made available worldwide, using which other countries can also predict their weather as well.

The NWS Supercomputers are one of the world's most powerful weather prediction systems.

AWIPS

As per the weather.gov, the Advanced Weather Interactive Processing System (AWIPS), is a technically sophisticated system for information processing, display, and communication. AWIPS is the foundation of the National Weather Service modernization and restructuring.

AWIPS is a computer system that uses satellite data and radar pictures in conjunction with hydrological and meteorological systems. As a result, the forecaster is better able to prepare and give more accurate forecasts and warnings in a timely manner.

WEATHER SENSORS FOR THE INTERNET OF THINGS

Weather sensor's job is to monitor the weather at a specified location and make the data public to everyone in the world. The technology underpinning this might be the Internet of Things (IoT), which is a sophisticated and cost-effective way of linking things to the internet and connecting the entire world of things. These are the devices that fundamentally sense temperature, pressure, humidity, intensity, rain worth, and so on. They react to changes in the environment and gather data from a wide variety of sources. The gathered data is shared with other connected devices and management systems to develop application-specific analyses and insights into valuable data trends. There are multiple types of sensors available within the paradigm that aid in the measurement of the aforementioned parameters.

According to campbellsci.com following are the IoT weather sensors that are used nowadays:

1. ClimaVUE50 - Digital Weather Sensing Element in a Compact Size

Figure 5. ClimaVUE50 [Source: campbellsci.com]

This sensor uses SDI-12 to record barometric pressure, wind, solar radiation, precipitation, air temperature, relative humidity, vapor pressure, and lightning strike (count and distance). *ClimaVUE50 will* do this without moving its parts, and also takes very less power. Data integrity will be promised by a pre-tilt sensor. This unique device is best suitable for quick development, in remote locations, for large networks, as part of a more complex system.

Temperature, air pressure, average lightning distance, number of lightning strikes, precipitation, ratio, radiation, tilt, wind direction, and wind speed are all factors to consider among the measurements made by this diverse product.

2. METSENS200 Wind and Compass Compact Weather Sensing Element

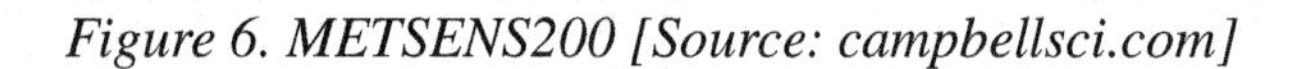

Figure 6. METSENS200 [Source: campbellsci.com]

METSENS200 measures wind speed associated with degreed direction via an ultrasonic sensing element. An associate degree unified electronic compass permits the MetSENS200 to produce correct, comparative wind direction measurements without being aligned during an definite means, making installation easier. WMO average wind speed and direction and current of air knowledge area unit provided.

The MetSENS200 may be easily linked with the MeteoPV Solar Resource Platform and Campbell Scientific data loggers that employ SDI-12, ModbusRS-485, RS-485, or NMEA RS-232.

Measurements made by this diverse product are wind direction and wind speed.

3. METSENS300 is a weather sensor that monitors temperature, relative humidity, and air pressure in a small package.

METSENS300 is a single, integrated instrument with no mechanical parts that monitors air temperature, relative humidity, and pressure in three double-louvered, naturally aspirated radiation shields. The data includes temperature, relative humidity, air pressure, absolute humidity, air density, and wet bulb temperature. The MetSENS300 works with any Campbell Scientific knowledge feller that supports SDI-12, RS-485, ModbusRS-485, or NMEA RS-232, as well as the MeteoPV solar Resource Platform. Temperature, pressure, and relative humidity are only a few of the measurements that were taken.

Figure 7. METSENS300 [Source: campbellsci.com]

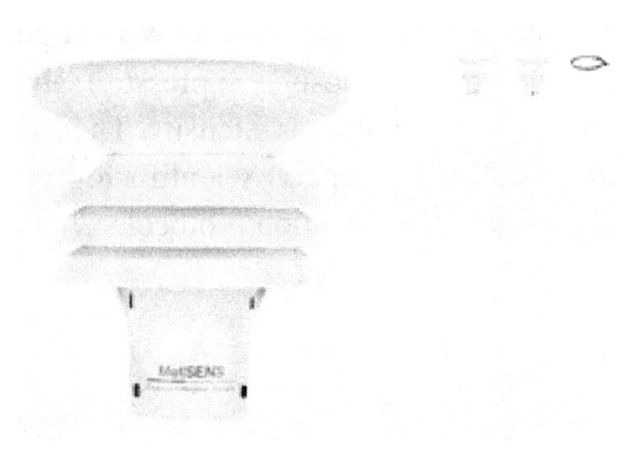

4. METSENS500 Compact Weather Sensing Element with Compass for Temperature, Relative Humidity, Atmospheric Pressure, and Wind.

METSENS500 uses an ultrasonic device to measure wind speed and direction, along with that, air temperature, relative humidity, and air pressure are measured within three double-louvered, radiation shields with natural aspiration. The internal electronic compass of MetSENS500 produces accurate, relative wind direction data without having to be set in a very specific way, making installation easier and are all available from the WMO. It works with the MeteoPV star Resource Platform as well as any Campbell Scientific expertise that supports SDI-12, RS-485, ModbusRS-485, or NMEA RS-232. Temperature, atmospheric pressure, ratio, wind direction, and wind speed are among the variables that are measured.

Figure 8. METSENS500 [Source: campbellsci.com]

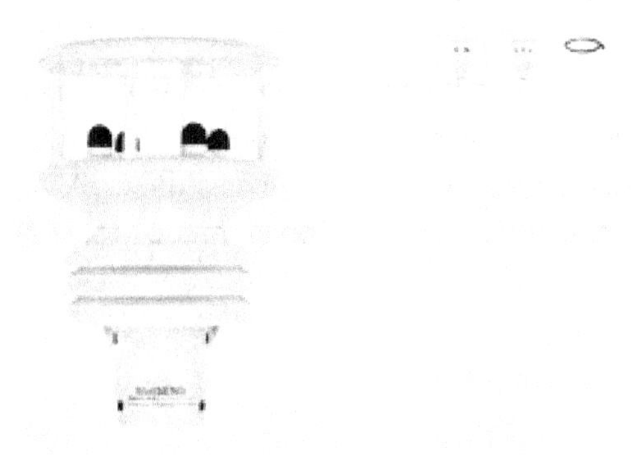

5. METSENS550 Compact Weather Sensing Element with Precipitation Connective and Compass for Temperature, RH, Atmospheric Pressure, and Wind.

This detector employs an ultrasonic detector to measure wind speed and direction with air temperature, ratio, and atmospheric pressure, all in a single, integrated instrument enclosed within three naturally aspirated radiation shields with no moving parts. A tipping bucket rain gauge can be connected to an incorporated aspect connection by users. The MetSENS550's integrated electronic compass allows it to provide accurate, relative wind direction measurements without the need for fine adjustment, making installation easier. The World Meteorological Organization (WMO) publishes data on average wind speed and direction, gusts, temperature, ratio, atmospheric pressure, absolute wetness, air density, wet bulb temperature, total precipitation, and precipitation intensity. The MetSENS550 is compatible with the MeteoPV solar Resource Platform and any Campbell Scientific logger that supports SDI-12, RS-485, ModbusRS-485, or NMEA RS-232 as a communication protocol.

Figure 9. METSENS550 [Source: campbellsci.com]

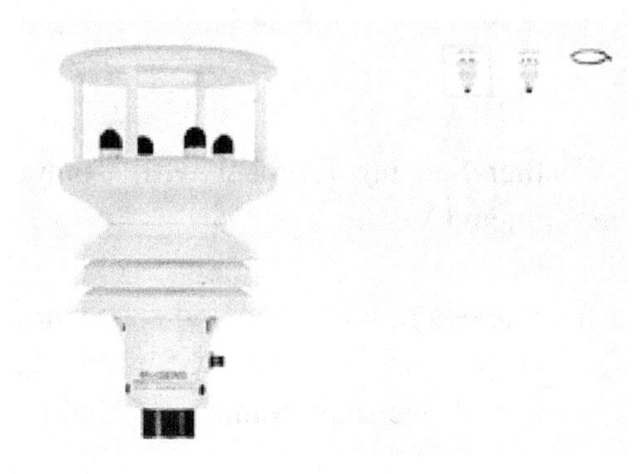

Measurements include air temperature, air pressure, ratio, wind direction, and wind speed. (A tipping bucket pluviometer can be connected to an inbuilt aspect connection.)

6. METSENS600 Compact Weather Sensing Element with Compass for Temperature, Relative Humidity, Atmospheric Pressure, Wind, and Precipitation

The above detector uses an ultrasonic detector to observe wind, speed and direction, as well as air temperature, ratio, and atmospheric pressure, all in one tool that is located within three double-louvered, logically articulated radiation protections with no moving machinery. A visual precipitation sensor detects small amounts of water. The combined electronic compass in the MetSENS500 allows it to offer precise,

Figure 10. METSENS600 [Source: campbellsci.com]

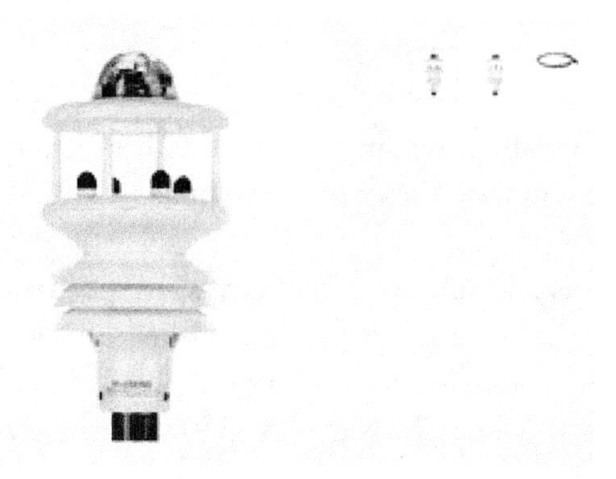

relative wind direction data without being focused in any precise way, making installation easier. The World Meteorological Organization (WMO) publishes data on average wind speed and direction, as well as blast, temperature, ratio, atmospheric pressure, absolute humidity, air density, wet bulb temperature, rain total, and rain intensity. The MetSENS600 may be simply linked with the MeteoPV star Resource Platform and any Joseph Campbell Scientific data logger that uses SDI-12, RS-485, ModbusRS-485, or NMEA RS-232 and uses SDI-12, RS-485, ModbusRS-485, or NMEA RS-232. This versatile device measures air temperature, air pressure, precipitation, ratio, wind direction, and wind speed, among other things.

CLASSIFICATION, AND APPLICATIONS OF IOT SENSORS IN WEATHER MONITORING

Weather sensors are divided into two groups based on their characteristics as Analog and Digital sensors. Analog sensors are those that produce a continuous output, while digital sensors are those that require conversion and transmission.

It's divided into Active and Passive sensors based on the supported input power supply. sensors that require external electricity to work are referred to as active sensors, whereas sensors that do not require external power to function are referred to as passive sensors. They get the electricity they need for their job from the number they're measuring.

Electrical sensors, chemical sensors, biosensors, electrochemical devices, optical devices, object data devices, and machine vision sensors are classified into four groups based on the detecting technique. Electrical sensors transform a sensed parameter into an electrical signal that can be used to measure or trigger a control device. Chemical sensors detect changes in chemical levels as well as chemical compounds in air and liquid mediums.

Biosensors are analytical sensors that aid in the identification of chemical components in biological samples. Electrochemical sensors are employed in a wide range of applications, including identifying vital gases, monitoring water quality (to name a few nitrogen compounds, alkali, and heavy metals), pharmaceutical, and biotechnological applications. The supply's solar beam is stopped by sensors that detect the target object, which is then thoroughly analyzed for its qualities. Temperature, vibrations, pH-value, pressure, strain, particle size, object material, field of force and other parameters are all measured by these sensors. Optical sensors include electrical conduction devices, photovoltaic cells (solar cells), and photodiodes. These are commonly used in analytical applications like environmental monitoring and human health, together with biosensors.

An object's information Sensors are particular to an object or application, such as CPU temperature observing and hydraulic oil temperature observing, or anywhere a temperature device is located within the system. Machine Vision Sensors can take and process photos, yielding important information. For real-time weather operations, a weather monitoring system (WMS) is employed. Below is an explanation of some of the WMS area units.

1. Air temperature measured by humidity and temperature sensors. The DHT11 is a sensor which is a widely used temperature and humidity sensor with a digital output. It makes use of a semiconductor device as well as an electrical phenomena humidification device. The DHT22 which is identical to

Figure 11. WM Sensors [Source E. N. S. S. Anjana,2021]

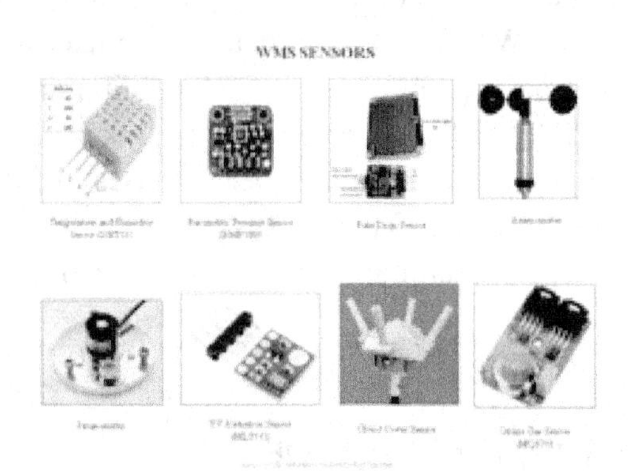

the DHT11, but DHT22 has a higher resolution and a wider measurement range. The receptiveness and exactness are both exceptional.

2. By detecting the differences in pressure or force applied per unit space by the atmosphere, atmospheric pressure sensors turn the information into electrical instincts. The BMP180 sensor is part of the BMP XXX series is a popular air pressure device with an I2C interface. The BMP changes with the altitude and weather. Because temperature changes cause pressure to drop, some BMP devices have a temperature sensor.
3. The rain board module in the rain drop sensor detects rain, and the control module analyses the analogue data before converting it to digital values for processing. The sensor's operation is controlled by these two modules. A sensitivity potentiometer and an LN393 comparator are included in the control module for testing Analog values. Variable resistors are included in the copper tracks on the rain board module. The status of the rain board determines the resistance.
4. The wind speed and direction detector an Anemometer: A weather plate is used to define the wind direction and it is measured using a potentiometer. Depending on the wind direction, the resistance changes. The output values from the analogue will be translated to degrees' notation and the directions up to 540 degrees are infrequently established, and these must be accurately categorized into 0–360-degree ranges of standards. It is possible to use a three-cup, three-wire wind gauge. The space between the device's cups and the stem switch shutting confirms the wind speed. The cups include a slight low magnetic body that helps this operation when the reed switch is missed once every spin. The Reed switch creates a digital output signal. The number of uprisings during a time period, typically 2.5 seconds, is increased to enhance wind speed.
5. Pyranometer for Solar Radiation: The worldwide irradiance is electromagnetic energy with a wavelength range of 300-3000nm, also known as broadband radiation. The pyranometer is used to determine the hemisphere's radiation compactness above the wavelength range. Incoming radiation is converted into electrical signals of various voltages by pyranometers. The voltages are resurrected as the global irradiance parameter, which influences the sensitivity of the device. The wavelength range of a Solar's spectrum is 0.15 to 4.0m, with a pyranometer used to measure a portion of it. The cosine response, or angle of radiation, can also be measured using the pyranometer. Solid-state/thermopile sensors or silicon photocells are used with excellent precision for light detection. LDR, photodiodes, and phototransistors can be used for detection functions, although their precision is

reduced.A thermopile pyranometer is used to measure sun radiation using the CMP11 instrument. It's commonly utilised in greenhouses and beneath plant canopies because of its spectral range of 280-2800nm. SQ110 is another extensively used sensor label for measuring sun radiation. In reaction to the intensity of sunlight in the visible range of the irradiance spectrum, this generates an output voltage.

6. Ultraviolet radiation: It detects UV that has been emitted. The device generates ultraviolet-intensity analog output voltages. The ultraviolet index is the international standard for gauging UV radiation from the sun. The UV index is calculated by dividing the output analogue voltage by 0.1V. The VEML6075 is a popular UV sensor that can detect both UVA (tanning rays) and UVB (burning rays) light bands. It uses a serial interface for device interaction. UV light sensor ML8511 is utilized in weather monitoring systems. It can efficiently patrol the light-weight region of 280-390nm. The majority of the UVA spectrum and a tiny portion of the UVB spectrum are contained within this region.
7. Ozone sensor: MQ131 is a semiconductor gas device that includes a device circuit and a heater circuit. For prime ozone concentration, a metal device could be utilized, while for low ozone concentration levels, a blue plastic device might be used. It can detect concentrations ranging from 10ppm to 1000ppm. An inside pre-heater aids in optimal sensing, although it takes around twenty-four hours to pre-heat to achieve optimum accuracy. For prime strong sunshine, peak gas concentrations occur.

Optical sensors and biosensors are widely employed in analytical applications such as environmental monitoring and human health. (E. N. S. S. Anjana,2021)

IOT APPLICATION OF A FOREIGN WEATHER MONITORING AND SURVEILLANCE

Weather monitoring technology that enables for continuous weather monitoring and surveillance at the same time from a remote location. The Internet of Things-based Remote Weather Monitoring Station is a full-featured ASCII text file lookout that can correctly detect the humidity, light intensity, temperature, and the observed parameters data can be saved on the open cloud "Thing Speak." The system also includes a camera that streams live video of the world being studied.

The above-mentioned example is a fully functional Remote Weather Monitoring system. Many of these systems can be built depending on a specific requirement, either by the meteorological department or by the research department. Considering the last few years, there have been numerous climatic changes, including the formation of cyclones and a large number of depressions on the seabed. In light of this, the demand for remote weather monitoring systems is growing. IoT frameworks can aid here by making data monitoring easier, keeping devices healthy, doing stress tests, and maintaining performance. (Vermal et al., 2017).

CONCLUSION

In this chapter, we'll look at how AIoT- artificial intelligence in the Internet of Things, might benefit both businesses and consumers. Although AI will continue to play a significant role in weather forecasting, the rising use of sensors in a range of applications across different domains has proved the sector's capacity to attract AI expertise. AI analyses continuous streams of data from IoT devices, uncovering patterns that traditional gauges miss. Machine learning combined with AI can also predict operation scenarios and highlight parameters that need to be tweaked for best results.

Intelligent IoT will eventually indicate which procedures are useless and should be eliminated and increase the efficiency by fine tuning the responsibilities.

For instance, Google employs artificial intelligence (AI) to reduce data center cooling costs via the Internet of Things (IoT).

REFERENCES

Anjana, E. N. S. S., & Student, B. (n.d.). *Review IoT Sensors Classification and Applications in Weather Monitoring*. Academic Press.

Bauer, P., Thorpe, A., & Brunet, G. (2015). The quiet revolution of numerical weather prediction. *Nature*, *525*(7567), 47–55. doi:10.1038/nature14956 PMID:26333465

Brenowitz, N. D., & Bretherton, C. S. (2019). Spatially extended tests of a neural network parametrization trained by coarse-graining. *Journal of Advances in Modeling Earth Systems*, *11*(8), 2728–2744. https://onlinelibrary.wiley.com/doi/ abs/10.1029/2019MS001711

Chevallier, F., Chéruy, F., Scott, N. A., & Chédin, A. (1998). A neural network approach for a fast and accurate computation of a longwave radiative budget. *Journal of Applied Meteorology*, *37*(11), 1385–1397. doi:10.1175/1520-0450(1998)037<1385:ANNAFA>2.0.CO;2

Ebert-Uphoff, I., Thompson, D. R., Demir, I., Gel, Y. R., Karpatne, A., Guereque, M., ... Smyth, P. (2017, September). A vision for the development of benchmarks to bridge geoscience and data science. *17th International Workshop on Climate Informatics*.

Fletcher, S. J. (2017). *Data assimilation for the geosciences: From theory to application*. Elsevier.

Gagne, D. J. II, McGovern, A., & Xue, M. (2014). Machine learning enhancement of storm-scale ensemble probabilistic quantitative precipitation forecasts. *Weather and Forecasting*, *29*(4), 1024–1043. doi:10.1175/WAF-D-13-00108.1

Grönquist, P., Yao, C., Ben-Nun, T., Dryden, N., Dueben, P., Li, S., & Hoefler, T. (2021). Deep learning for post-processing ensemble weather forecasts. *Philosophical Transactions of the Royal Society A*, *379*(2194), 20200092.

Hamill, T. M., & Whitaker, J. S. (2006). Probabilistic quantitative precipitation forecasts based on reforecast analogs: Theory and application. *Monthly Weather Review*, *134*(11), 3209–3229.

Krasnopolsky, V. M., Fox-Rabinovitz, M. S., & Chalikov, D. V. (2005). New approach to calculation of atmospheric model physics: Accurate and fast neural network emulation of longwave radiation in a climate model. *Monthly Weather Review*, *133*(5), 1370–1383. doi:10.1175/MWR2923.1

Lagerquist, R., McGovern, A., & Smith, T. (2017). Machine learning for real-time prediction of damaging straight-line convective wind. *Weather and Forecasting*, *32*(6), 2175–2193.

Lazo, J. K., Morss, R. E., & Demuth, J. L. (2009). 300 billion served: Sources, perceptions, uses, and values of weather forecasts. *Bulletin of the American Meteorological Society*, *90*(6), 785–798.

LeCun, Y., Bottou, L., Bengio, Y., & Haffner, P. (1998). Gradient-based learning applied to document recognition. *Proceedings of the IEEE*, *86*(11), 2278–2324.

McGovern, A., Elmore, K. L., Gagne, D. J., Haupt, S. E., Karstens, C. D., Lagerquist, R., ... Williams, J. K. (2017). Using artificial intelligence to improve real-time decision-making for high-impact weather. *Bulletin of the American Meteorological Society*, *98*(10), 2073–2090.

Oord, A. V. D., Dieleman, S., Zen, H., Simonyan, K., Vinyals, O., Graves, A., . . . Kavukcuoglu, K. (2016). *Wavenet: A generative model for raw audio.* arXiv preprint arXiv:1609.03499.

Rasp, S., Dueben, P. D., Scher, S., Weyn, J. A., Mouatadid, S., & Thuerey, N. (2020). WeatherBench: a benchmark data set for data-driven weather forecasting. *Journal of Advances in Modeling Earth Systems, 12*(11).

Ronneberger, O., Fischer, P., & Brox, T. (2015, October). U-net: Convolutional networks for biomedical image segmentation. In *International Conference on Medical image computing and computer-assisted intervention* (pp. 234-241). Springer.

Russakovsky, O., Deng, J., Su, H., Krause, J., Satheesh, S., Ma, S., ... Fei-Fei, L. (2015). Imagenet large scale visual recognition challenge. *International Journal of Computer Vision*, *115*(3), 211–252.

Shi, X., Chen, Z., Wang, H., Yeung, D. Y., Wong, W. K., & Woo, W. C. (2015). Convolutional LSTM network: A machine learning approach for precipitation nowcasting. *Advances in Neural Information Processing Systems*, 28.

Taillardat, M., Mestre, O., Zamo, M., & Naveau, P. (2016). Calibrated ensemble forecasts using quantile regression forests and ensemble model output statistics. *Monthly Weather Review*, *144*(6), 2375–2393.

Verma, G., Gautam, A., Singh, A., Kaur, R., Garg, A., & Mehta, M. (2017). IOT Application of a Remote Weather Monitoring & Surveillance Station. *International Journal of Smart Home*, *11*(1), 131–140.

Vogel, P., Knippertz, P., Fink, A. H., Schlueter, A., & Gneiting, T. (2018). Skill of global raw and post-processed ensemble predictions of rainfall over northern tropical Africa. *Weather and Forecasting*, *33*(2), 369–388. doi:10.1175/WAF-D-17-0127.1

Section 2

Chapter 6
Advancements in Weather Forecasting With Deep Learning

Nidhi Tejas Jani
Pandit Deendayal Energy University, India

Rajeev Kumar Gupta
https://orcid.org/0000-0002-5317-9919
Pandit Deendayal Energy University, India

Santosh Kumar Bharti
https://orcid.org/0000-0002-0627-6433
Pandit Deendayal Energy University, India

Arti Jain
https://orcid.org/0000-0002-3764-8834
Jaypee Institute of Information Technology, India

ABSTRACT

The changes in the weather play a significant role in people's planning. It has attracted the attention of several study communities due to the fact that it has an impact on human life all over the world. But weather forecasting is a challenging task because it is dependent on a variety of factors such as wind speed, wind direction, global warming, etc. Deep learning-based solutions have seen a lot of success in the geospatial domain over the last few years. In the past few years, a variety of deep learning-based weather forecasting models have been proposed. The forecasting techniques used traditionally are highly parametric and so are complex. In this chapter, deep learning techniques which are used for weather forecasting, such as Multilayer Perceptron, Jordan Recurrent Neural Network, Elman Recurrent Neural Network, etc., are discussed in detail. This chapter presents a comparative analysis of various deep learning-based weather forecasting models that are currently available.

DOI: 10.4018/978-1-6684-3981-4.ch006

Figure 1. Weather forecasting working

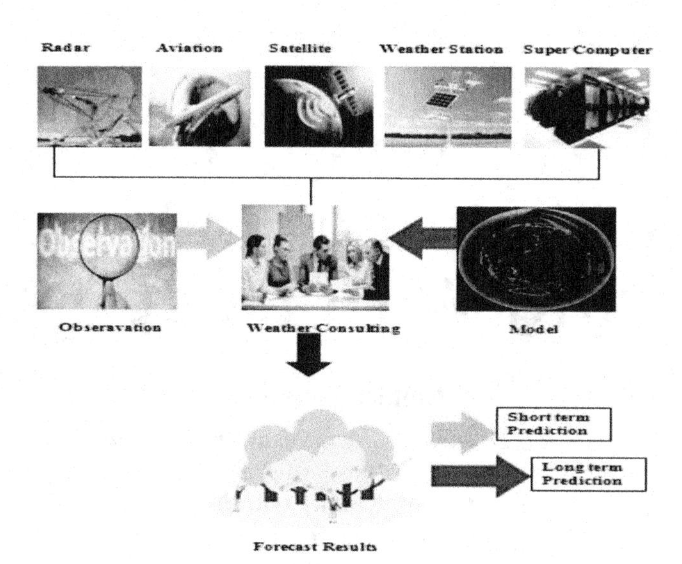

INTRODUCTION

Artificial Neural Networks (ANN) have grown in popularity in recent years, and are now being used in a variety of tasks such as pattern recognition, time-series analysis, classification, and regression. Weather prediction is considered as a procedure regarding science and technology that is predicting weather conditions for a specific time and location in advance. Accurate weather forecasts can aid individuals in their short-, medium-, and long-term planning efforts, as well as in emergency situations. As a result, before making a selection on where to go, it is vital to be aware of current weather patterns. Therefore, it increases the requirement of handiness of tools used for weather prediction accurately. This requirement is more noticeable if forecasting is considered for short-term forecasting also known as nowcasting. Prediction is on the basis of the sliding window algorithm. The outcome of every month is calculated to check accuracy.

The different atmosphere attributes like temperature, wind and humidity which affects different aspects of human livelihood. In the last few decades, a wide variety of weather prediction models have been deployed (Yadav,2013). The majority of these models are based variants of fuzzy logic and artificial neural networks. This chapter discusses some of the existing weather forecasting models along with their pros and cons. Makhamisa et al. (Makhamisa, 2020) discusses three neural network models, namely Elman recurrent neural network (ENN), Jordan recurrent neural network (JNN) and Multi-layer perceptron (MLP) for rainfall forecasting which is also discussed in (Goodfellow,2016, Nielsen,2015 and Lewis,2016). Using these methods, it is possible to compare the model's performance over time which helps us to determine the best model for forecasting the weather (Yadav,2013). ENN, JNN and MLP models are used to describe the rise of rainfall and solar irradiations. Therefore, work here can be summed up as prediction of weather using deep learning techniques. But the question that needs to be addressed is regarding the accuracy of forecasting techniques.

The weather forecasting process is illustrated in Figure 1. First, data is gathered and submitted for observation, then a perfect model is selected and submitted to weather consulting, and last, findings are announced, which provide short- and long-term predictions, respectively.

The chapter here is organized as follows. Section 2 discusses some existing deep learning-based methods used for the weather forecasting. Section 3 includes various regression models used for the weather forecasting. Section 4 present the comparative analysis of different weather forecasting models and Section 5 highlights the conclusion and future work.

DIFFERENT DEEP LEARNING APPROACHES USED FOR WEATHER FORECASTING

The number of weather forecasting approaches have been introduced over the last few years, majority of these approaches are based on Machine Learning and Deep Learning. This section discusses some existing Deep Learning based approaches that may be used weather forecasting. Major ways through which we can predict weather is by observing the current condition of weather, like temperature, humidity, pressor, max. win speed, wind speed and wind direction, motion of air and clouds present in the sky, looking at the pattern of previous weather that represents current condition or by observing a switch in pressure of air. All of these are features which help to understand weather conditions in a better way. But eventually the sun is accountable for the weather. Data is collected with the use of radar, barometer and thermometer and weather is predicted. All the features such as humidity, weather patterns, wind direction and others are necessary for weather forecasting. The output is predicted to help for long-term and short-term planning. Weather prediction on the basis of deep learning is considered as a powerful supplement for the standard method and thus many researchers have accomplished results by introducing data driven deep learning approaches for weather prediction.

Artificial Neural Network

Artificial Neural Networks which are also simply known by Neural networks come under clusters of units or nodes that are connected and loosely model the neurons in the biological mind. ANN falls into the category of supervised machine learning which means that data is labeled and trains algorithms to predict results with accuracy (Wasserman,1993, Christopher, 2016, Russell,2016). ANN extracts encouragement that explains work of biological neurons in the brain. As ANN is based on supervised learning, it absorbs from the sample by formulating a mapping between input and output (Ghanbarzadeh, 2009). Artificial neural networks include input layer, at least one hidden layer and output layer. Every layer contains nodes that indicate neurons and it is associated with weights. The node which is internal contains two functions i.e transfer and activation function (Christopher,2016, Russell,2016). In the transfer function shown in equation 1 where input represents "Xi" and weights are represented by "Wi" and is as follow:

$$f\left(x\right)=\sum w_i x_i + b_i \tag{1}$$

Here, b_i represent the bias value. Activation function has the responsibility of modeling the relationship and they are non-linear. Activation function is represented by "$\varnothing$". The activation function in

Figure 2. ANN Architecture

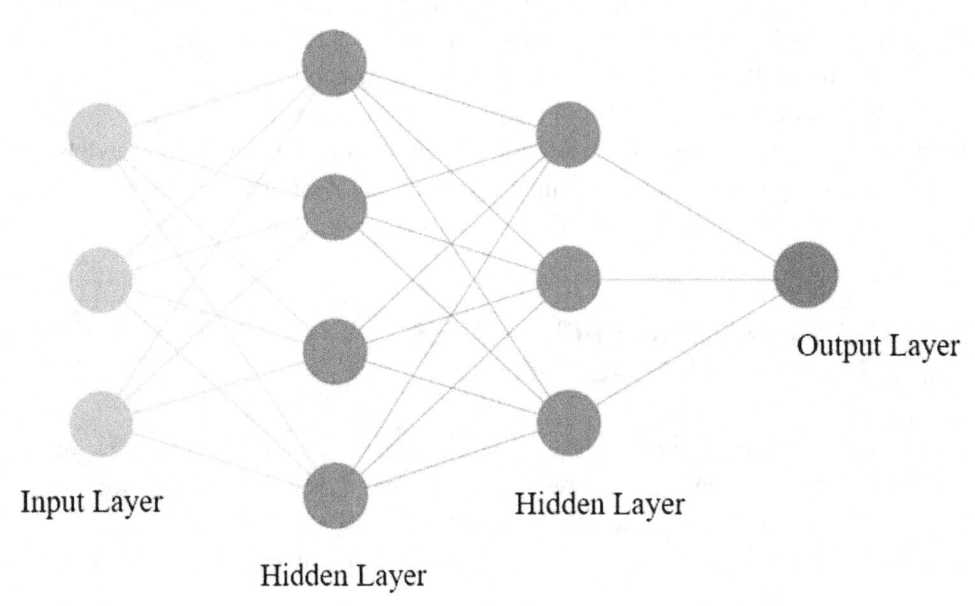

addition is differentiable (Ghanbarzadeh,2009). The output of the activation function in given by equation 2:

$$y_i = \varnothing\left(f\left(x\right)\right) \tag{2}$$

Figure 2 illustrates the architecture of ANN which includes input, output and hidden layers. In the input layer there are three nodes, output layer has one node while two hidden layer includes four and three nodes respectively. Artificial Neural Network here is utilized for binary classification, also known as two-class classification because output layer includes one node. The directed arrows indicate the wights. In the input layer each node matches the functionality that is used for prediction and so it has 3 features. The node of hidden layer receives weighted input as a set which is described by transfer function in Eq1 and it will give output which is given by Eq 2.

Makhamisa et al. (Makhamisa, 2020), discussed the comparative analysis of three deep learning-based models namely MLP, ENN and JNN. All three models are trained on six features i.e., temperature, humidity, pressor, max. win speed, wind speed and wind direction. Figure 3, shows the features used to trained the model for weather forecasting.

All three models were initially tested on the basis of hourly meteorological data from Lesotho from the 1st January 2016 to 26th day of March 2021. This dataset contains 2045 instances with 6 features which helps for short term forecasting. The Figure 4 illustrates the distribution of dataset for all six features as independent predictors.

Figure 4 describes features used for forecasting that establishes input layer's nodes for deep learning architectures like MLP, Jordan RNN and ENN. The models are introduced with the help of R statistical programming language (Verzani,2005, Kohl,2015, Stowell,2014). Artificial neural networks are implemented using RSNNS package (Bergmier,2012). The model uses Elman RNN, Multilayer perceptron

Figure 3. Features used for weather forecasting

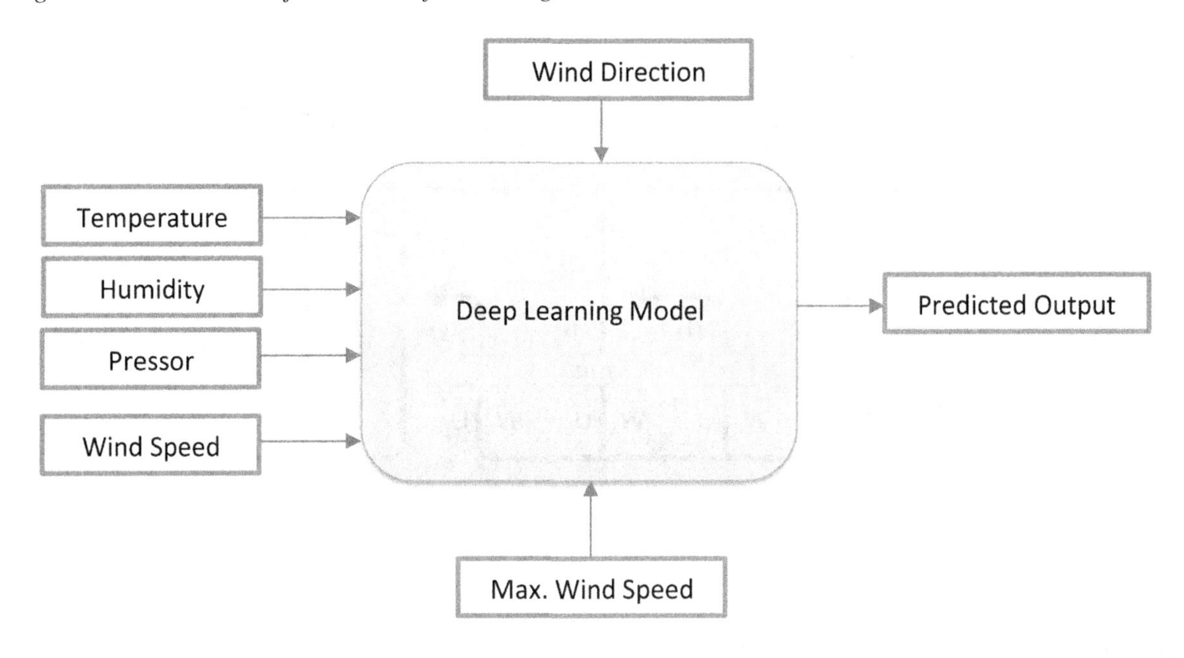

and Jordan RNN. Weather forecasting is done using the three modes and it performs tasks like sunshine forecasting and precipitation forecasting. The model is trained and tested and thus 80% of the data is trained, other 10% data for validation and 10% is tested. The accuracy for sunshine forecasting using multilayer perceptron model is 91% while for Jordan RNN achieves accuracy 97% for sunshine forecasting. Elman RNN has the accuracy 96% for sunshine forecasting.

Figure 4. Features for weather forecasting (Makhamisa,2020)

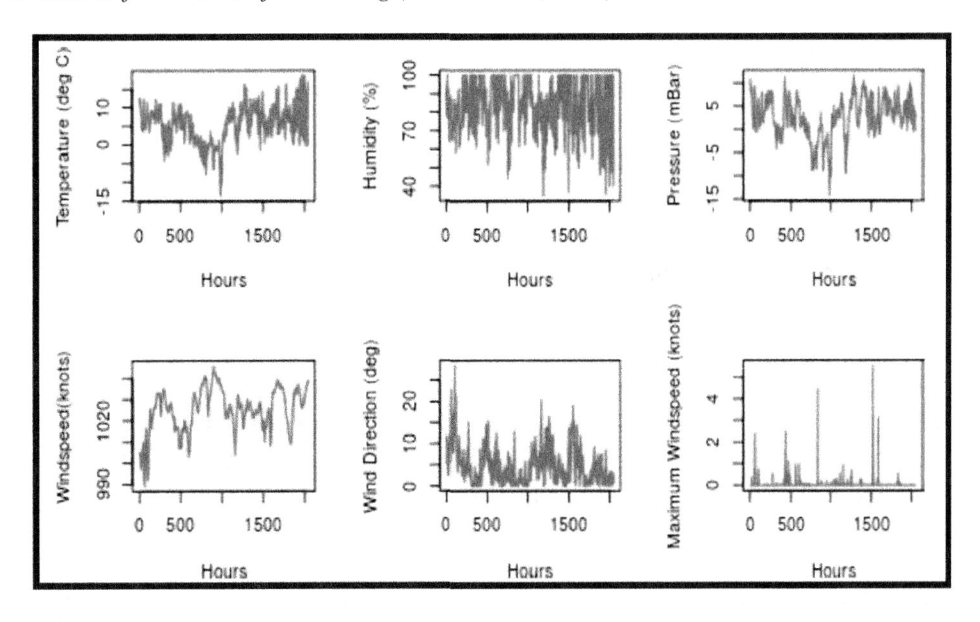

Figure 5. LSTM Architecture (Marco,2020)

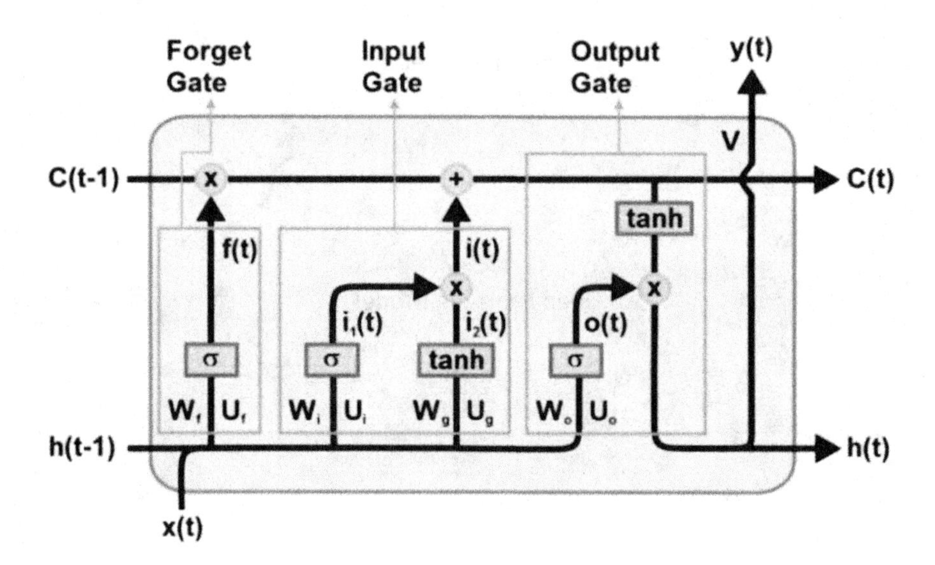

Weather Prediction Using LSTM Method

Now let's see LSTM method is and how it can be used for weather forecasting. Long short-term memory, also known as LSTM, is introduced to defeat the problem of vanishing gradient that is present in recurrent neural networks. In the network gradient flow was enhanced. Instead of having a hidden later, an LSTM unit consists of cell state, input gate, forget gate and output gate. Cell state gives information with the complete sequence and also represents network memory. The things which are applicable from the time steps of the previous ones should be added is taken care by forget gate. Output gate determines the output values of current time step. Figure 5 illustrates the architecture of LSTM.

LSTM cell consisting of 't' time with U as a weight matrix is associated with the input vector having 't' time. The cell of LSTM is combined to t+1 and t-1 time LSTM cell and with Was weight matrix. LSTM cell is also associated with 't' time output vector and V weight matrix. U and W are the matrices that consist of submatrices and they are attached to various components of LSTM units. Time is shared between every weight matrix.

At the processing time the cell state sends related information and thus the previous time step's information appears at every time step which decreases the consequences of short-term memory. While training every time steps, the important information is learnt by gates. It will then decide to add or forget the information and then include them into cell state or remove from cell state. Thus LSTM solves the problem of vanishing gradient as it permits the data recovery that are sent in the memory. LSTM are useful for processing, classifying and predicting time series such as weather forecasting.

Forget gate is the first gate and it takes care whether the information to be saved or to be deleted. The prior hidden state information and current input information are given through a function named sigmoid. If the output is near to 0 then that means we can forget the information is to be saved. Forget gate can be express mathematically by the equation 3.

$$f(x) = \sigma\left(x(t)U_f + h(t-1)W_f\right) \tag{3}$$

Here, $x(t)$ shown the input to the LSTM, σ represents the activation function and h(t-1) represent the input from the previous timestamp. Next gate is the input gate which is used to upgrade the state of the cell. The input of the sigmoid function is the current input and previous hidden state. The sigmoid function says that the output close to 1 is considered as important. Input gate also sends current input and hidden state to tanh function to compress values that are between 1 and -1 which will enhance tuning of the network. After that the output of both sigmoid and tanh functions are multiplied. The sigmoid function's output determines which important information to keep from the output of the tanh function. Equation 4 shows the mathematically equation for input gate.

$$i_1(t) = \sigma\left(x(t)U_i + h(t-1)W_i\right)$$

$$i_2(t) = tanh\left(x(t)U_g + h(t-1)W_g\right)$$

$$i(t) = i_1(t) * i_2(t) \tag{4}$$

Cell state is calculated after the input gate is activated. Firstly, the cell state having the previous time step is multiplied element by element with the forget gate's output. Thus, the probability of ignoring values of cell state when multiplied by value near to 0. After that input gate's output is added element by element to the cell state. Thus, the output obtained will be the latest cell state. Equation 5 shows the mathematically equation for cell state.

$$\mathrm{C(t)} = \sigma\left(f(t) * c(t-1) + i(t)\right) \tag{5}$$

Output gate is the last gate that determines the value of approaching a hidden state that holds previous input information. Firstly, current input and currently hidden state are added and given to sigmoid. Then the latest cell state is given to tanh. Lastly, the output of tanh function and sigmoid function are then multiplied to determine information hidden state should include. The output will be the latest hidden state. After that the latest hidden state and latest cell state are taken over for the other step. Equation 6 shows the mathematically equation for output gate.

$$o(t) = \sigma\left(x(t)U_o + h(t-1)W_o\right)$$

$$h(t) = tanh\left(C(t) * o(t)\right) \tag{6}$$

In this section we have learnt about the LSTM method and so it is considered one of the best approaches for real time prediction of weather. Also, prediction of rainfall can be done using intensified LSTM based recurrent neural network.

Support Vector Method for Weather Prediction

As in the earlier section, we have learnt the LSTM method, likewise SVM is also one of the methods of deep learning that can be used for forecasting weather. Support vector machine, shortly known as SVM is a useful classification, pattern recognition and regression technique. It comes under the category of supervised learning with corresponding algorithms. SVM are built on the decision planes concept that determine decision boundaries. X which is the input vector is mapped with the Ø(x) in feature space. SVM can be shown by equation 7:

$$p\left(x\right)= q\varnothing\left(x\right)+r \tag{7}$$

Here r is considered as bias and the weight vector is represented as q.

Advantage of a support vector machine is it works well without any knowledge prior to it. SVM needs a linear function to categorize a dataset that is linearly separable and for a dataset which is non- linear. It needs two hyper planes to be separated. Support vector machines work well when multi- dimensional hyper planes are developed for separating a data set given into two different categories optimally. The method, support vector regression was used to show prediction of rainfall in order to observe data of rainfall in Bangladesh (Hasan,2015).

DIFFERENT REGRESSION METHODS FOR WEATHER FORECASTING

Weather forecasting has various methods that are available with respect to time. For the duration of 48 hours a short-range weather forecasting is available. Similarly medium range forecasting is for the duration from 3 to 7 days. While for the duration of more than 7 days we have long range forecasting but still there is no such limit for the duration. The models of regression are the one broadly used for estimation of the future happenings. In this section we will learn more about the different regression methods.

Logistic Regression

Logistic regression is a technique used when variables are dependent and it is binary. Logistic regression is considered to be a predictive analysis technique like other regression analysis. It is applied in order to define data and it also describes the association of binary dependent variables with one or more than one nominal, interval, ordinal or any independent variable at ratio-level. A multi predictor model of logistic regression originated for the probabilistic prediction of an average rainfall which is on the timescale of month (Prasad,2010). The study is for the regions like the whole India, Orissa which is on east and Gujarat which is on west (i.e these two are homogeneous subdivisions of India).

Bayesian Regression

Bayesian Regression is a method which undertakes statistical study and it has a background of Bayesian inference. For prediction of the tropical cyclone during the high season of hurricane(July-September) in central north pacific, a poison linear regression technique is applied and it has a bayesian framework (Chu and Zhao,2007).

Principal Component Regression

This regression technique is built on principal component analysis. Here in principal component regression rather than regression, dependent variables which are on explanatory variables and are utilized as predictors. Multivariate principal component regression and canonical correlation analysis are the two models that are useful for forecasting regular rainfall in the region of USAPI, specifically in spring and winter (Yu et al,1997).

Multiple Linear Regression

MLR technique utilizes various variables that are explanatory which are responsible in predicting the result of the response variable. Main aim of this model is to build relation between the response variable and explanatory variables. MLR was applied in order to originate a model for predicting parameters of weather (Paras,2012). It was also discovered that the model proposed was very much capable for weather condition forecasting for a specific station with the use of locally collected data.

COMPARATIVE ANALYSIS OF DIFFERENT DEEP LEARNING TECHNIQUES FOR WEATHER FORECASTING

This section gives a brief analysis regarding different models used (Ayman,2021). Table 1 consists of different features for weather forecasting and uses a model for testing. The table also has accuracy reported that gives results after training and testing the model accordingly.

CONCLUSIONS AND FUTURE WORK

This chapter describes weather forecasting and how prediction is performed regarding the weather. It also includes features responsible for weather prediction which helps with long-term and short-term planning. The chapter also discusses various deep learning methods that are useful for weather forecasting. Application for short-term weather forecasting using deep learning and is based on Lesotho's weather is described. The three models MLP, Jordan RNN and Elman RNN are used for sunshine prediction. Results of high accuracy are provided from the work by combining techniques having more advantages than other forecasting methods. We can summarize weather forecasting as useful as well as highly accurate which is beneficial for planning. Understanding patterns of weather is considered to be very helpful when we are making decisions. It is not possible to analyze all the methods of regression. Forecasting a weather still has limitations. For future work the focus will be on having better accuracy for weather forecasting by using a group of stated deep learning models rather than using those individual models.

Table 1. Different deep learning methods

Features	Model	Real Time	Evaluation Parameter over Dataset	Scale
Temperature (Santhosh et al., 2010)	**Training** BP algorithm.	No	Training dataset: 200 data records Root Mean square: 2.16%	Short term as well as Local
Rainfall (Afan et al., 2015)	RNN	No	Dataset: ENZO (2 Experiments) 1)Training: 75% data, Testing: 25% data R^2 = 84.8%, SME = 125 2)Training: 50% data, Testing: 50% data R^2 = 59.9%, Root mean square error = 55.29%	Short term as well as Local
Temperature Humidity Dew point Pressure (Daris et al., 2018)	RNN+LSTM	No	Dataset: NCDC (50:50, Training and Testing) Accuracy= 96.06%	Short term as well as Local
Temperature Humidity Pressure (Nitin et al., 2019)	The Random Forest Classification	Yes	Dataset: Delhi (Training and Testing, 75: 25) Accuracy = 87.90%	Short term as well as Local
Temperature Humidity Rainfall (Sanjay et al., 2012)	**Training** BP algorithm	Yes	Sensors are used for recording various pattern	Short term/Local
Speed of wind (Amir et al., 2017)	RNN+LSTM	Yes	Dataset: METAR Data is accessed every 'h' hours for training Mean Absolute Error = 1.18% Root Mean Square Error = 1.62%	Short term/Regional
Temperature (Al-Matarneh et al., 2014)	**Training** BP algorithm	No	Datasets: Anman and Taipei Anman: VAF = 92.484 and Mean Absolute Error = 1.462 Taipei: VAF = 98.2926 and Mean Absolute Error = 0.0542	Short term/Local
Temperature wind speed (Siamak, 2019)	3d-CNN	Yes	Dataset: NCDC Wind forecasting: Training and Testing: 90:10. Temperature Forecasting: Training and Testing: 97: 03 Esbjerg Location (Mean Absolute Error) = 1.71% Odense Location (Mean Absolute Error) = 0.79% Roskilde Location (Mean Absolute Error) = 1.84%	Long term as well as Regional
Temperature Wind speed Dew point Geopotential height (Aditya et al., 2015)	DBN+GP	Yes	Dataset: IGRA 1)Root mean square error for duration of 6 hrs is 2.11%, 2)Root mean square error for duration 12 hrs is 1.03%, 3)Root mean square error for duration of 24 hrs is 1.01%	Short term, long term as well as Regional
Tropical Cyclone Extratropical Cyclones Tropical depression Atmospheric rivers (Evan et al., 2017)	CNN and Encoder-Decoder	No	Dataset: Simulation Output Small Level Data: Training and Testing: 67:33 Medium Level data: Training and Testing: 80:20 Large Level Data: Training and Testing: 81.5:18.5 3D-CNN Results: Tropical Cyclones: 56.40% Extra-Tropical Cyclones: 27.74% Atmospheric Rivers: 33.95% Tropical Depression: 96.57%	Long term as well as Local
Temperature Dew point Wind speed (James et al., 2014)	SVR and Auto Encoder	No	Dataset: HKO Training and Testing: 90:10 Temperature Normalized: Mean Square Error = 0.8117% R^2 = 0.915. Wind Speed Normalized: Root Mean Square Error = 0.2522% R^2 = 0.891 Dew Point Normalized: Mean Square Error = 0.9552% R^2 = 0.901	Long term as well as Local

REFERENCES

Afan, Salman, Kanigoro, & Heryadi. (2015). Weather forecasting using deep learning techniques. Proc. Int'l. Conf. Advanced Computer Science and Info. Systems (ICACSIS), 281-285.

Grover, Kapoor, & Horvitz. (2015). A deep hybrid model for weather forecasting. *Proceedings of the 21th ACM SIGKDD Int'l. Conf. Knowledge Discovery and Data Mining*, 379-386. 10.1145/2783258.2783275

Ghaderi, Sanandaji, & Ghaderi. (2017). *Deep forecast: Deep learning-based spatio-temporal forecasting*. arXiv preprint arXiv:1707.08110

Abdalla, A. M., Ghaith, I. H., & Tamimi, A. A. (2021). Deep learning weather forecasting survey. *International Conference on Information Technology (ICIT).*

Bergmier, C., & Benitez, J. (2012). Neural networks in R using the stuttgart neural network simulator: RSNNS. *Journal of Statistical Software*, *46*(7), 1–26.

Christopher, M. B. (2016). *Pattern Recognition and Machine Learning*. Springer-Verlag.

Chu, P. S., & Zhao, X. (2007). A Bayesian Regression Approach for Predicting Seasonal Tropical Cyclone Activity Over the Central North Pacific. *Journal of Climate*, *20*(15), 4002–4013. doi:10.1175/JCLI4214.1

Daris, Fente, & Singh. (2018). Weather forecasting using artificial neural networks. *2018 2nd Int'l. Conf. on Inventive Communication and Computational Technologies (ICICCT),* 1757-1761.

Racah, Beckham, Maharaj, Kahou, & Pal. (2016). *Extreme Weather: A large-scale climate dataset for semi-supervised detection, localization, and understanding of extreme weather events*. arXiv preprint arXiv:1612.02095.

Ghanbarzadeh, A., Noghrehabadi, A., Assareh, E., & Behrang, M. (2009). Solar radiation forecasting based on meteorological data using artificial neural networks. *Proc. 7th IEEE International Conference on Industrial Informatics*, 227-231. 10.1109/INDIN.2009.5195808

Goodfellow, I., Bengio, Y., & Courville, A. (2016). *Deep Learning*. MIT Press. Available from http://www.deeplearningbook.org

Hasan, N., Nath, N. C., & Rasel, R. I. (2015). A support vector regression model for forecasting rainfall. *Proc. 2nd International Conference on Electrical Information and Communication Technology (EICT)*, 554-559. 10.1109/EICT.2015.7392014

Liu, Hu, You, & Chan. (2014). Deep neural network-based feature representation for weather forecasting. Proc. on the Int'l. Conf. on Artificial Intelligence (ICAI), 1-7. doi:10.1109/ICACSIS.2015.7415154

Kohl, M. (2015). *Introduction to Statistical Analysis with R*. Bookboon.

Al-Matarneh, Sheta, Bani-Ahmad, Alshaer, & Al-Oqily. (2014). Development of temperature-based weather forecasting models using neural networks and fuzzy logic. *Int'l. J. of Multimedia and Ubiquitous Engineering, 9*(12), 343-366.

Lewis, N. (2016). *Deep Learning Made Easy with R: A Gentle Introduction for Data Science*. CreateSpace Independent Publishing Platform.

Lewis, N. (2017). *Neural Networks for Time Series Forecasting with R: Intuitive Step by Step for Beginners*. CreateSpace Independent Publishing Platform.

Senekane, Mafu, & Taele. (2020). Weather Nowcasting Using Deep Learning Techniques. In *Data Mining - Methods, Applications and Systems*. . doi:10.5772/intechopen.84552

Del Pra. (2020). *Time series forecasting with deep learning and attention mechanism*. Towards Data Science. https://towardsdatascience.com/time-series-forecasting-with-deep-learning-and-attention-mechanism-2d001fc871fc

Nielsen, M. (2015). *Neural Networks and Deep Learning*. Determination Press. https://static.latexstudio.net/article/2018/0912/neuralnetworksanddeeplearning.pdf

Singh, Chaturvedi, & Akhter. (2019). Weather Forecasting Using Machine Learning Algorithms. 2019 Int'l. Conf. on Signal Processing and Communication (ICSC), 171-174.

Paras, A., & Mathur, S. (2016). Simple Weather Forecasting Model Using Mathematical Regression. Indian Res. *J. Extension Educ*, *1*(Special Issue), 161–168.

Prasad, Dash, & Mohanty. (2009). A logistic regression approach for monthly rainfall forecasts in meteorological subdivisions of India based on DEMETER retrospective forecasts. *Int. J. Climatology, 30*, 1577-1588. . doi:10.1002/joc.2019

Russell, S. J., & Norvig, P. (2016). *Artificial Intelligence: A Modern Approach*. Pearson Education Limited. https://zoo.cs.yale.edu/classes/cs470/materials/aima2010.pdf

Sawaitul, Wagh, & Chatur. (2012). Classification and prediction of future weather by using back propagation algorithm-an approach. *International Journal of Emerging Technology and Advanced Engineering*, *2*(1), 110–113.

Mehrkanoon, S. (2019). Deep shared representation learning for weather elements forecasting. *Knowledge-Based Systems*, *179*, 120–128. doi:10.1016/j.knosys.2019.05.009

Baboo, S. S., & Shereef, I. K. (2010). An efficient weather forecasting system using artificial neural network. *International Journal of Environmental Sciences and Development*, *1*(4), 321–326. doi:10.7763/IJESD.2010.V1.63

Stowell, S. (2014). *Using R for Statistics*. Apress.

Verzani, J. (2005). *Using R for Introductory Statistics*. Chapman & Hall.

Wasserman, P. D. (1993). *Advanced Methods in Neural Computing*. John Wiley & Sons, Inc.

Yadav, A. K., & Chandel, S. (2013). Solar radiation prediction using artificial neural network techniques: A review. *Renewable & Sustainable Energy Reviews*, *33*, 772–781. doi:10.1016/j.rser.2013.08.055

Chapter 7
Multivariate Time Series Forecasting of Rainfall Using Machine Learning

Shilpa Hudnurkar
https://orcid.org/0000-0001-7854-2019
Symbiosis Institute of Technology, Symbiosis International University (Deemed), India

Vidur Sood
Symbiosis Institute of Technology, Symbiosis International University (Deemed), India

Vedansh Mishra
Symbiosis Institute of Technology, Symbiosis International University (Deemed), India

Manobhav Mehta
Symbiosis Institute of Technology, Symbiosis International University (Deemed), India

Akash Upadhyay
Symbiosis Institute of Technology, Symbiosis International University (Deemed), India

Shilpa Gite
Symbiosis Centre for Applied Artificial Intelligence, Symbiosis International University (Deemed), India

Neela Rayavarapu
Symbiosis International University (Deemed), India

ABSTRACT

Predicting rainfall is essential for assessing the impact of climatic and hydrological changes over a specific region, predicting natural disasters or day-to-day life. It is one of the most prominent, complex, and essential weather forecasting and meteorology tasks. In this chapter, long short-term memory network (LSTM), artificial neural network (ANN), and 1-dimensional convolutional neural network LSTM (1D CNN-LSTM) models are explored for predicting rainfall at multiple lead times. The daily weather parameter data of over 15 years is collected for a station in Maharashtra. Rainfall data is classified into three classes: no-rain, light rain, and moderate-to-heavy rain. The principal component analysis (PCA) helped to reduce the input feature dimension. The performance of all the networks are compared in terms of accuracy and F1 score. It is observed that LSTM predicts rainfall with consistent accuracy of 82% for 1 to 6 days lead time while the performance of 1D CNN-LSTM and ANN are comparable to LSTM.

DOI: 10.4018/978-1-6684-3981-4.ch007

INTRODUCTION

Time series forecasting has now become a significant area of interest as it has gained economic importance. Vast amounts of data are being generated in various domains and most of this is in the form of time-series data. The analysis and study of time series data hold great importance due to the need for forecasting, in fields such as weather and stock markets. It is very difficult to adapt traditional linear methods for solving multivariate, multi-input forecasting problems which makes them unfit for time series forecasting (Liu et al., 2019). Rainfall prediction is crucial because the lives and livelihoods of people depend on it. In the agricultural sector, farming activities largely depend on rainfall and weather conditions. Prediction of rainfall is challenging and complex due to its dynamic nature (Srinivas et al., 2013). Large variability in rainfall during the rainy season is also observed (Hrudya et al., 2021). This necessitates rainfall prediction over a small geographical area to increase crop yield and prevent the farmers from incurring losses due to incorrect forecasting (Singh et al., 2021).

In India, India Meteorological Department (IMD) issues weather forecasts. The prediction models used by IMD for daily rainfall prediction are dynamic. Sikka (2009) narrated how numerical weather prediction (NWP) models (dynamic models) were developed and how National Centre for Medium-Range Weather Forecasting (NCMRWF), India, evolved over the years in medium-range predictions (3 to 10 days lead time prediction)(Sikka, 2009). NWP models are based on thermodynamic equations that model the current state of the atmosphere (Laing & Evans, 2011). These models require supercomputers to solve the thermodynamic equations on various spatial resolutions and are very complex as they process a huge amount of data.

With the advances in technology and by having the satellites dedicated to weather data and images, a lot of data related to weather variables, on various temporal scales, is being generated, analyzed, and recorded. Machine learning (ML) algorithms have been found suitable to process high-volume data, detect the pattern of the data, to learn from examples. Supervised machine learning (ML) algorithms require a large amount of data to train the artificial intelligent network. With the availability of the data, various ML algorithms such as long short-term memory networks (LSTM), artificial neural networks (ANN), decision trees, and support vector machines (SVM) have been studied over the previous decade to find out the most optimal algorithm for rainfall prediction (Brereton & Lloyd, 2010; Rathnayake et al., 2011; Saha et al., 2016; Shrivastava et al., 2012). Intelligent models of neural networks like LSTM and ANN can efficiently solve multiple input variable problems (Hudnurkar & Rayavarapu, 2021; Wahyono et al., 2020). For rainfall prediction with ML or deep learning methods, weather variable data is needed to be fed to the model as training data. The data must be preprocessed to accomplish better training of the models. The trained network then requires testing for the unseen data.

This chapter discusses the development of an intelligent model for rainfall forecasting with multiple lead times and with better accuracy. The significant contributions of this study are as follows:

1. Development of a robust model that can predict rainfall using multivariate time series data.
2. A comparative study of three artificial intelligence-based prediction models is carried out.
3. A model for predicting rainfall accurately for multiple lead times from 1 to 6 days has been developed, trained, and tested.

The chapter is organized in sections namely, Background, Proposed Model, Dataset, Experimentation, Evaluation and Result Analysis, and Summary. The Background section reviews ML algorithms used for time series applications; the Proposed Model section details the methodology of the present work. The material used for the study is described in the Dataset section. The experiments carried out are detailed in the Experimentation section. The results are discussed in the Evaluation and Result Analysis section and the last section summarizes the chapter and provides a brief discussion on the scope of future work.

BACKGROUND

In this section, the use of ANN, LSTM, and 1D CNN-LSTM models has been reviewed. ANN, the most basic version (Hudnurkar & Rayavarapu, 2018) of neural networks, has been used in various fields for forecasting using univariate and multivariate data. It has been applied to problems like financial time series prediction (Ankenbrand & Tomassini, 1996), which involved univariate data at its core.

Hung et al. used hourly data, spanning four years, to develop a neural network capable of forecasting rainfall, 1 to 6 hours into the future (Hung et al., 2009). Darji et el.; surveyed the use of ANN for rainfall forecasting (M. P. Darji et al., 2015). Mishra et al. identified ANN as a promising data mining technique for rainfall forecasting (Mishra et al., 2017). Mishra et al. trained a feed-forward neural network (FFNN) to forecast rainfall over North India (Soni et al., 2018). Abhishek et al. tested three different neural network algorithms for weather forecasting (Abhishek et al., 2012). Lee et al. used ANN for late spring early summer rainfall forecasting (Lee et al., 2018). Darji used various architectures of ANN for rainfall prediction to find monthly and yearly rainfall patterns (M. Darji, 2019). Benevides et al. used a nonlinear autoregressive network with exogenous inputs (NARX) for forecasting intense hourly rainfall (Benevides et al., 2019). For flood prediction, Chai et al. classified rainfall and compared backpropagation and Radial Basis Function ANNs (Chai et al., 2017). Hudnurkar and Rayavarapu used ANN and SVM for the binary classification of rainfall time series for a station in Maharashtra state, India (Hudnurkar & Rayavarapu, 2022). According to literature ANN using the regression approach has been used for prediction more often than the classification approach.

Long short-term memory (LSTM), a kind of recurrent neural network (RNN), is well suited for multivariate time series rainfall forecasting (Sood et al., 2021; Sri Rahayu et al., 2020). It has been used in various fields to understand multiple types of forecasting problems, such as cloud resources (Tran et al., 2018), wind turbines, and grid interaction (Y. S. Wang et al., 2020). Along with other classical models, Yu et al. applied an LSTM model (short-term prediction) to predict solar irradiance (Yu et al., 2019). The authors compared LSTM with classical models such as ARIMA, SVR, and concluded that LSTM performed better under most weather conditions. The authors concluded that the LSTM approach was able to overcome the typical problem of vanishing gradient common to typical RNNs. Wojtkiewicz et al., tested two types of RNNs (Wojtkiewicz et al., 2019) to predict solar irradiance. It was found that LSTM had slightly less error compared to the multivariate Gated Recurrent Unit. Haq et al. used El Nino and Indian Ocean Dipole as inputs to LSTM for rainfall prediction (Haq et al., 2021). Out of the univariate and multivariate approaches explored by the authors, they found multivariate prediction results better than univariate prediction. Haq et al. also varied hyperparameters to optimize the network for better prediction results.

Wu(2020) applied the principal component analysis (PCA) method, an unsupervised ML technique, to their LSTM model to reduce the dimension of input data (Wu, 2020). Results showed that with the increase in time scale, a combination of PCA and LSTM models achieved better results than the autoregressive moving average and backpropagation models. Wang et al. tested CNN-LSTM (Y. S. Wang et al.,

2020) for predicting the output of wind power plants. They used a time sliding window algorithm and constructed the neural network input dataset, following which CNN-LSTM was applied to this dataset. The model reached a high prediction accuracy. Li et al. tested a hybrid CNN-LSTM model for forecasting the next 24 hours' PM 2.5 concentration (Li et al., 2020). Several models were compared, and the results detailing parameters such as mean absolute error (MAE) and root mean square error (RMSE) scores showed that CNN-LSTM outperformed all other models, despite its short training time. In most of the approaches, the performance of the network was evaluated using MAE, RMSE, mean squared error (MSE), mean absolute percentage error (MAPE). Jang et al. (2020) used Bi-LSTM to increase the accuracy in text classification. A Bi-LSTM component in conjunction with an attention mechanism was used for sentiment classification (Jang et al., 2020). Xu et al. used 1D CNN-LSTM for the recognition of epileptic seizures. The EEG signal data were preprocessed and then used in the developed 1D CNN-LSTM model (Xu et al., 2020). This model also reached a high prediction accuracy. *Table 1* summarizes artificial intelligence (AI) models with performance metrics used for various prediction applications.

Table 1. Some of the existing practices and model, region, and performance metrics used therein

Author	Region	Model used	Performance Metrics
Hung et al	Bangkok	MLP	Efficiency index
Lee et al.	Korea	ANN	RMSE, MSE
Darji et al.	India	ANN	MSE, MAE
Abhishek et a.	Karnataka	BPFNN, BPA	MSE
Mishra et al.	North India	FFNN	Regression analysis
Tran et al.	Vietnam	LSTM	MAE
Ankenbrand et al.	Switzerland	ANN	MAE, MSE
Wang et al.	China	PCA-LSTM	MAPE
Liu et al.	China	3 phase LSTM	MAE, RMSE
Wojtkiewicz et al.	USA	GRU v LSTM	MAPE
Yu et al.	China	LSTM	MAPE
Wu	China	PCA-LSTM	RMSE, MAE
Wang et al.	China	CNN-LSTM	Prediction accuracy
Li et al.	China	CNN-LSTM	MAE, RMSE
Benevides et al.	Portugal	NARX	Net.RMS, F1 score

PROPOSED MODEL

Time series forecasting using multivariate data is much more suitable for real-world applications, although training the developed model using this multivariate data is a complex task. However, training the developed model with multivariate data for rainfall forecasting is reasonable since the complexity of the model will be in proportion to its importance in the real world. Multivariate data is noisy, i.e., it always contains some variables showing significant randomness, these variables are difficult to analyze (Liu et al., 2019), but it also contains variables that significantly affect the response variables. Predicting rainfall is a tedious task as it depends on various parameters like wind direction, visibility, wind speed, maximum and minimum temperature, etc. Now, determining the most relevant parameters that affect the response variable is challenging because it is important to strike a perfect balance between the complexity and efficiency of the model. In this chapter, three prediction models, with multivariate time series data as input, are proposed and compared based on their performance in predicting rainfall.

The three prediction models proposed in this chapter are ANN, LSTM, and 1D CNN-LSTM. The general flow of the process for the development of these prediction models is presented in *Figure 1*. It is the multivariate time series forecasting involving the acquisition of weather parameter data from a reliable source. After data acquisition, preprocessing of the data was carried out. The data cleaning and sorting were followed by a PCA application to transform the large dataset into a smaller one. PCA was employed for the selection of important features. The variance specified for feature selection was 80%. The dataset of selected features obtained by PCA was divided into training and testing datasets for prediction purposes. The three models, namely, ANN, LSTM, and 1D CNN-LSTM, were then developed. Hyperparameters for each model were experimented with for better prediction accuracy. Trained models were tested separately for the test dataset.

Figure 1. Process flow

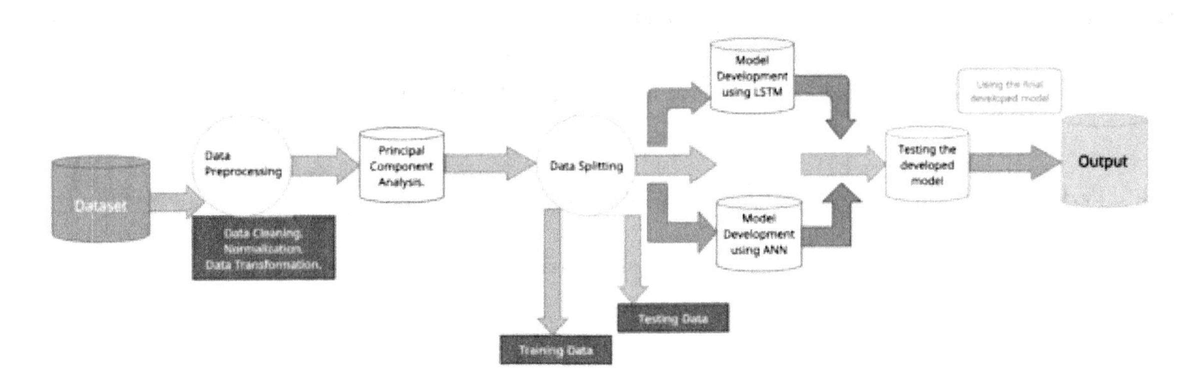

The input dataset comprised various parameters such as wind speed, wind direction, visibility, minimum and maximum temperature. The dataset was cleaned to get a dataset with no missing records. MS excel and ML techniques were used for the same. The response variable namely, rainfall, was then classified into three categories as follows:

- No rain: 0 mm
- Very Light to Light: 0.01mm to 18mm
- Moderate to Heavy: > 18mm

The PCA technique and the prediction models used in the study are briefly described in the following subsections.

Principal Component Analysis (PCA)

PCA is an unsupervised ML technique that is used to reduce the dimensions of huge datasets such that it transforms the dataset consisting of a large number of features (or variables) into a smaller number of components (Hasan, S.B.M. & Abdulazeez, 2021).

PCA aims to reduce the dimensionality as well as identify meaningful features from the dataset (Hasan, S.B.M. & Abdulazeez, 2021). Reducing the number of features (or variables) of a dataset might take a bit of a toll on the accuracy, but after performing PCA, a smaller set of variables is obtained that holds the majority of information of the pre-transformed dataset. The amount of information carried by the post PCA dataset always depends on the specified variance.

Artificial Neural Networks (ANN)

The term "artificial neural network" refers to a sub-field of artificial intelligence influenced by the brain. A computational network based on biological neural networks that construct the structure of the human brain is known as an ANN (S. Wang, 2003). ANNs, like neural networks in the human brain, have neurons that are coupled to each other in various layers of the networks. They consist of nodes (also known as artificial neurons) that are connected; the weights associated with these nodes are adjusted during the training of the model (Sivanandam, S. N., & Deepa, 2007). Many different architectures and techniques have evolved from the underlying foundation of neural networks, such as ANN, LSTM, CNN, etc. *Figure 2* presents an ANN with two hidden layers, an input layer, and an output layer.

Figure 2. A simple architecture of Artificial Neural Network

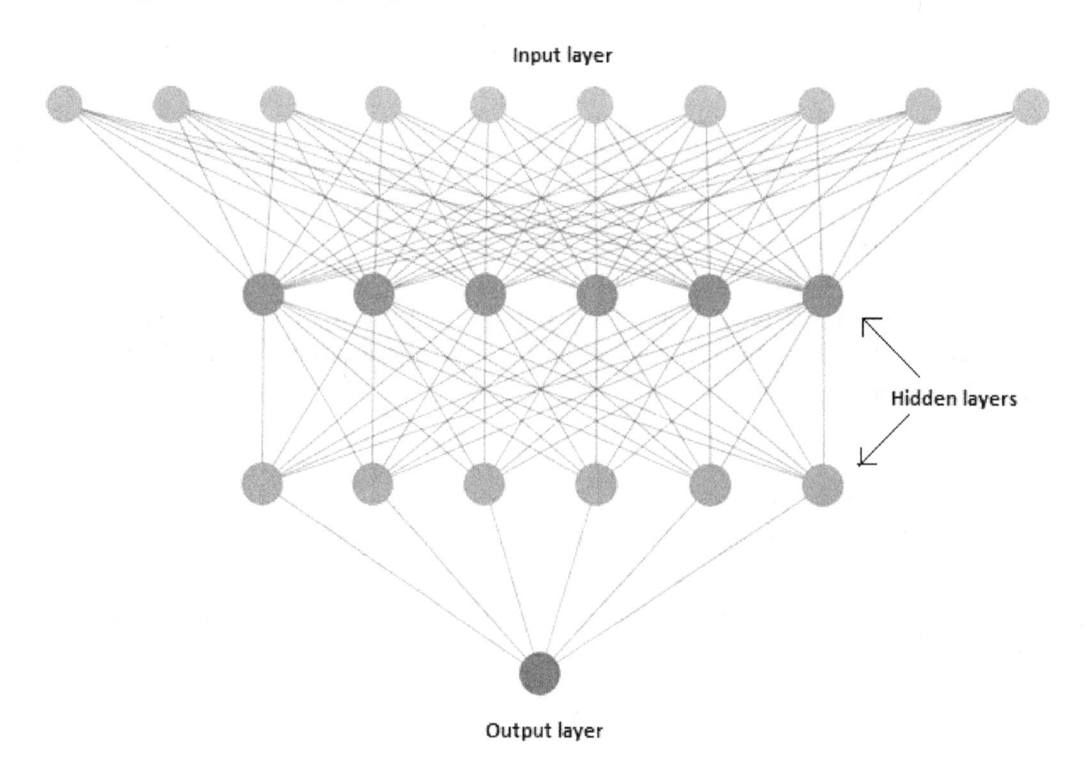

Long Short-Term Memory (LSTM) Networks

LSTM networks are a category of RNN capable of learning long-term dependency in sequence prediction problems (Wahyono et al., 2020). The most important factor in designing LSTMs is to solve the long-term dependency problem which usually arises in basic neural networks. The ability of LSTM to remember information for long periods, by default, and learn without struggle serves as the novelty and the biggest taking point of this network (Liu et al., 2020). All RNNs have the chain form of repeating modules of neural networks. LSTM differs in repeating modules; however, they retain a chain-like structure. A basic LSTM architecture is shown in Figure 3.

An LSTM module has three gates and a cell state. These provide them the power to learn, forget or remember information from each of the units. A cell unit consists of an input gate, output gate, and forget gate. One can add or remove the information to the cell state through this. The forget gate improves the accuracy of the module by deciding which information from the previous cell should be retained by using a sigmoid function.

LSTM's ability to learn and remember long sequences of input data makes it perfect for time series forecasting applications.

Figure 3. The architecture of long short-term memory network (Charco et al., 2021)

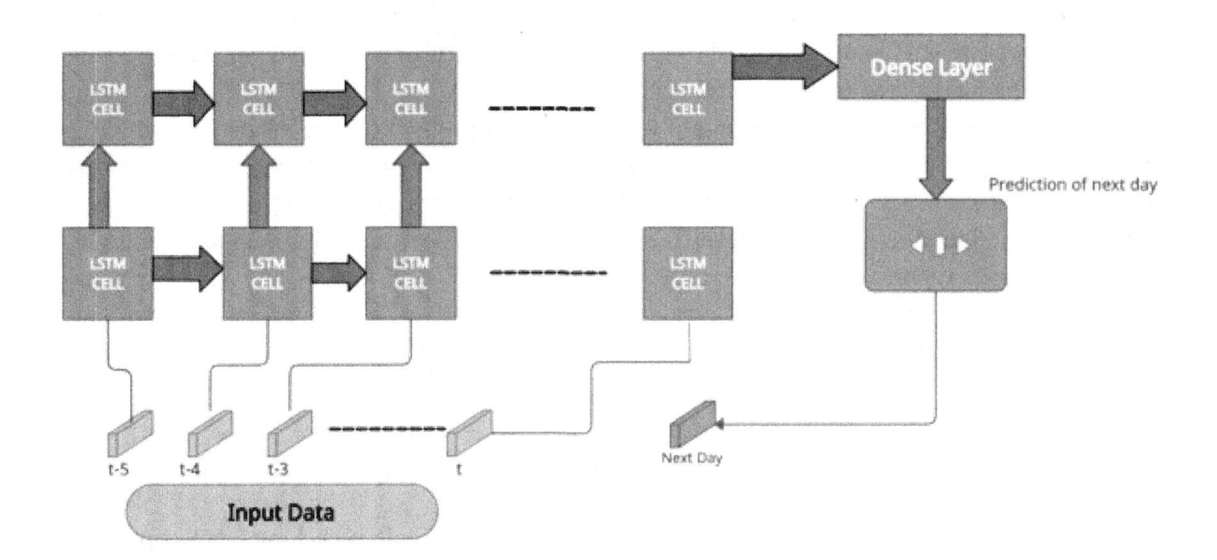

1D Convolution Neural Network-LSTM (1D CNN-LSTM)

By executing one-dimensional convolution operations with various filters, the 1D CNN can extract the effective and representative features of 1D time-series sequence data. The 1D CNN's convolutional filters and feature maps are all one-dimensional in this research, allowing it to match the one-dimensional character of rainfall data. The architecture of the 1D CNN-LSTM network is shown in *Figure 4*.

Figure 4. The architecture of the 1D CNN-LSTM network

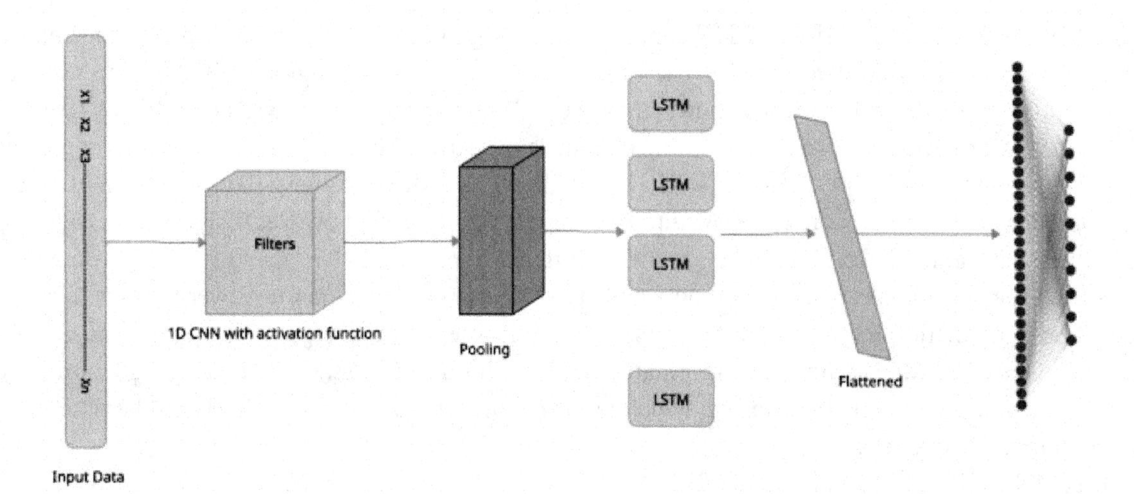

DATASET

Data of various weather parameters such as maximum temperature, minimum temperature, humidity, and sea level pressure are observed, collected, and maintained by IMD. The daily data of the Kolhapur station of Maharashtra, India, was collected from the National Data Centre of IMD, Pune. *Figure 5* shows the map of Kolhapur, located in the state of Maharashtra, India.

The daily surface data of Kolhapur station with station index 43157, were available from the year 2000 to the year 2019. The data comprised the daily data of over 30 parameters, including temperature, dew point temperature, dry bulb temperature, wet bulb temperature, relative humidity, average wind speed, wind direction, rainfall, station level pressure, mean sea level pressure, number of clouds and their direction, and evaporation.

Few of the parameters, namely temperature, average wind speed, and sea level pressure were continuous, while few such as wind direction and visibility were categorical.

EXPERIMENTATION

As mentioned in the methodology, the dataset was preprocessed. It included checking for missing records, converting the response variables into classes, and checking for categorical variables, if any.

Data Preprocessing and Analysis

There were missing records that were removed from the dataset during the cleaning stage. The total records available after the cleaning stage were over 5500. For the prediction at multiple lead times, three classes were considered. These classes are detailed in *Table 2* The class-wise number of records for the response variable available in the dataset is shown in *Table 3*.

Figure 5. Map of Kolhapur (Modified from the source: https://stategisportal.nic.in/stategisportal/Maharastra_BharatMaps/map.aspx)

Table 2. Rainfall classes

Rainfall (in mm)	Class
No rain	0 (No rain)
0.01 to 18	1 (Very light to light rain)
>18	2 (Moderate to heavy)

Table 3. Class wise number of records

Class	Number of records
No rain (0)	3338
Very light to light rain (1)	2226
Moderate to heavy rain (2)	259

Feature Selection

The dataset consists of 30 independent features and one dependent variable. The values of the dependent variables for the preceding days also become essential for forecasting; hence, in total, the model would need to learn the relationship between 60 variables and the corresponding output. These dimensions will increase in multiples of 30 if values of more than one previous day are considered. PCA has been applied to the dataset to reduce this dimensionality, which should help the model do a better job at generalizing. With the explained variance of 0.80, eleven PCA components were selected. *Figure 6* presents the selected features.

Figure 6. Most important features by Principal Component Analysis

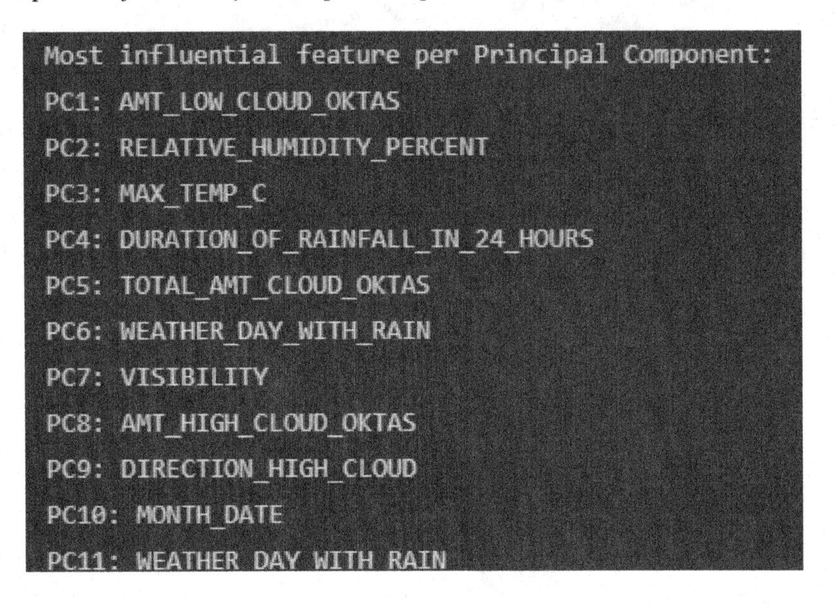

Model Parameters

Of total daily records, the dataset was divided into training and test datasets as 80% and 20%, respectively. Three prediction models, ANN, LSTM, and 1D CNN-LSTM were developed, trained, and tested. After each model was trained, it was tested for over 1100 records. The model parameter selection for LSTM is presented in *Table 4*.

Table 4. Main Parameters of single-layer LSTM

Parameter	Value
Hidden layer (s)	1 (LSTM)
Nodes in Hidden Layer(s)	12
Optimizer	Adam
Learning Rate	0.0001
Training Epochs	Dynamic Configuration (Early Stopping)
Activation function(s)	Linear, SoftMax
Batch Size	128

The competing ANN model was designed with the same parameters, the only difference being the type of hidden layer employed – Dense. The input is a PCA array, each row of which consists of the values of each of the 12 features on a given day and those for the preceding five days, including the rainfall class of the previous days. The output obtained is the probability score corresponding to each of the three output classes. *Figure 7* shows the single-layer architecture of the LSTM network for rainfall prediction.

Figure 7. Single-layer LSTM model architecture

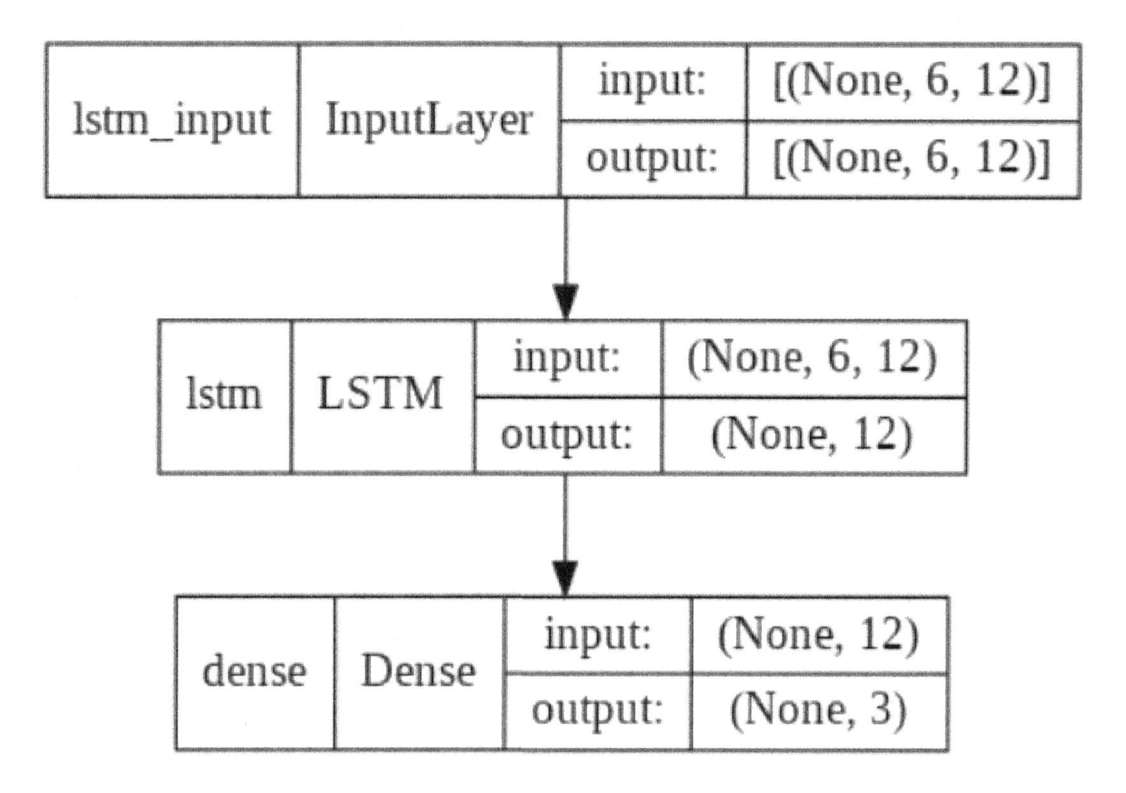

After training all the three models, each one was tested for one day- ahead rainfall prediction. The same model was then used for rainfall prediction for multiple lead times by systematically increasing the lead time of prediction by one day at a time. This procedure was repeated for getting the prediction for a lead time of up to six days. The results of training and testing of all the models are discussed in the next section.

EVALUATION AND RESULT ANALYSIS

Figure 8 shows the change in the accuracy with the number of epochs while training the LSTM model and *Figure 9* shows the variation in training and validation loss with the number of epochs. As can be seen from the graph, training continued till 303 epochs. It can also be observed that more than 80% accuracy was achieved during the training stage. Test results of all the models were then obtained. The false-positive results are shown in Table 5. False-positive results are considered errors as they show positive when observed values are negative. In this case, when there is no rain, the model classified it as a rainy day. As seen in Table 5, the developed ANN model gave more false positives than the LSTM and 1D CNN-LSTM models for all three classes of the response variable.

Figure 8. Graph of the number of epochs versus accuracy during LSTM training

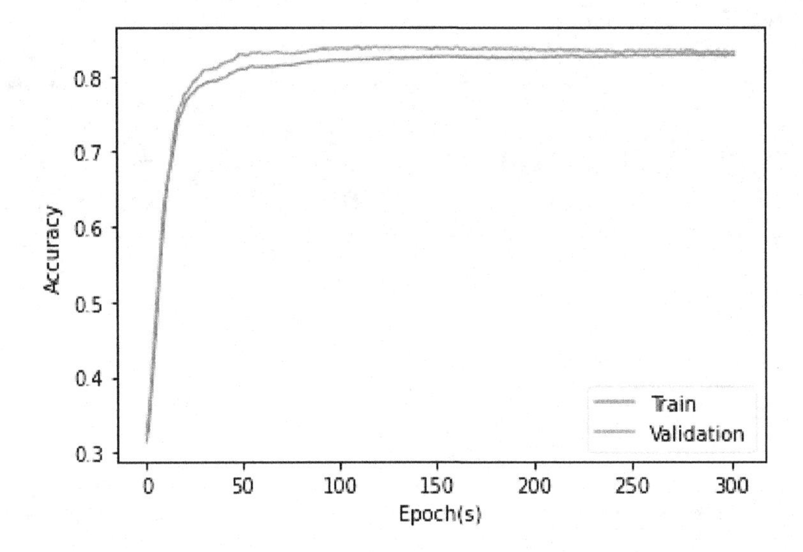

Figure 9. Graph of the number of epochs versus loss during LSTM training

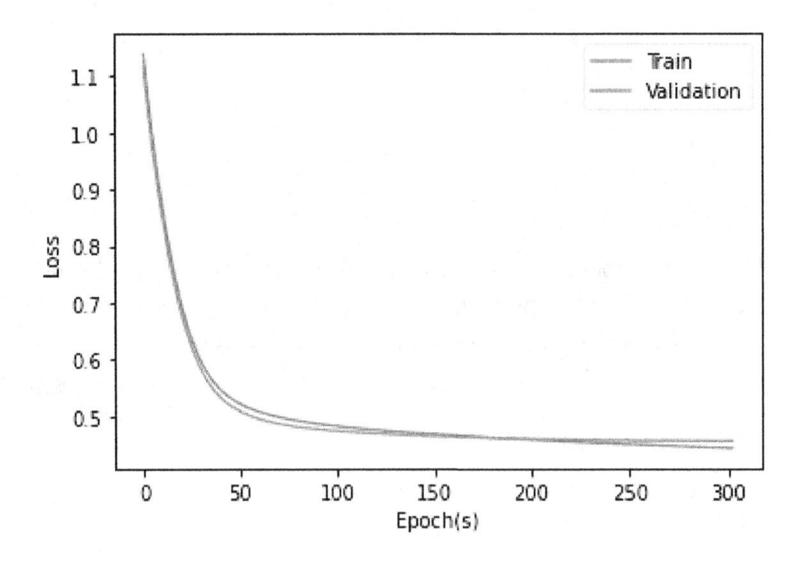

The performance of the three networks was then compared for their validation accuracy and F1 score. As shown in *Table 6*, LSTM outperformed with the highest accuracy, and the F1 score of the LSTM and 1D CNN-LSTM model was better than that of the ANN model.

The test performance of these three models for multiple lead time rainfall prediction in terms of accuracy was then compared and is shown in *Table 7*. The representation of output during testing the network is presented in *Figure 10*. The multiclass classification results can be better understood from the confusion matrix, and the same is presented in *Figure 11*. A confusion matrix visualization provides a quick summary of the prediction results of a model and is used to gauge the performance of the prediction models. *Figure 11* represents test results obtained for one-day lead time rainfall prediction by ANN, LSTM, and 1D CNN-LSTM.

Table 5. Model wise False positives (FPs) for the forecasting

CLASS	ANN	LSTM	1D CNN-LSTM
No rain	121	99	97
Very light to light rain	112	101	102
Moderate to heavy rain	4	0	1

Figure 10. Output format

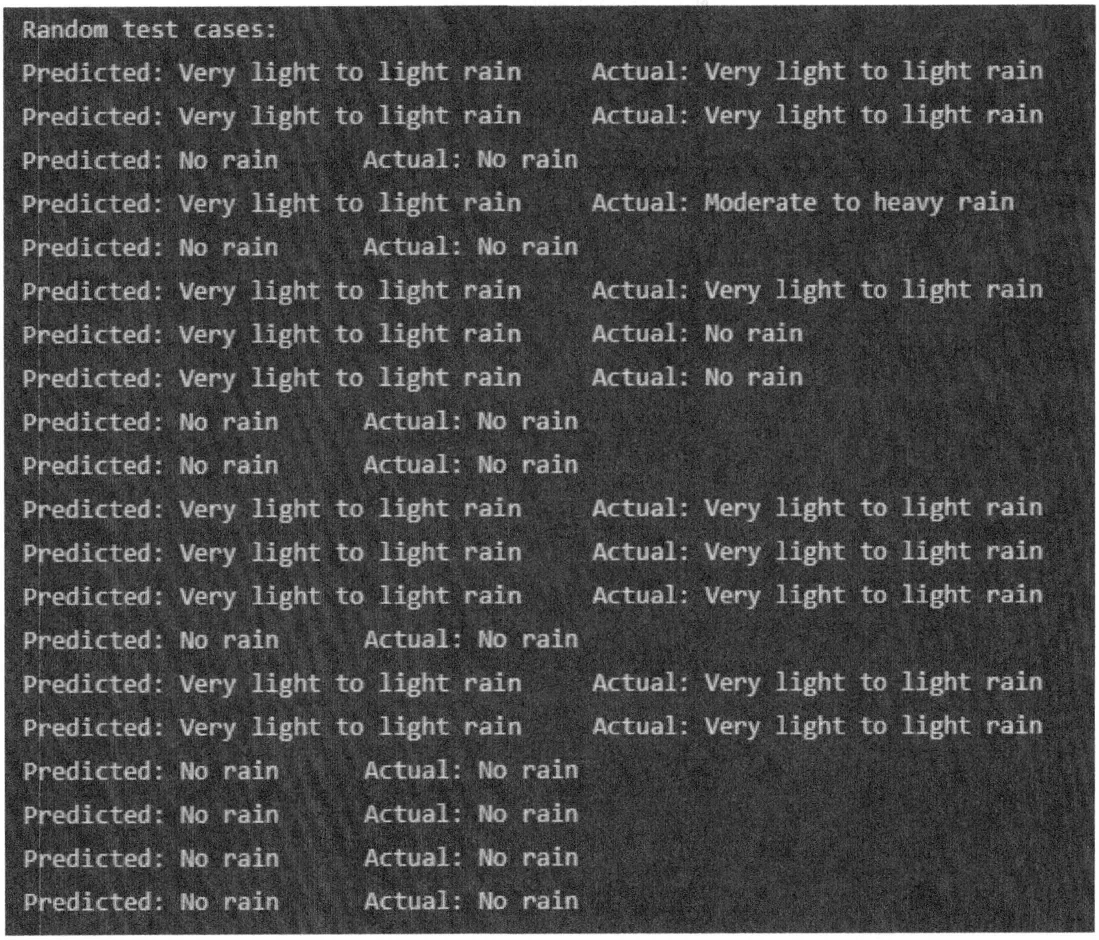

```
Random test cases:
Predicted: Very light to light rain    Actual: Very light to light rain
Predicted: Very light to light rain    Actual: Very light to light rain
Predicted: No rain      Actual: No rain
Predicted: Very light to light rain    Actual: Moderate to heavy rain
Predicted: No rain      Actual: No rain
Predicted: Very light to light rain    Actual: Very light to light rain
Predicted: Very light to light rain    Actual: No rain
Predicted: Very light to light rain    Actual: No rain
Predicted: No rain      Actual: No rain
Predicted: No rain      Actual: No rain
Predicted: Very light to light rain    Actual: Very light to light rain
Predicted: Very light to light rain    Actual: Very light to light rain
Predicted: Very light to light rain    Actual: Very light to light rain
Predicted: No rain      Actual: No rain
Predicted: Very light to light rain    Actual: Very light to light rain
Predicted: Very light to light rain    Actual: Very light to light rain
Predicted: No rain      Actual: No rain
Predicted: No rain      Actual: No rain
Predicted: No rain      Actual: No rain
Predicted: No rain      Actual: No rain
```

Figure 12 shows the confusion matrix of six-day lead time rainfall prediction by ANN, LSTM, and 11D CNN-LSTM. From *Table 7*, it can be observed that the LSTM model performed consistently for all the lead times, whereas 1D CNN-LSTM was better than LSTM for up to two days lead time prediction. Although the ANN model performed consistently, the accuracy was lower than that of the other two models. The employed LSTM model was able to do a good job with classes 0 and 1, which are 'No rain' and 'Very light to light rain,' respectively. The model, however, was not able to generalize class 2 accurately, which may be due to, among other reasons, the fact that there is an inherent label imbalance in the dataset, as shown in *Table 3*

Figure 11. Confusion Matrix: t+1 - ANN, LSTM and 1D CNN-LSTM respectively

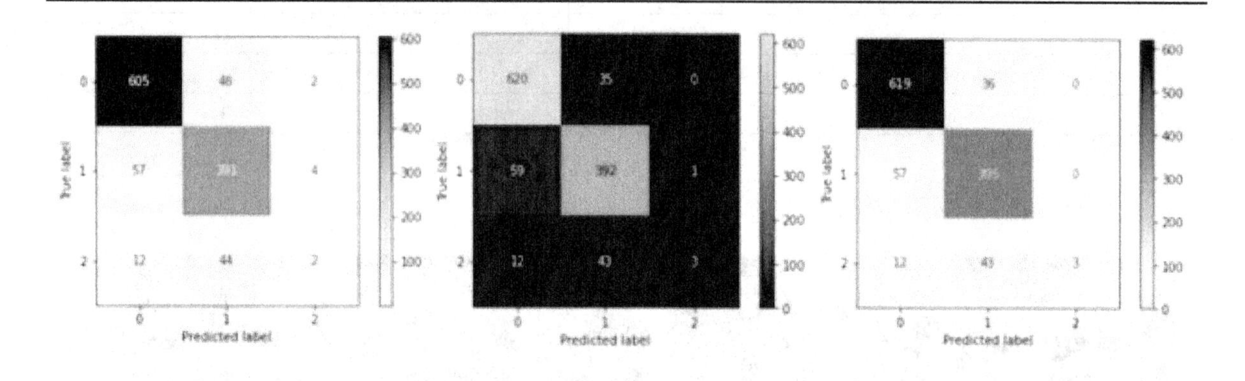

Figure 12. Confusion Matrix: t+6 - ANN, LSTM and 1D CNN-LSTM, respectively

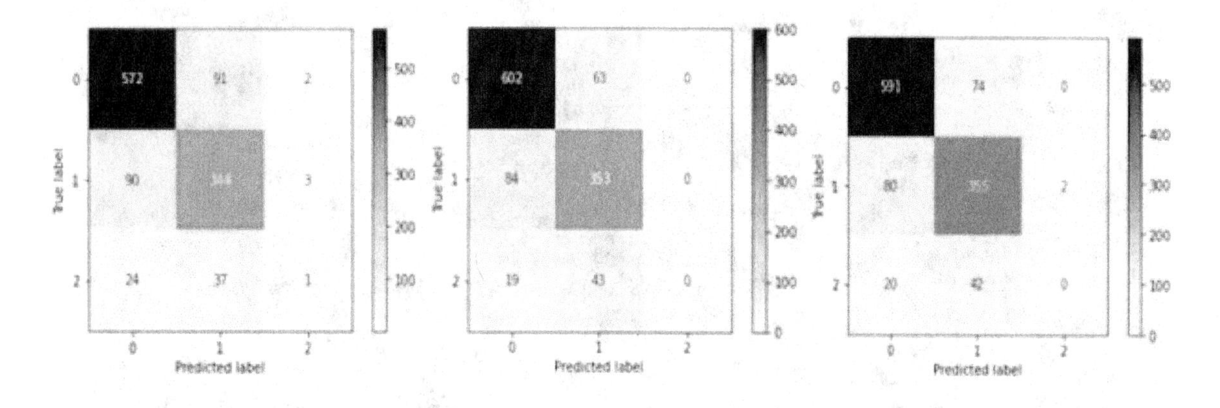

Table 6. Performance evaluation of ANN, LSTM and 1D CNN-LSTM

Parameter	ANN	LSTM	1D CNN-LSTM
F1-Score	0.78	0.81	0.81
Validation set accuracy (%)	76.81	82.82	80.01

Table 7. Accuracy for one to six days ahead rainfall prediction obtained with ANN, LSTM, 1D CNN-LSTM

Model	Lead time					
	t+1	**t+2**	**t+3**	**t+4**	**t+5**	**t+6**
The accuracy obtained with the ANN model (%)	81.12	76.87	75.63	75.11	77.54	75.06
The accuracy obtained with the LSTM model (%)	87.21	83.18	83.52	82.83	82.32	81.53
The accuracy obtained with the 1D CNN-LSTM model (%)	88.05	84.60	81.23	80.83	82.97	80.88

From *Figure 11*, it can be seen that for the lead time of (t+1), all the three models – namely, ANN, LSTM, and 1D CNN-LSTM performed well, with LSTM and 1D CNN-LSTM doing slightly better than the ANN.

The result in *Table 7* shows that the false positives in LSTM are less than ANN, and this single aspect demonstrates the usefulness of LSTM over ANN for multivariate data. Considering different evaluation parameters like accuracy and F1-Score for both ANN and LSTM, as shown in *Table 6*, it can be observed that LSTM performs better than ANN. There is a label imbalance in the dataset as the number of labels having a value of 2(moderate to heavy rain) is much smaller than the number of the other two labels (refer to *Table 3*).

The overall results reveal that for multiple lead time rainfall prediction, LSTM performs consistently with higher accuracy as compared to ANN and 1D CNN-LSTM networks.

FUTURE RESEARCH DIRECTIONS

The work described in the chapter was done for a station in Maharashtra six days ahead rainfall prediction. If developed as an application, this could be used as a rainfall alert system for farmers of a specific region. This work can be further extended to study other stations to form a cluster of homogeneous regions based on the rainfall. Another aspect in which this work can be extended is to fine-tune the hyperparameters of the network. This study has used numerical data, however, AI techniques such as deep learning need to explore image data. For example, the satellite images taken at a very short interval can be processed for feature detection and prediction.

CONCLUSION

Rainfall prediction is a complex task as it is dependent on many interrelated weather variables. Various tools of AI have been found promising for the solution of multivariate time series forecasting problems. Supervised techniques require appropriate and ample data to train the model. In recent times, because of advanced information and technology, vast amounts of weather parameter data are available. However, all the available data may not be suitable or may not be strongly correlated with the predictand. In this chapter, a study was carried out to predict rainfall at multiple lead times to compare the performance of three ML algorithms, namely, ANN, LSTM, and 1D CNN-LSTM.

Kolhapur station from the state of Maharashtra, India was selected as a case study. The dataset obtained from IMD, Pune was preprocessed. From the available weather variable data, important variables were selected by employing PCA. This helped to identify the 11 most significant predictors out of 30. ANN, LSTM, and 1D CNN-LSTM models were developed, trained, and tested for a day-ahead rainfall prediction. The lead time was then increased up to six days. The performance of these models was compared based on prediction accuracy and F1 score. The experiments revealed that the results obtained using the LSTM network were consistent in predicting rainfall at all the lead times and were more accurate than the other two models. The accuracy of ANN was about 75% for all the lead times whereas the 1D CNN-LSTM network showed comparable performance with about 80% accuracy at all the lead times.

Such prediction models can be utilized for farmers to know the rainfall pattern in advance, for common people who need to commute frequently, and for water resource managers for short-term water management.

ACKNOWLEDGMENT

The authors would like to thank the National Data Center of India Meteorological Department, Pune for providing the data for this study.

REFERENCES

Abhishek, K., Kumar, A., Ranjan, R., & Kumar, S. (2012). A rainfall prediction model using artificial neural network. *Proceedings - 2012 IEEE Control and System Graduate Research Colloquium, ICSGRC 2012,* 82–87. 10.1109/ICSGRC.2012.6287140

Ankenbrand, T., & Tomassini, M. (1996). Forecasting financial multivariate time series with neural networks. *International Symposium on Neuro-Fuzzy Systems, Proceedings*. 10.1109/ISNFS.1996.603826

Benevides, P., Catalao, J., & Nico, G. (2019). Neural network approach to forecast hourly intense rainfall using GNSS precipitable water vapor and meteorological sensors. *Remote Sensing, 11*(8), 966. Advance online publication. doi:10.3390/rs11080966

Brereton, R. G., & Lloyd, G. R. (2010). Support Vector Machines for classification and regression. *Analyst (London), 135*(2), 230–267. doi:10.1039/B918972F PMID:20098757

Chai, S. S., Wong, W. K., & Goh, K. L. (2017). Rainfall Classification for Flood Prediction Using Meteorology Data of Kuching, Sarawak, Malaysia: Backpropagation vs Radial Basis Function Neural Network. *International Journal of Environmental Sciences and Development, 8*(5), 385–388. doi:10.18178/ijesd.2017.8.5.982

Charco, J. L., Roque-Colt, T., Egas-Arizala, K., Pérez-Espinoza, C. M., & Cruz-Chóez, A. (2021). Using Multivariate Time Series Data via Long-Short Term Memory Network for Temperature Forecasting. *Advances in Intelligent Systems and Computing, 1273*(November), 38–47. doi:10.1007/978-3-030-59194-6_4

Darji, M. (2019). Rainfall forecasting using neural networks. *International Journal of Research and Analytical Reviews*.

Darji, M. P., Dabhi, V. K., & Prajapati, H. B. (2015). Rainfall forecasting using neural network: A survey. *Conference Proceeding - 2015 International Conference on Advances in Computer Engineering and Applications, ICACEA 2015,* 706–713. 10.1109/ICACEA.2015.7164782

Haq, D. Z., Novitasari, D. C. R., Hamid, A., Ulinnuha, N., Farida, Y., Nugraheni, R. D., Nariswari, R., Rohayani, H., Pramulya, R., & Widjayanto, A. (2021). Long short-term memory algorithm for rainfall prediction based on El-Nino and IOD data. *Procedia Computer Science*, *179*, 829–837. doi:10.1016/j.procs.2021.01.071

Hasan, S. B. M., & Abdulazeez, A. M. (2021). A Review of Principal Component Analysis Algorithm for Dimensionality Reduction. *Journal of Soft Computing and Data Mining*, *02*(01), 20–30. doi:10.30880/jscdm.2021.02.01.003

Hrudya, P. H., Varikoden, H., & Vishnu, R. (2021). A review on the Indian summer monsoon rainfall, variability and its association with ENSO and IOD. *Meteorology and Atmospheric Physics*, *133*(1), 1–14. doi:10.100700703-020-00734-5

Hudnurkar, S., & Rayavarapu, N. (2018). Performance of Artificial Neural Network in Nowcasting Summer Monsoon Rainfall: A case Study. *1st International Conference on Data Science and Analytics, PuneCon 2018 - Proceedings*. 10.1109/PUNECON.2018.8745413

Hudnurkar, S., & Rayavarapu, N. (2021). Predicting summer monsoon rainfall in the Indian Context: Impact and Challenges. In B. Veress & J. Szigethy (Eds.), *Horizons in Earch Science Research* (pp. 1–32). Nova Science Publishers, Inc.

Hudnurkar, S., & Rayavarapu, N. (2022). Binary classification of rainfall time-series using machine learning algorithms. *Iranian Journal of Electrical and Computer Engineering*, *12*(2), 1945–1954. doi:10.11591/ijece.v12i2.pp1945-1954

Hung, N. Q., Babel, M. S., Weesakul, S., & Tripathi, N. K. (2009). An artificial neural network model for rainfall forecasting in Bangkok, Thailand. *Hydrology and Earth System Sciences*, *13*(8), 1413–1425. doi:10.5194/hess-13-1413-2009

Jang, B., Kim, M., Harerimana, G., Kang, S. U., & Kim, J. W. (2020). Bi-LSTM model to increase accuracy in text classification: Combining word2vec CNN and attention mechanism. *Applied Sciences (Switzerland)*, *10*(17), 5841. Advance online publication. doi:10.3390/app10175841

Laing, A., & Evans, J.-L. (2011). *Introduction to Tropical Meteorology* (2nd ed.). The COMET® Progra. https://ftp.comet.ucar.edu/memory-stick/tropical/textbook_2nd_edition/index.htm

Lee, J., Kim, C. G., Lee, J. E., Kim, N. W., & Kim, H. (2018). Application of artificial neural networks to rainfall forecasting in the Geum River Basin, Korea. *Water (Switzerland)*, *10*(10), 1448. Advance online publication. doi:10.3390/w10101448

Li, T., Hua, M., & Wu, X. U. (2020). A Hybrid CNN-LSTM Model for Forecasting. *IEEE Access: Practical Innovations, Open Solutions*, *8*, 26933–26940. doi:10.1109/ACCESS.2020.2971348

Liu, F., Cai, M., Wang, L., & Lu, Y. (2019). An Ensemble Model Based on Adaptive Noise Reducer and Over-Fitting Prevention LSTM for Multivariate Time Series Forecasting. *IEEE Access: Practical Innovations, Open Solutions*, *7*, 26102–26115. doi:10.1109/ACCESS.2019.2900371

Liu, F., Lu, Y., & Cai, M. (2020). A hybrid method with adaptive sub-series clustering and attention-based stacked residual LSTMs for multivariate time series forecasting. *IEEE Access: Practical Innovations, Open Solutions*, *8*, 62423–62438. doi:10.1109/ACCESS.2020.2981506

Mishra, N., Soni, H. K., Sharma, S., & Upadhyay, A. K. (2017). A comprehensive survey of data mining techniques on time series data for rainfall prediction. *Journal of ICT Research and Applications*, *11*(2), 167–183. doi:10.5614/itbj.ict.res.appl.2017.11.2.4

Rathnayake, V. S., Premaratne, H. L., & Sonnadara, D. U. J. (2011). Performance of neural networks in forecasting short range occurrence of rainfall. *Journal of the National Science Foundation of Sri Lanka*, *39*(3), 251–260. doi:10.4038/jnsfsr.v39i3.3629

Saha, M., Pabitra, M., & Ravi, N. (2016). Autoencoder-based identification of predictors of Indian monsoon. *Meteorology and Atmospheric Physics*, *128*(5), 613–628. doi:10.100700703-016-0431-7

Shrivastava, G., Karmakar, S., Kumar Kowar, M., & Guhathakurta, P. (2012). Application of Artificial Neural Networks in Weather Forecasting: A Comprehensive Literature Review. *International Journal of Computers and Applications*, *51*(18), 17–29. doi:10.5120/8142-1867

Sikka, D. R. (2009). *Two decades of medium-range weather forecasting in India: National Centre for Medium-Range Weather Forecasting*. Center of Ocean-Land-Atmosphere Studies.

Singh, T., Mohadikar, M., Gite, S., Patil, S., Pradhan, B., & Alamri, A. (2021). Attention Span Prediction using Head-pose Estimation with Deep Neural Networks. *IEEE Access: Practical Innovations, Open Solutions*, *9*, 1–1. doi:10.1109/ACCESS.2021.3120098

Sivanandam, S. N., & Deepa, S. N. (2007). *Principles of soft computing*. Academic Press.

Soni, H. K., Mishra, N., Upadhyay, A. K., & Sharma, S. (2018). Development and Analysis of Artificial Neural Network Models for Rainfall Prediction by Using Time-Series Data. *International Journal of Intelligent Systems and Applications*, *10*(1), 16–23. doi:10.5815/ijisa.2018.01.03

Sood, V., Mehta, M., Mishra, V., Upadhyay, A., Hudnurkar, S., Gite, S., & Rayavarapu, N. (2021). A Bibliometric Survey on the Use of Long Short-Term Memory Networks for Multivariate Time series forecasting. *Library Philosophy and Practice*, *2021*, 1–21.

Sri Rahayu, I., Djamal, E. C., Ilyas, R., & Bon, A. T. (2020). Daily temperature prediction using recurrent neural networks and long-short term memory. *Proceedings of the International Conference on Industrial Engineering and Operations Management,* 2700–2709.

Srinivas, C. V., Hariprasad, D., Bhaskar Rao, D. V., Anjaneyulu, Y., Baskaran, R., & Venkatraman, B. (2013). Simulation of the Indian summer monsoon regional climate using advanced research WRF model. *International Journal of Climatology*, *33*(5), 1195–1210. doi:10.1002/joc.3505

Tran, N., Nguyen, T., Nguyen, B. M., & Nguyen, G. (2018). A Multivariate Fuzzy Time Series Resource Forecast Model for Clouds using LSTM and Data Correlation Analysis. *Procedia Computer Science*, *126*, 636–645. doi:10.1016/j.procs.2018.07.298

Wahyono, T., Heryadi, Y., Soeparno, H., & Abbas, B. S. (2020). Enhanced lstm multivariate time series forecasting for crop pest attack prediction. *ICIC Express Letters*, *14*(10), 943–949. doi:10.24507/icicel.14.10.943

Wang, S. (2003). Artificial Neural Network. In *Interdisciplinary Computing in Java Programming* (pp. 81–100). Springer US. doi:10.1007/978-1-4615-0377-4_5

Wang, Y. S., Gao, J., Xu, Z. W., Luo, J. D., & Li, L. X. (2020). A prediction model for ultra-short-term output power of wind farms based on deep learning. *International Journal of Computers, Communications & Control*, *15*(4), 29–38. doi:10.15837/ijccc.2020.4.3901

Wojtkiewicz, J., Hosseini, M., Gottumukkala, R., & Chambers, T. L. (2019). Hour-ahead solar irradiance forecasting using multivariate gated recurrent units. *Energies*, *12*(21), 1–13. doi:10.3390/en12214055

Wu, S. (2020). Research on Wind Power Ultra-short-term Forecasting Method Based on PCA-LSTM. *IOP Conference Series: Earth and Environmental Science*, *508*(1), 0–8. doi:10.1088/1755-1315/508/1/012068

Xu, G., Ren, T., Chen, Y., & Che, W. (2020). A One-Dimensional CNN-LSTM Model for Epileptic Seizure Recognition Using EEG Signal Analysis. *Frontiers in Neuroscience*, *14*(December), 1–9. doi:10.3389/fnins.2020.578126 PMID:33390878

Yu, Y., Cao, J., & Zhu, J. (2019). An LSTM Short-Term Solar Irradiance Forecasting under Complicated Weather Conditions. *IEEE Access: Practical Innovations, Open Solutions*, *7*, 145651–145666. doi:10.1109/ACCESS.2019.2946057

KEY TERMS AND DEFINITIONS

1-Dimensional Convolution Neural Network Long Short-Term Network: A long short-term memory network that uses the properties of a convolutional neural network to extract features, is called a 1D CNN LSTM network.

Artificial Neural Network: The network that uses some functions and tries to mimic the function of the human brain is called an artificial neural network.

Confusion Matrix: The way to present the output of the classifier is called as confusion matrix. It enables the user to understand how the classifier performed in terms of various evaluation parameters such as accuracy, precision, and F1 score.

Epochs: While using supervised Machine Learning techniques, the network is trained by presenting some examples. To reduce the error between the predicted and observed examples, the training data is repeated and presented to the network. These repetitions are called epochs.

Long Short-Term Memory Network: An artificial intelligence network that can retain past information by using a gate-like structure built using functions is called a long short-term memory network.

Multiclass Classification: In artificial intelligence, dividing the response variable into more than two classes and determining the original class using AI techniques is called multiclass classification.

Principal Component Analysis: When the dataset contains many features, a technique that returns the most important set of features without much loss of information contained in the dataset, is called principal component analysis.

Chapter 8
Deep Learning Solutions for Analysis of Synthetic Aperture Radar Imageries

Nimrabanu Memon
Pandit Deendayal Energy University, India

Samir B. Patel
https://orcid.org/0000-0002-4280-6446
Pandit Deendayal Energy University, India

Dhruvesh P. Patel
https://orcid.org/0000-0002-2074-7158
Pandit Deendayal Energy University, India

ABSTRACT

The potential of Synthetic Aperture Radar (SAR) to detect surface and subsurface characteristics of land, sea, and ice using polarimetric information has long piqued the interest of scientists and researchers. Traditional strategies include employing polarimetric information to simplify and classify SAR images for various earth observation applications. Deep learning (DL) uses advanced machine learning algorithms to increase information extraction from SAR datasets about the land surface, as well as segment and classify the dataset for applications. The chapter highlights several problems, as well as what and how DL can be utilized to solve them. Currently, improvements in SAR data analysis have focused on the use of DL in a range of current research areas, such as data fusion, transfer learning, picture classification, automatic target recognition, data augmentation, speckle reduction, change detection, and feature extraction. The study presents a small case study on CNN for land use land cover classification using SAR data.

DOI: 10.4018/978-1-6684-3981-4.ch008

INTRODUCTION

One of the best examples of an active microwave remote sensing system is Synthetic Aperture Radar (SAR), which includes radiation with 1mm longer wavelengths. Another benefit is the longer wavelength. Objects ten times smaller in size than the microwave wavelength are almost transparent in this region, allowing it to penetrate clouds and operate in all weather conditions.

It has revolutionized many earth observation areas such as agriculture, hydrology, and the study of surface deformation, urban planning, marine applications, space technology, and others. Table 1 shows the applications of SAR at various wavelengths. The ability of the microwaves to penetrate through clouds, atmospheric constituents, and even subsurface penetration has made this technology irreplaceable. SAR sensors in space have improved their ability to discern individual radar characteristics on the earth's surface. In contrast, aerial sensors can create very high-resolution images.

Table 1. SAR wavelengths and their applications (Podest, 2018)

Frequency	Wavelength	Application
VHF	300 kHz – 300 MHz	Biomass, Foliage, and Ground penetration
P	0.3 GHz – 1 GHz	Soil moisture and Biomass
L	1 GHz – 2GHz	Agriculture sector, Forestry, and Soil moisture
C	4 GHz-8 GHz	Ocean, and agriculture
X	8 GHz – 12 GHz	Agriculture, ocean, high-resolution radar
Ku	14GHzZ – 18 GHz	Glaciological applications, snow cover mapping
Ka	27GHzZ – 47 GHz	High-resolution radars

However, it is not easy to deal with SAR images. SAR images are more than two-dimensional images, like in the case of optical data. They are associated with many distortions, and grainy textured speckles affect their quality.

CHALLENGES WITH SAR IMAGERIES

Radar backscattering response strongly depends on the orientation of targets. Due to slant range geometry, SAR imageries foreshortening, layover, and shadow effects dominate undulating or hilly terrain. These effects limit the interpretation of SAR data.

Also, the scattering mechanism and speckle in the SAR image make it difficult to understand and are visually different from optical imageries. Searching a tiny target of interest in such imageries by visual inspection is very time consuming and requires lots of expertise and a good amount of knowledge of the area under study)

(Wang et al., 2015). Without prior experience, it is almost impractical for non-experts (students, researchers willing to work on SAR datasets for their research interest) to gain useful information from SAR imagery and promote the development of automatic techniques for information retrieval (automatic target recognition).

The trend shows a need for an increase in the information content of SAR images. Traditional methods involve multiple operations like multi polarisation, multi-frequency, multi-temporal SAR images, and enhanced azimuth and range resolution. However, interpreting a single SAR image is a complex and tedious task. Combining several images for enhanced information retrieval can further increase the nonlinearity and complexity in the image analysis.

As a result, there is limited ground truth information available with SAR data. Thus, it is difficult for researchers working in this field to effectively classify the data and utilize entire information from SAR data. We will discuss the modern techniques (deep learning techniques) widespread nowadays in remote sensing and how these techniques can aid some of the challenges associated with SAR data.

There are three types of SAR data: polarimetric SAR, interferometric SAR, and differential interferometry. To cover SAR data analysis of each type is beyond the scope of this paper. Thus, the article will only include polarimetric SAR data analysis.

POLARIMETRIC SAR (POLSAR) DATA ANALYSIS

Polarimetric information is essential to both passive and active microwave remote sensing. It is the most challenging aspect of microwave remote sensing. One can derive qualitative and quantitative physical information of urban, land, ocean, ice, snow, etc., utilizing polarimetric information from artificial and natural targets.; hence it is widely used in various earth observation applications. In this context, the researchers' primary goal is to gain as much information as possible from the image by exploiting the electric vector's polarimetric information (in both vertical and horizontal directions) of the incident E.M. wave. To study the polarization, researchers mainly use scattering information of the E.M. wave in the image.

Scattering Mechanisms

The specular behavior of the earth's surface is the leading cause of single scattering; unique interaction between the E.M. wave and operating environments accompanied such dispersion.

Single Bounce Scattering

1. Preserves the orientation angle of polarization
2. The first and fourth elements, i.e., SHH and S.V. V of the scattering matrix, are equal.

Double Bounce Scattering

Even (or double) bounce scattering is the two successive surface scattering that occurs due to the interaction of E.M. waves with two locally orthogonal surfaces. In other words, the incident EMR reflected by the first specular surface becomes an incident wave for the second surface. The second surface is spatially orthogonal to the first scattering surface before reaching the receiver. The resulting scattering phenomenon will be a dihedral or double-bounce scattering figure 1 as the receiver receives the maximum information reflected from the two surfaces. Such type of scattering mechanism is found mainly in urban areas. Thus urban areas have a very high backscatter value in all polarization states. Hence, it appears bright in all the images.

Texture Analysis

Growing crops contains a wide range of canopy with different geometrical structures, shapes, and sizes. Some plants or at least some of their components are sensitive to the orientation of the SAR sensor. Before going into detail about texture analysis and texture feature extraction, let us first understand what texture means in the image and why it is necessary to identify the objects and features in the satellite image. Any image, satellite imagery, aerial photograph, or image used for medical purposes such as x-ray, etc.,is associated with two crucial characteristics: tone and texture. The image tone is the varying tone of the gray shades of the resolution cells, such as pixels. The texture is the spatial or statistical distribution of the gray tones (Haralick et al., 1973). The relationship between texture and tone is intertwined, much like a wave and a particle.

Figure 1. Different scattering mechanisms (Pottier and Ferro-Famil, 2008)

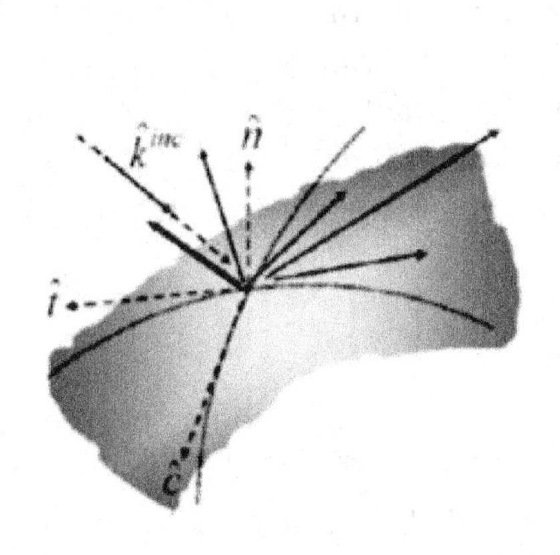

(a) Single bounce scattering from surface

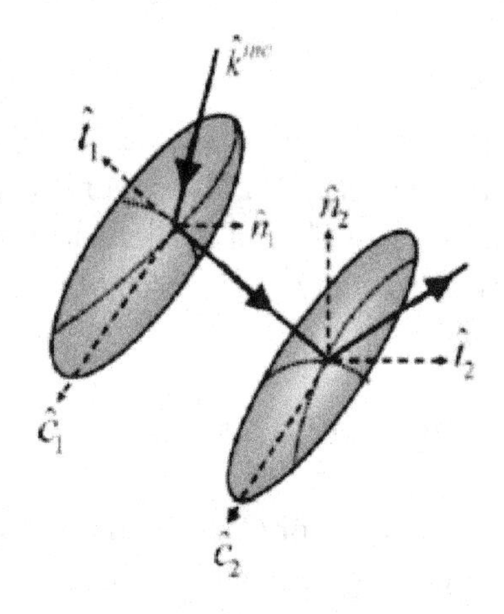

(b) Dihedral or double bounce scattering from two orthogonal surfaces

The texture detects land surface features in an image based on topographical characteristics. Surface texture is examined and measured using smoothness, coarseness, and regularity. It is a regional descriptor of the information present in the image for image analysis (Zhang et al., 2009). Aside from that, it contains critical information on the structure of surface characteristics and their link to the Environment. Haralick et al. (1973) classified it as fine, coarse, smooth, rippling, molled, lineated, or irregular. Consider an example of rice crop monitoring and mapping in SAR. Water covers the rice fields during the sowing period. As a result, the image appears smooth with a black tone. However, as time passes and the crop grows, the backscatter value rises, as does the level of roughness in the image. Thus field appears brighter as the level of roughness on the surface rises. Thus, one can use a texture with phenology knowledge to identify and discriminate different crops under study. The texture is a vital characteristic of an image that helps categorize the surface's type and land use under investigation. Thus, it can be beneficial in various remote sensing applications.

STATE OF THE ART TECHNIQUES FOR SAR IMAGE ANALYSIS

Machine learning algorithm learns from data and enables us to tackle and fix problems too difficult to solve with the classic methods or programs written and designed by humans. Deep learning is an advanced machine learning technique that is helpful for image recognition and classification.

The most important advantage of using deep learning over traditional visual recognition methods is that they learn automatically from the data.

Deep learning can learn valuable features (information) from an image for new specific tasks. It does not require much expertise and effort to design features by hand.

It consists of neural networks with more than two layers and thus is called deep. Also, it makes better use of big data. It provides a comprehensive end-to-end learning framework that integrates learning feature transformations and classification through the backpropagation process (Zhang et al., 2016). Internet companies, such as Microsoft, Google, Facebook, etc., use DL in various image analysis activities such as face recognition, image classification, object detection, etc.

Deep learning provides better presentations of the feature of remote sensing-related applications. Many papers include using deep learning algorithms that analyze SAR images, such as automatic target detection, parameter inversion, terrain categorization, despeckling SAR images, data fusion. For example, many authors have demonstrated the ability of "Convolutional neural networks (CNN)" to monitor the intermediate and high level of quality from raw images through convolutional and pooling layers. Recent research has shown that features presented by CNN are beneficial for large scale image recognition (Krizhevsky et al., 2012, He et al., 2016, Simonyan and Zisserman, 2014), object detection (Redmon et al., 2016, Girshick et al., 2016), and semantic segmentation (Long et al., 2015, Noh et al., 2015).

SCOPE OF MACHINE LEARNING IN SAR DATA ANALYTICS

For Inadequate Labeled Data

Deep learning needs a good amount of training data to converge. In other words, the higher the amount of ground truth data, the higher the algorithm's accuracy, but simultaneously it becomes expensive due to an increase in computational cost. However, one can get high-resolution SAR data with temporal variability but limited labeled data (training data with ground truth information). This problem is not restricted to SAR but is also associated with remotely sensed datasets.

Again, most deep learning techniques face a big challenge of data inadequacy to yield accurate results. So the question arises of what modern techniques are available to deal with a limited amount of labeled datasets for training. Here are some solutions we have identified through literature (Ball et al., 2017), which can be helpful in the absence of abundant labeled data:

Using Transfer Learning: Train algorithm on one imagery, extract some mid-level features from the input data, and use it to predict or classify (for instance, image classification problem). The transfer learning approach may be advantageous when working with temporal datasets. It allows the algorithm to be trained on a single date dataset with ground truth and predict the response on different date datasets (change detection, crop phenology monitoring). In their paper, the authors of (Pan and Yang, 2010) used dual CNN with a transfer learning approach to improve the network's learning.

Data Augmentation: Data augmentation is a strategy to gradually increase diversity in a training database without collecting new data. Standard methods of augmenting data are scaling, translation, 90o rotation, finer angles, flipping or adding salt and pepper noise, and visual effects. Many studies use this technique to detect a target using a SAR database automatically. For example, (Ding et al., 2016) used a combination of CNN and three data augmentation methods to detect a SAR target. They register test data on training data by using translation based on some criteria. They individually applied translation with CNN and linear SVM. Experiments showed that translation with CNN outperforms translation with linear SVM since the CNN method can handle 2D images directly and exploit each pixel's context through convolution. (Lewis et al., 2019) investigated three machine learning networks to fill the synthetic and measured data research gap. They used two generative adversarial networks and modified a convolutional autoencoder to address the disparity between synthetic and measured data. They did this by generating new, realistic labeled data obtained by translating data from two domains (synthetic and measured) and joining the manifold into an intermediate representation.

An Improvement Over Model-Based Techniques (For Nonlinear Data)

SAR systems such as crop monitoring require external background soil information, plant height, crop density, plant rows and cropping, crop sensitivity, moisture content, etc. All this information together makes crop monitoring very difficult and highly nonlinear. Models using all these parameters can be time-consuming by manual approach or inaccurate due to inadequate data collection or error in handling the interrelationships among the input variables.

In (Shanmugapriya et al., 2019), authors studied the multiparametric SAR data to identify bajra (pearl millet) in mixed crop conditions. Among the parameters under study, the backscattering coefficient was highly sensitive to the vegetation canopy's dielectric constant and other factors such as row orientation of plantation, density, etc. These parameters change for different crops. Thus, the interconnectivity of each parameter made the problem (i.e., discrimination of pearl millet from other crops) highly complex and nonlinear. Deep learning (DL) models such as ANN, CNN, RNN, etc., can handle complex and nonlinear data and thus may be helpful to ease the complexity associated with such input data. DL algorithms deal with pixel-level processing, but they also consider the spatial correlation of each pixel with its surrounding pixels and objects. Also, DL algorithms have shown their potential in learning hierarchical features from the input data and from learning smaller features to learning complex & more abstract features in the deeper portions. Thus deep learning algorithms can aid the complexity & nonlinearity of the data and provide excellent results.

Despite these benefits, before incorporating any DL algorithms into a study, a better thorough understanding of the deep learning architectures yields better results because deeper architectures are time-consuming, can be computationally expensive, and may require high computational graphics such as GPU to run the algorithms. Also, optimizing DL architectures is essential for every researcher or practitioner to consider accurate and reliable results obtained through DL algorithms.

Feature Extraction Using Deep Learning

Image features are extracted based on the type and information required for a particular application. For example, feature extraction may be pixel-based, object-based, or structural. Conversely, deep learning can represent and organize many levels of information. It can reveal complex relationships within data (Cire¸san et al., 2010). Deep learning can map different levels of data extraction and integrate them from low to high levels (Riesenhuber and Poggio, 1999). For instance, deep learning can represent the scene as a unitary transformation in scene recognition problems using local spatial variations and structural patterns captured by low-level features. Here no external segmentation stage or separate object extraction is needed (Zhang et al., 2016).

DL for SAR Image Classification

PolSAR image classification is an essential application of radar remote sensing. Traditionally, decomposition techniques based on pixel-wise target decomposition parameters classify PolSAR images. Literature showed that they hardly consider spatial patterns that contain rich information in PolSAR imageries (Xu et al., 2016). Deep learning uses spatial and polari-metric information to automatically extract efficient features from the imagery for classification.

In (Chen et al., 2003), the author combined "Fuzzy c-means (FCM)" and the fast-paced practice "Dynamic Learning Neural Network" (DLNN) to classify fully polarimetric SAR data (AIRSAR) acquired by JPL-Airborne SAR System. It used the statistical properties of the image and tried to take ad- vantage of a fuzzy neural network. Complex Wishart Distribution computes the distance measure used in this method. Instead of preselecting the polarized channels, the author used all [C] matrix elements to form a feature vector. FCM used during the training phase effectively handled the mixed pixels. The fuzzy neural network used fuzzy partitioning in a supervised manner, and the complex Wishart distance assisted in exploring the full polarimetric feature for classification. The authors (Lardeux et al., 2009) employed SVM on AIRSAR polarimetric data for tropical vegetation cartography. They derived and evaluated different polarimetric indicators' contributions from full-pol data. (Krizhevsky et al., 2012) trained a deep convolutional network to classify about 1.2 million high-resolution imageries and gained top-1 & top-5 error rates on the test data in the ImageNet LSVRC-2010 contest. (Zhou et al., 2016) used CNN for polarimetric SAR image classification using six-real channeled data of the covariance matrix. (Lv et al., 2015) used a deep belief network (DBN) to classify PolSAR data for urban land use and land cover.

[Geng et al., 2015] used CNN and A.E. to create a deep convolutional A.E. to classify high-resolution SAR images. In another paper by [Hou et al., 2015], Restricted Boltzmann Machines (RBM) filters the noisy SAR image, and a three-layer deep belief network (DBN) was given the resultant image as input to perform the classification task.

Multi-Sensor Data Fusion Based on DL Algorithms

The progress in remote sensing technology and the addition of different sensors such as biosensors, chemical sensors, and onboard satellite sensors also increases the volume of data from all these sensors. So, a need arises to combine data from sensors to extract optimal information from them. Data fusion can be an effective way to utilize the enormous volume of data from numerous sources. In other words, data fusion means combining different datasets from multiple sensors and multiple sources to get more useful information from the data. The fusion of information from various sensors having different physical characteristics can enhance the level of understanding of our surroundings. Fusion can provide the basis for planning, decision-making, and control autonomous and intelligent machines (Hall and Llinas, 1997).

Before deep learning, Intensity-Hue-Saturation (IHS), High Pass Frequency (HPF), Principal Component Analysis (PCA), and two-dimensional Discrete Wavelet Transform (DWT) were employed in the field of data fusion for the fusion of two images (Abdikan et al., 2008). For instance, in [Pohl and Van Genderen, 1998], the authors applied data fusion to pattern recognition, object detection, visual enhancement, and area surveillance. The paper (Kussul et al., 2016) proposed advanced intelligent methods to fuse data acquired by multiple heterogeneous sources using machine learning techniques.

In (Schmitt et al., 2018) paper, authors proposed "SENI-2", deep learning-based SAR data fused with optical imagery contains 282,384 corresponding image patches of c-band Sentinel -1 and optical image Sentinel-2. It was collected over the globe and through all the meteorological seasons. It showed promising results for applications like SAR image colorization, creating artificial optical imageries from SAR datasets, and SAR-Optical image matching. The researchers proposed the fused data to support the R.S. community's advancement in deep learning.

Authors (Waske and Benediktsson, 2007) exploit multi-temporal SAR images with optical images to classify multisensor datasets. A multisensor dataset is a fusion of two datasets acquired from two different sensors. In this paper, first, the image was separately classified using SVM and was used in subsequent fusion with the optical dataset, which was again classified using SVM. Also, they used voting schemes to prepare the final classification of the fused image. It was then compared with existing traditional classification techniques like maximum likelihood (XML) classifier & decision tree and found that the adopted approach was best among these and significantly improved the classification accuracy of the multisensor dataset.

SOME DEEP LEARNING TECHNIQUES AND THEIR ARCHITECTURE FOR SAR DATA CLASSIFICATION

This section provides brief information on important and popular deep learning techniques for SAR image segmentation and classification.

Neural Networks

SAR images can be parametric, semi-parametric, and nonparametric approaches. Neural networks lie in semi-parametric models between parametric and nonparametric extremes for probability density estimation (Sugawara and Nikaido, 2014). They do not make strong assumptions on probability distributions, and hence they are flexible on the system's complexity. They, therefore, are useful to build models. A neural network classifier can be single as well as multi-layered. Single-layer networks can fully solve problems in which hyperplanes linearly separable training observations. On the other hand, in networks with just one additional layer (hidden layer), input data is entirely transformed by the hidden layer into a higher dimension where the problem becomes linearly separable (Montavon et al., 2012). The same technique is useful in designing SVM for the classification of data.

Learning

It is required to teach every neural network how to perform specific tasks. This process is iterative. Generally, there are two ways to perform these tasks: supervised and unsupervised learning. Supervised cases require predictors (x1, x2, x3,…., xn) and corresponding responses (y1, y2, y3, ..., yn). The predictor data feed the model, and it produces outputs (y1′, y2′, y3′, …., yn′). In this type of learning, the model guided by the given output response (y1, y2,.., yn) is used to predict the output (y1′, y2′, y3′, …., yn′) and thus is said to be a supervised approach, by minimizing an objective function commonly known as the cost function. Regression and classification problems work on the principle of supervised learning. In contrast, only the input data is provided for the unsupervised learning approach. The algorithm discovers the natural patterns and structures in the data. Clustering and principal component analysis (PCA) uses the unsupervised learning approach.

Neural Network (N.N.) Models

N.N. model is dependent on the following factors:

1. Layers in the network
2. The number of Neurons within the layer
3. Neuron to layer and layer to layer connection
4. Selection of activation function
5. Weighting function on the input layers
6. Training of network

The model's accuracy depends on these parameters, so these parameters should be optimized during the design of the network model. They vary with respect to input data and the type of application for which the network is applied (classification/segmentation).

Network Architecture

Usually, the connection between the neurons is unidirectional. However, there are some networks where there is a bidirectional connection. These connections fall into two categories: interlayer and intra layer, which describe the connection between layers and within. There are several types of N.N.; the most commonly used N.N. type is "feed-forward" N.N (figure 2). The signal in the model only proceeds forward from input layers, passing through any intermediate layers and reaching to output layer without any feedback. In other words, if the network can take the input layer only from the forward direction, then the network is called a feed-forward network. Whereas if the output is sent back to the preceding layer as input, i.e., the output is fed back to each neuron in the same layer, the network is called feedback or lateral feed connection. Moreover, if the excellent consists of connection of loops, then it is referred to as recurrent network [Alawlaki, 2006].

Figure 2. Neuron in Artificial Neural Network

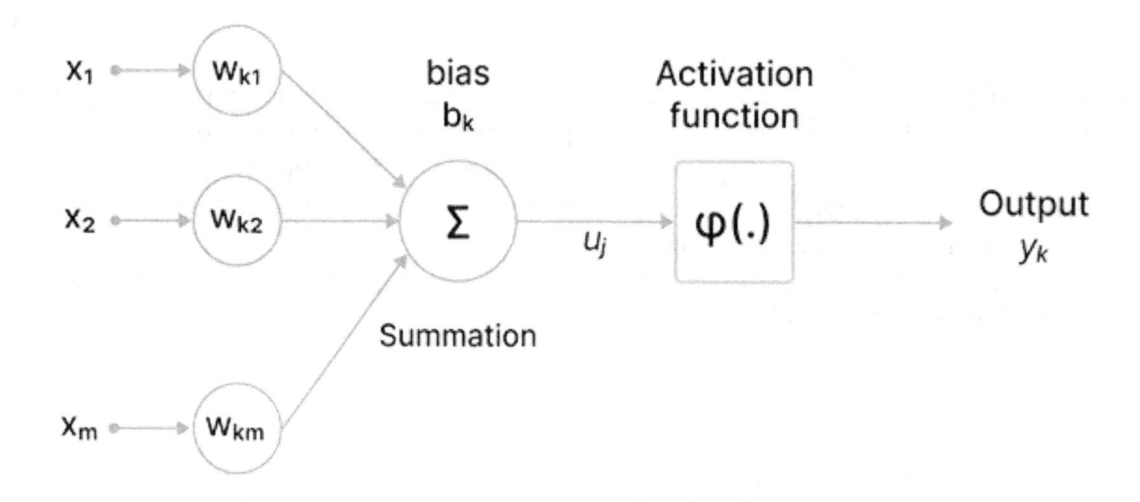

Artificial Neural Networks (ANN)

ANN is a popular tool to analyze remotely sensed data. Many researchers have proven the potential of an artificial neural network to be a powerful and self-adaptive method in pattern recognition compared to traditional linear models and simple nonlinear analysis (Alawlaki, 2006, Keiner and Yan, 1998). From the early 1990s, artificial neural networks have been applied to analyze the R.S. images with good results (Atkinson and Tatnall, 1997). Figure 2 depicts the ANN architecture.

The advantage of using ANN:

1. It can learn complex functional relationships between input and output data by employing a non-linear response function, iterating many times in particular network architecture (Lek and Gu'egan, 1999).
2. It can generalize in a noisy environment and thus is more robust for incomplete or imprecise data (Hewitson and Crane, 1994).
3. To incorporate a priori information and realistic physical parameters into the analysis (Foody, 1995b, Foody, 1995a).

Additionally, ANN provides a supervised learning approach with less training data than the maximum probability. The rules to recognize a category depend on both this category and other class categories (Paola and Schowengerdt, 1995). Moreover, artificial neural networks also allow fuzzy classifications by considering the activation values to measure fuzzy membership that belongs to a class (Civco, 1993, Foody, 1995b Carpenter et al., 1999, Mas, 2004). Due to these advantages, ANNs perform tasks more accurately than any other statistical classification techniques, particularly for the complex feature space and data that follows different statistical distributions (Schalkoff, 1992, Benediktsson et al., 1993). Comparatively, many studies proved that ANN could classify remotely sensed data more precisely than maximum likelihood [Civco, 1993, Paola and Schowengerdt, 1995, Children, 1997, Gopal et al., 1999, Kavzoglu and Mather, 1999].

Convolutional Neural Networks

Neural networks have shown great potentiality to learn complex, high dimensional, nonlinear mappings in pattern recognition applications. However, there are certain limitations when applied to image processing. Large image dimensions can create training complexity as the number of interconnected networks and weights increases. Networks are fully connected, so there is no translation or local invariance. At the same time, the input can be in any order, but the satellite images have strongly 2D correlations and local structures. Thus one of the most important and popularly used operations in image processing, pattern recognition, and computer vision, as proposed by Yann LeCun, is the convolution Neural Network. It is defined in two.

Figure 3. Linear Systems

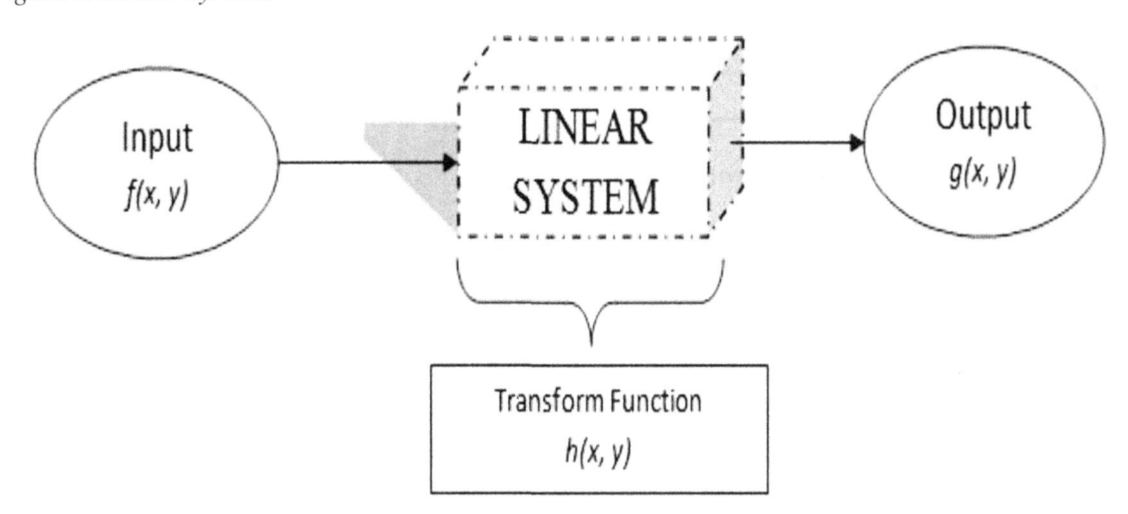

input functions producing another function by modifying one of the inputs. It is popularly known for two reasons, one for forming the basis for spatial image processing and another as it forms the mathematical foundation for linear systems.

Spatial Image processing: It modifies the values of the input image f(x,y) based on the neighboring pixels h(x,y), which is mathematical, defined as shown in equation 1:

$$g\left(x,y\right)=\int\int_{-\infty}^{\infty}f\left(x,y\right)h\left(\tau-x,\gamma-y\right)d\tau d\gamma$$

Here the output image g(x,y) is the modified f(x,y).

Linear systems: It is used in the super-resolution process generating a high-resolution image from a set of low-resolution images. Figure 3 shows a block diagram of linear systems. Image restoration utilizes the linear system in conjunction with the image acquisition process representing linear system to ideal world-scene transformed into a digital image after convolution with the transform function of the camera or any image input device. CNN can also be used as an Edge detector and hence can be useful for urban applications of remote sensing.

Figure 4. CNN architecture

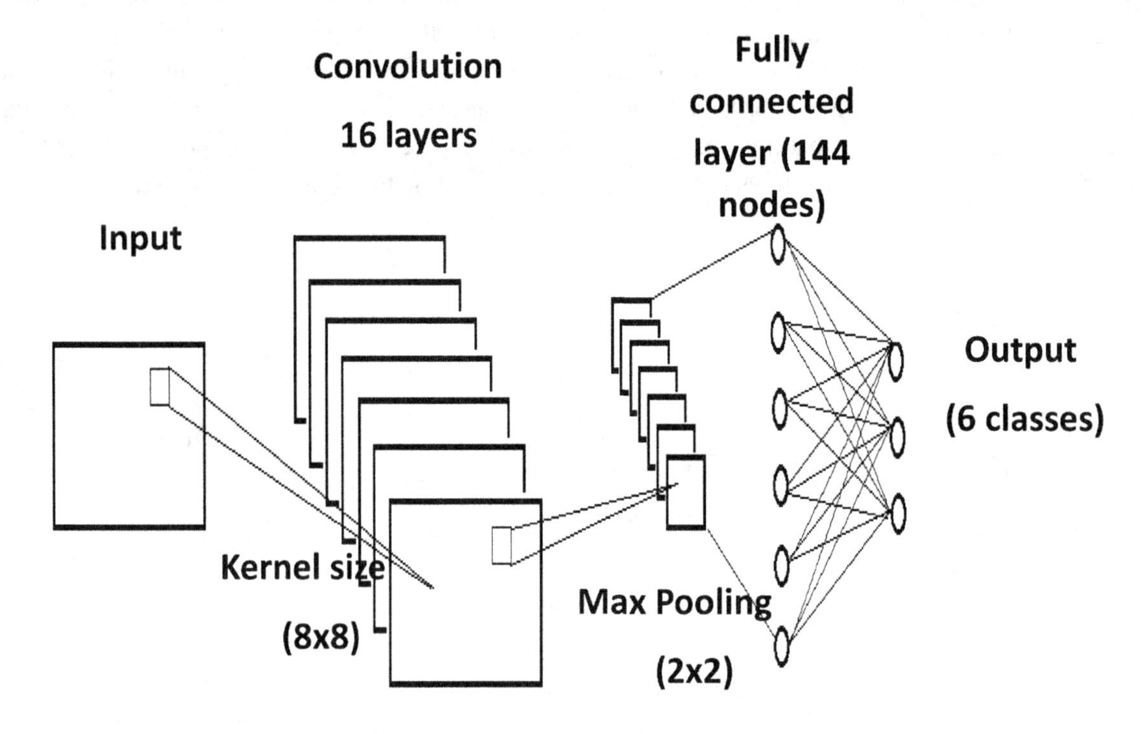

The Convolution method suits especially for image processing due to the following reasons:

1. It has Local Receptive fields where each node is connected to a limited set of nodes and not fully connected as in the case of neural networks.
2. Secondly, it provides translation invariance by sharing weights and reducing the number of training weights.
3. It is used for spatial sub-sampling, making the convolution robust against local distortions.
4. It can learn features. Thus features are learned and not trained by us.

Moreover, these features can be interpreted, which is an added advantage over ANN, where features are only weights that are hard to interpret.

CNN (figure 4) is one of the important algorithms, widely used in many fields. The method can easily recognize objects in the image based on morphology and structure. However, CNN has a major limitation. When applied for pixel-level classification, due to its translational invariance characteristics, the land cover boundaries and outlines of the features in the classified image get distorted. Thus it fails to classify the image accurately. To overcome this difficulty [Pan and Zhao, 2017] have proposed a central point enhanced convolutional neural network (CE-CNN) which can classify high-resolution satellite images. Compared with K-NN, MLP, CNN, SVM, maximum likelihood classifier, and classification & regression tree (CART), the proposed method improves the classification accuracy with better discrimination of boundaries and fewer distortions in the classified image found.

- Hardware for Deep Learning: The GPU is the heart of deep learning applications. It increases the processing speed significantly. However, when deploying a neural network, there are alternatives to GPUs (Graphics Processing Units), namely the FPGA (Field Programmable Gate Array). FPGAs can implement custom data types, whereas GPUs are architecturally limited. With neural networks transforming in various ways and expanding into new industries, the adaptability provided by FPGAs is advantageous. Table… describes the hardwares helpful in processing deep learning algorithms.
- Processor Abbreviation Description: Application-specific integrated circuits ASICs Custom circuits, such as Google Tensor Processor Units (TPU), provide excellent performance with the highest efficiency.
- Field-programmable gate arrays: (FPGAs) FPGAs provide performance close to ASICs. They are also flexible and adaptable over time to apply a new mind.
- Graphics processing units: (GPUs) GPUs are popular for A.I. computations. It offers compatible processing capabilities, making them faster in rendering images than CPUs.
- Central processing units: (CPUs) CPUs are general-purpose processors, with a performance not ideal for graphics and video processing. (https://docs.microsoft.com/en-us/azure/machine-learning/how-to-deploy-fpga-web-service)

Tools and Libraries Useful in Deep Learning Implementation

TensorFlow

An end-to-end open-source platform for machine learning (ML) with a comprehensive, flexible ecosystem of tools, libraries, and community resources allows researchers to push the state-of-the-art ML. (https://www.tensorflow.org/).

It has a flexible architecture that allows developers to deploy computation on multiple CPUs or GPUs in a desktop, mobile device, or server.

Keras

It is a high-level API of TensorFlow and runs on top of either TensorFlow or Theano. It also enables experimentation with speed. (https://keras.io/).

1. Official high-level API of TensorFlow
2. Multi-backend, multi-platform
3. Allows easy and fast prototyping
4. It runs seamlessly on CPU and GPU

5. It supports both CNN and RNN

Pytorch

Pytorch is a scientific computing framework with GPU optimizations (https://pytorch.org/). Easy to implement It is more flexible due to its dynamic computational graph. It is C-based and thus has high execution efficiency.

CAFFE

CAFFE refers to Convolution Architecture for Feature Extraction, an open-source framework for deep learning (https://caffe.berkeleyvision.org/).

1. C++/CUDA architecture
2. Supports command-line, python, and Matlab interfaces
3. Includes fast and well-tested codes
4. Tools, reference modules, demos, and recipes
5. Seamless switch between CPU & GPU

Google Colab

A free cloud service developed on Jupyter notebooks provides free GPU and Pytorch, TensorFlow, Keras, and OpenCV frameworks (https://colab.research.google.com).

EXPERIMENTAL ANALYSIS

RISAT is an Indian microwave satellite with multiple resolutions, polarizations, and modes that can operate during the day, night, and in all weather conditions. The launch of RISAT from Shriharikota, Nellore Dist., A.P., INDIA on April 26, 2012, marks the beginning of a new class of Earth observation imaging products and services. The RISAT High-Resolution SAR will operate at 5.35 GHz in the C-band.

The literature survey indicates the potential efficiency of neural networks for SAR image classification (figure 5). Hence, we experimented with two algorithms based on neural networks (artificial neural network and convolutional neural network) to classify the SAR image. The image is a dual-pol Risat cFRS image covering some parts of the Mumbai region. The data was multilooked into 7x10 looks in azimuth and range direction, respectively, and Raney decomposition was applied to make it more interpretable. We then classified the decomposed image into five land cover classes, namely water, forest, urban, wetland/saltpans, and mangrove forests, based on visual interpretation using ANN and CNN.

We chose 46392 sample pixels with reference to google earth's image, out of which 80% of the pixels (37113 pixels) were used for training. The rest, 20% (9279 pixels), were kept for testing/validation. The hyperparameters for ANN were tuned using the GridSearchCV method. The best (optimal) parameters with respect to our data were selected and used to train the network. The difference between both networks is that ANN takes pixel-based input. At the same time, in CNN, a patch-based input is fed as it

Figure 5. Summary of model rates in percentage

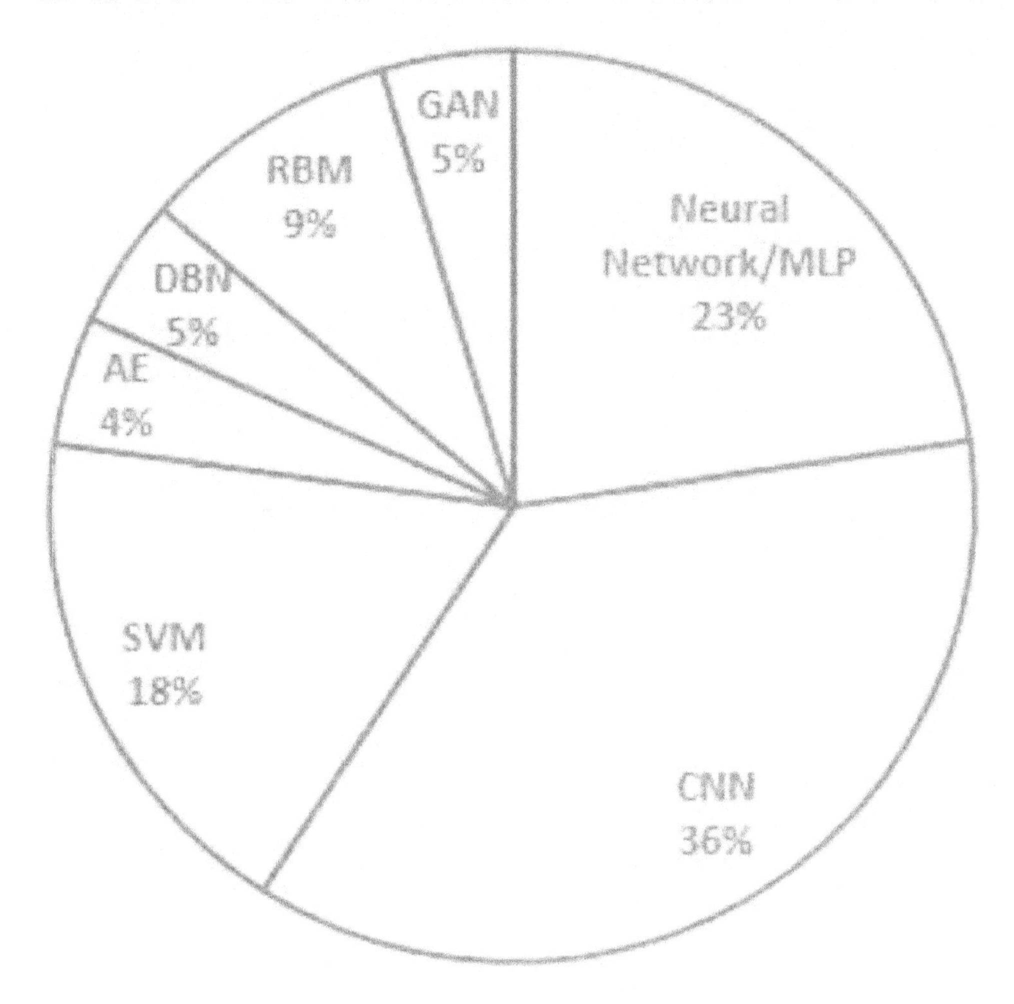

also considers the spatial information of the input pixel. CNN parameters such as kernel size, number of layers for convolution and pooling, patch size were chosen based on a manual experiment.

Table 2 shows the accuracy obtained by both algorithms, and the graph compares this accuracy in Figure 6.

As per Table 2, both the algorithms performed well in the classification task and generalized well on complex data such as dual-PolSAR (here Risat-1). The visual inspection in figure 7 and quantitative analysis as per table 1 proved that deep learning could be useful for classifying SAR data. In addition, CNN gave better accuracy than ANN. It may be because it takes the pixel information and considers the surrounding spatial information, which allows the algorithm to discriminate well between two classes and thus perform superior to ANN during the experiment. Both the tasks performed on high computational GPU gave faster results while handling a large amount of data.

Figure 6. Comparison of obtained model accuracies

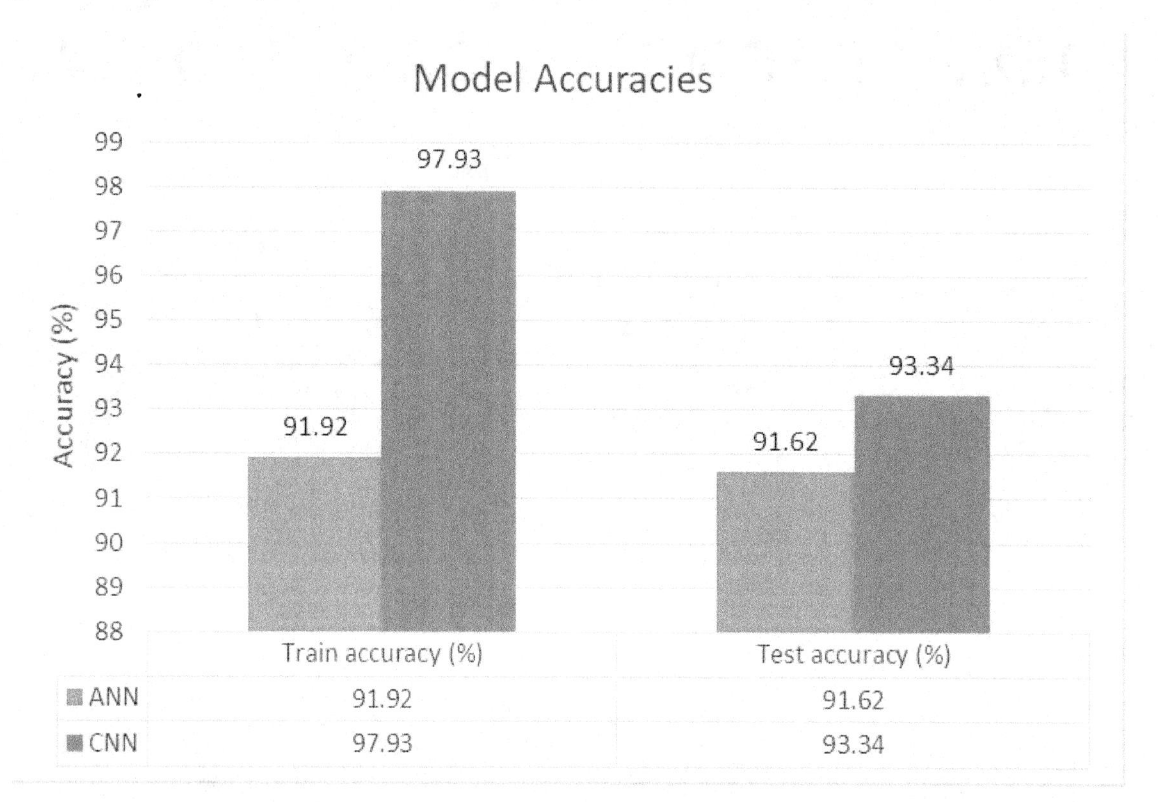

Table 2. Accuracies obtained through experiment analysis

Models	Train accuracy (%)	Test accuracy (%)
ANN	91.92	91.62
CNN	97.93	93.34

ADVANTAGES OF PROPOSED FRAMEWORK

In this literature survey, we surveyed 52 articles, out of which 16 articles were on SAR technology, and 36 were on deep learning techniques. Out of these 36 articles, 20 review DL architectures, and 16 for SAR data processing. Figures 8 and 9 show the number of articles for each application.

As shown in figure 5, 36% of these articles have used CNN for applications like feature extraction, automatic target recognition (ATR), PolSAR image classification, speckle reduction, and transfer learning. 23% used neural networks or multi-layer perceptron for classification, speckle reduction, and change detection. 18% have used support vector machines for SAR data classification, data fusion, and transfer learning. While rest includes RBM (restricted Boltzmann machines), DBN (deep belief networks), A.E. (autoencoders),

GAN (generative adversarial network) for similar tasks (the percentage obtained for each model from the papers surveyed and the figures may vary as per one's interest in literature). Thus, overall analysis

Figure 7. Results of experimental analysis

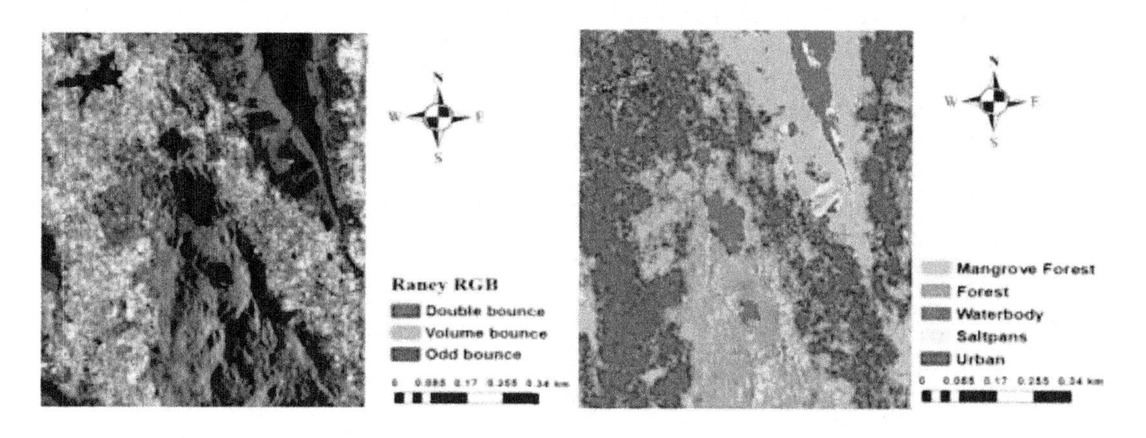

(a) Raney RGB for reference

(b) Artificial neural network

(c) Convolutional Neural network

shows that convolutional neural networks are the most potent and popular DL architecture used in SAR remote sensing for almost every application, from feature extraction to transfer learning.

SUMMARY

This paper reviews Synthetic Aperture Radar remote sensing data analysis. The paper includes a comprehensive survey of the classical techniques used for SAR data analysis with modern deep learning techniques. Classification of SAR image plays a vital role for any Earth Observation (E.O.) applications whether it is agriculture, hydrology, urban landscape, or disaster management; all of these applications emphasize real-time analysis of remotely sensed data from spaceborne and well airborne data. SAR data provide timely and accurate data, and for those periods, the optical remote sensing data limits the data acquisition due to various atmospheric factors like cloud, fog, etc.

Figure 8. Total DL models as per their applications for SAR datasets

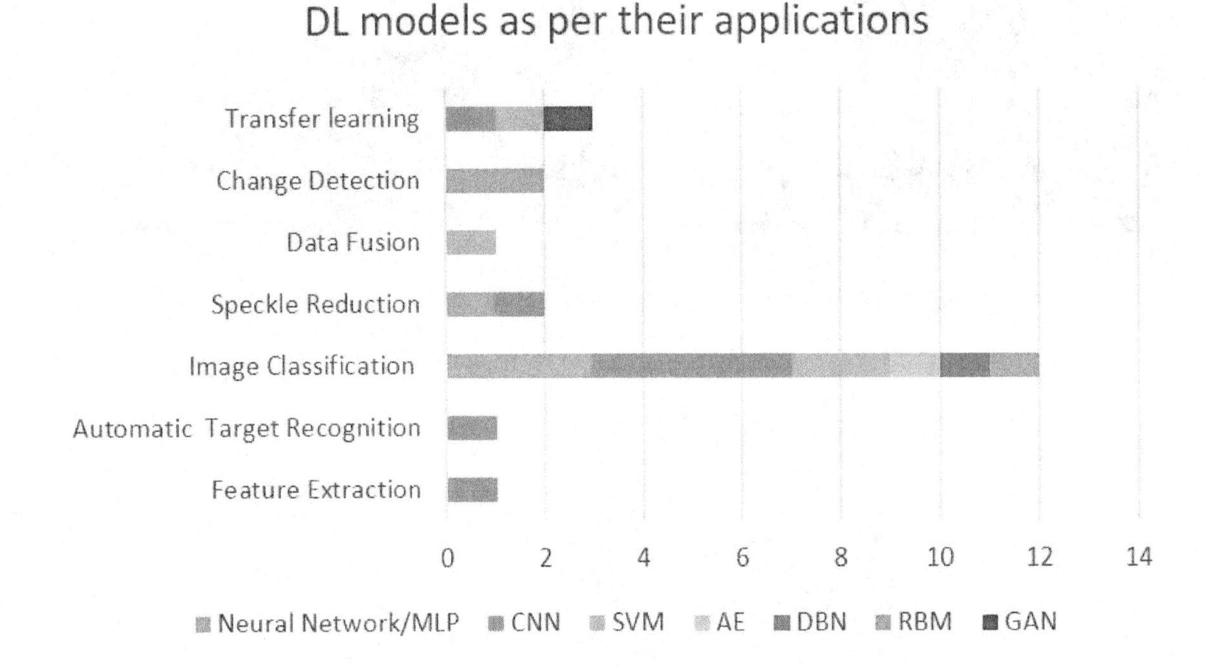

Figure 9. Deep Learning models for SAR data analysis

Models	SAR Data Analysis						
	Feature Extraction	Automatic Target Recognition	Image Classification	Speckle Reduction	Data Fusion	Change Detection	Transfer learning
NN/MLP			3	1		1	
CNN	1	1	4	1			1
SVM			2		1		1
AE			1				
DBN			1				
RBM			1			1	
GAN							1

Moreover, SAR data provides both high spatial and scattering information simultaneously. However, the SAR data acquisition is more expensive due to its active sensor (self-illumination) to sense the earth's surface. It requires good expertise to interpret SAR datasets compared to other remotely sensed datasets due to the complex phenomena involved in this technology. The literature survey also throws light on various complexities and the dependencies of the SAR sensor with the target (surface) properties. The study focuses on the classical techniques for SAR polarimetric analysis, such as decomposition and texture analysis. These techniques use the physical scattering phenomenon between the target and the sensor to understand & extract useful information contained in the imagery. Moreover, radar backscatter depends upon various target properties such as dielectric constant, surface roughness, local incidence angle, moisture content, height, orientation, etc. all these make SAR data more complex and nonlinear. Thus SAR data analysis needs expert-level knowledge to utilize the data and is hard to interpret.

The paper describes the challenges associated with SAR data have and how the use of deep learning techniques can be an advantage to overcome these difficulties. Deep learning techniques can easily handle high dimensional (big data), complex, nonlinear data, and it automatically learns essential features from the data without explicit programming. It thus can be useful in SAR data analysis. It can benefit many applications such as removing speckle noise from the image, multisensor fusion of two datasets, image segmentation and classification, temporal data analysis, and automatic target recognition. The main drawback of deep learning algorithms is the requirement of a good amount of data for training and associated labeled information (ground truth). The higher the amount of ground truth information, the higher the algorithm's accuracy. However, we can find a massive amount of SAR data for training but less ground truth information. This unavailability of properly labeled data can lead to different outcomes, or the algorithms may be difficult to converge. The paper also covers techniques like transfer learning and data augmentation, which will be helpful when there is inadequate availability of such datasets. The article also includes essential tools and libraries for implementing deep learning algorithms.

In SAR data analysis, the parameters such as speckle reduction, the effect of incidence angle, sun illumination, soil moisture conditions, etc., significantly contribute to various E.O. applications. Hence changing any of these parameters has a different impact on the dataset. These parameters need to be explored further for a particular application. Furthermore, the use of high processing GPUs for implementing deep learning algorithms can enhance the capabilities of the algorithms to handle high dimensional, large datasets while speeding up the computational process compared to the regular CPUs. Many other deep learning techniques like Auto Encoder, Deep Belief Networks, Sparse Auto Encoders, etc., can further be explored.

ACKNOWLEDGMENT

The authors are thankful to SAC ISRO for providing Risat cFRS data for experiment analysis. The research is a part of the funded project by Space Applications Center, Ahmedabad at Pandit Deendayal Energy University (PDEU), Gandhinagar, under the grant "ORSP/R&D/ISRO/2016/SPXX".

REFERENCES

Abdikan, S., Balik Sanli, F., Bektas Balcik, F., & Goksel, C. (2008). Fusion of SAR images (pulsar and radarsat-1) with multispectral spot image: A comparative analysis of resulting images. *International Archives of the Photogrammetry, Remote Sensing and Spatial Information Sciences - ISPRS Archives, 37*, 1197–1202.

Alawlaki, A. A. (2006). *Statistical methods of classification in GIS*. HTTP: //hdl.handle.net/10603/126331

Atkinson, P. M., & Tatnall, A. (1997). Introduction neural networks in remote sensing. *International Journal of Remote Sensing, 18*(4), 699–709. doi:10.1080/014311697218700

Ball, J. E., Anderson, D. T., & Chan, C. S. (2017). Comprehensive survey of deep learning in remote sensing: Theories, tools, and challenges for the community. *Journal of Applied Remote Sensing, 11*(04), 1. doi:10.1117/1.JRS.11.042609

Benediktsson, J. A., Swain, P. H., & Ersoy, O. K. (1993). Conjugate-gradient neural networks in multi-source and very-high- dimensional remote sensing data classification. *International Journal of Remote Sensing, 14*(15), 2883–2903. doi:10.1080/01431169308904316

Carpenter, G. A., Gopal, S., Macomber, S., Martens, S., Woodcock, C. E., & Franklin, J. (1999). A neural network method for efficient vegetation mapping. *Remote Sensing of Environment, 70*(3), 326–338. doi:10.1016/S0034-4257(99)00051-6

Chen, C.-T., Chen, K.-S., & Lee, J.-S. (2003). The use of fully polarimetric information for the fuzzy neural classification of SAR images. *IEEE Transactions on Geoscience and Remote Sensing, 41*(9), 2089–2100. doi:10.1109/TGRS.2003.813494

Chiuderi, A. (1997). Multisource and multitemporal data in land cover classification tasks: The advantage offered by neural networks. In Geoscience and Remote Sensing, 1997. IGARSS'97. Remote Sensing-A Scientific Vision for Sustainable Development., 1997 IEEE International, 4, 1663–1665.

Cire¸san, D. C., Meier, U., Gambardella, L. M., & Schmidhuber, J. (2010). Deep, big, simple neural nets for handwritten digit recognition. *Neural Computation, 22*(12), 3207–3220. doi:10.1162/NECO_a_00052 PMID:20858131

Civco, D. L. (1993). Artificial neural networks for land-cover classification and mapping. *International Journal of Geographical Information Science, 7*(2), 173–186.

Ding, J., Chen, B., Liu, H., & Huang, M. (2016). Convolutional Neural Network with Data Augmentation for SAR Target Recognition. *IEEE Geoscience and Remote Sensing Letters, 13*(3), 364–368. doi:10.1109/LGRS.2015.2513754

Foody, G. (1995a). Using prior knowledge in artificial neural network classification with a minimal training set. *Remote Sensing, 16*(2), 301–312. doi:10.1080/01431169508954396

Foody, G. M. (1995b). Land cover classification by an artificial neural network with ancillary information. *International Journal of Geographical Information Systems, 9*(5), 527–542. doi:10.1080/02693799508902054

Geng, J., Fan, J., Wang, H., Ma, X., Li, B., & Chen, F. (2015). High- Resolution SAR Image Classification via Deep Convolutional Autoencoders. *IEEE Geoscience and Remote Sensing Letters, 12*(11), 2351–2355. doi:10.1109/LGRS.2015.2478256

Girshick, R., Donahue, J., Darrell, T., & Malik, J. (2016). Region- based convolutional networks for accurate object detection and segmentation. *IEEE Transactions on Pattern Analysis and Machine Intelligence, 38*(1), 142–158. doi:10.1109/TPAMI.2015.2437384 PMID:26656583

Gopal, S., Woodcock, C. E., & Strahler, A. H. (1999). Fuzzy neural network classification of global land cover from a 1 avhrr data set. *Remote Sensing of Environment*, *67*(2), 230–243. doi:10.1016/S0034-4257(98)00088-1

Hall, D. L., & Llinas, J. (1997). An introduction to multisensor data fusion. *Proceedings of the IEEE*, *85*(1), 6–23. doi:10.1109/5.554205

Haralick, R. M., & Shanmugam, K. (1973). Textural features for image classification. *IEEE Transactions on Systems, Man, and Cybernetics*, *3*(6), 610–621.

He, K., Zhang, X., Ren, S., & Sun, J. (2016). Deep residual learning for image recognition. *Proceedings of the IEEE conference on computer vision and pattern recognition*, 770–778.

Hewitson, B. C., & Crane, R. G. (1994). *Neural nets: applications in geography: applications for geography* (Vol. 29). Springer Science & Business Media. doi:10.1007/978-94-011-1122-5

Hou, B., Luo, X., Wang, S., Jiao, L., & Zhang, X. (2015). Polarimetric SAR images classification using deep belief networks with learning features. *International Geoscience and Remote Sensing Symposium (IGARSS)*, (2), 2366–2369. 10.1109/IGARSS.2015.7326284

Kavzoglu, T., & Mather, P. M. (1999). Pruning artificial neural networks: An example using land cover classification of multisensor images. *International Journal of Remote Sensing*, *20*(14), 2787–2803. doi:10.1080/014311699211796

Keiner, L. E., & Yan, X.-H. (1998). A neural network model for estimating sea surface chlorophyll and sediments from thematic mapper imagery. *Remote Sensing of Environment*, *66*(2), 153–165. doi:10.1016/S0034-4257(98)00054-6

Krizhevsky, A., Sutskever, I., & Hinton, G. E. (2012). Imagenet classification with deep convolutional neural networks. Advances in neural information processing systems, 1097–1105.

Kussul, N., Shelestov, A., Lavreniuk, M., Butko, I., & Skakun, S. (2016). Deep learning approach for large scale land cover mapping based on remote sensing data fusion. In *Geoscience and Remote Sensing Symposium (IGARSS), 2016 IEEE International* (pp. 198–201). IEEE. 10.1109/IGARSS.2016.7729043

Lardeux, C., Frison, P. L., Tison, C., Souyris, J. C., Stoll, B., Fruneau, B., & Rudant, J. P. (2009). Support vector machine for multifrequency SAR polarimetric data classification. *IEEE Transactions on Geoscience and Remote Sensing*, *47*(12), 4143–4152. doi:10.1109/TGRS.2009.2023908

Lek, S., & Gu'egan, J.-F. (1999). Artificial neural networks as a tool in ecological modelling, an introduction. *Ecological Modelling*, *120*(2-3), 65–73. doi:10.1016/S0304-3800(99)00092-7

Lewis, B., DeGuchy, O., Sebastian, J., & Kaminski, J. (2019). Realistic SAR data augmentation using machine learning techniques. *Proceedings of the Society for Photo-Instrumentation Engineers*, •••, 10987.

Long, J., Shelhamer, E., & Darrell, T. (2015). Fully convolutional net-works for semantic segmentation. *Proceedings of the IEEE Conference on Computer Vision and Pattern Recognition*, 3431–3440. doi:10.1117/12.2518452

Lv, Q., Dou, Y., Niu, X., Xu, J., Xu, J., & Xia, F. (2015). Urban land use and land cover classification using remotely sensed SAR data through deep belief networks. *Journal of Sensors, 2015*. doi:10.1155/2015/538063

Mas, J. (2004). Mapping land use/cover in a tropical coastal area using satellite sensor data, gis and artificial neural networks. *Estuarine, Coastal and Shelf Science*, *59*(2), 219–230. doi:10.1016/j.ecss.2003.08.011

Montavon, G., Orr, G., & Mu¨ller, K.-R. (2012). *Neural networks: Tricks of the trade* (vol. 7700). Springer.

Noh, H., Hong, S., & Han, B. (2015). Learning deconvolution network for semantic segmentation. *Proceedings of the IEEE International Conference on Computer Vision*, 1520–1528.

Pan, S. J., & Yang, Q. (2010). A survey on transfer learning. *IEEE Transactions on Knowledge and Data Engineering*, *22*(10), 1345–1359. doi:10.1109/TKDE.2009.191

Pan, X., & Zhao, J. (2017). A central-point-enhanced convolutional neural network for high-resolution remote-sensing image classification. *International Journal of Remote Sensing*, *38*(23), 6554–6581. doi :10.1080/01431161.2017.1362131

Paola, J. D., & Schowengerdt, R. A. (1995). A detailed comparison of backpropagation neural network and maximum-likelihood classifiers for urban land use classification. *IEEE Transactions on Geoscience and Remote Sensing*, *33*(4), 981–996. doi:10.1109/36.406684

Podest, E. (2018). *SAR for Mapping Land Cover*. Academic Press.

Pohl, C., & Van Genderen, J. L. (1998). Review article multisensor image fusion in remote sensing: Concepts, methods and applications. *International Journal of Remote Sensing*, *19*(5), 823–854. doi:10.1080/014311698215748

Pottier, E., & Ferro-Famil, L. (2008). Advances in SAR Polarimetry applications exploiting polarimetric spaceborne sensors. *2008 IEEE Radar Conference*, 1–6. 10.1109/RADAR.2008.4720872

Redmon, J., Divvala, S., Girshick, R., & Farhadi, A. (2016). You only look once: Unified, real-time object detection. *Proceedings of the IEEE conference on computer vision and pattern recognition*, 779–788. 10.1109/CVPR.2016.91

Riesenhuber, M., & Poggio, T. (1999). Hierarchical models of object recognition in cortex. *Nature Neuroscience*, *2*(11), 1019–1025. doi:10.1038/14819 PMID:10526343

Schalkoff, R. J. (1992). *Pattern recognition*. Wiley Online Library.

Schmitt, M., Hughes, L. H., & Zhu, X. X. (2018). The sen1-2 dataset for deep learning in SAR-optical data fusion. *ISPRS Annals of the Photogrammetry. Remote Sensing and Spatial Information Sciences*, *4*(1), 141–146.

Shanmugapriya, S., Haldar, D., & Danodia, A. (2019). Opti-mal datasets suitability for pearl millet (Bajra) discrimination using multiparametric SAR data. *Geocarto International*, 6049.

Simonyan, K., & Zisserman, A. (2014). *Very deep convolutional networks for large-scale image recognition*. arXiv preprint arXiv:1409.1556v6.

Sugawara, E., & Nikaido, H. (2014). Properties of adeabc and adeijk efflux systems of Acinetobacter baumannii compared with those of the acrab-tolc system of escherichia coli. *Antimicrobial Agents and Chemotherapy*, *58*(12), 7250–7257. doi:10.1128/AAC.03728-14 PMID:25246403

Wang, H., Chen, S., Xu, F., & Jin, Y. Q. (2015). Application of deep- learning algorithms to MSTAR data. *International Geoscience and Remote Sensing Symposium (IGARSS)*. 10.1109/IGARSS.2015.7326637

Waske, B., & Benediktsson, J. A. (2007). Fusion of support vector machines for classification of multi-sensor data. *IEEE Transactions on Geoscience and Remote Sensing*, *45*(12), 3858–3866. doi:10.1109/TGRS.2007.898446

Xu, F., Jin, Y. Q., & Moreira, A. (2016). A Preliminary Study on SAR Advanced Information Retrieval and Scene Reconstruction. *IEEE Geoscience and Remote Sensing Letters*, *13*(10), 1443–1447. doi:10.1109/LGRS.2016.2590878

Zhang, L., Xia, G.-S., Wu, T., Lin, L., & Tai, X. C. (2016). Deep learning for remote sensing image understanding. *Journal of Sensors*, *2016*, 2016. doi:10.1155/2016/7954154

Zhang, L., Zou, B., Zhang, J., & Zhang, Y. (2009). Classification of polarimetric SAR image based on support vector machine using multiple-component scattering model and texture features. *EURASIP Journal on Advances in Signal Processing*, *2010*(1), 960831. doi:10.1155/2010/960831

Zhou, Y., Wang, H., Xu, F., & Jin, Y. Q. (2016). Polarimetric SAR Image Classification Using Deep Convolutional Neural Networks. *IEEE Geoscience and Remote Sensing Letters*, *13*(12), 1935–1939. doi:10.1109/LGRS.2016.2618840

Chapter 9
Harnessing Artificial Intelligence for Drought Management

Ved Prakash Singh
https://orcid.org/0000-0002-2281-5687
Ministry of Earth Sciences, Bhopal, India & Indian Institute of Technology, Patna, India

Shirish Khedikar
Ministry of Earth Sciences, Pune, India

Jimson Mathew
Indian Institute of Technology, Patna, India

Tanvi Garg
Medi-Caps University, Indore, India

ABSTRACT

Unlike other natural hazards, drought is a slow-onset, creeping natural hazard. Its effects often accumulate slowly over a considerable period and may linger for years after the termination of the drought events. Absence of precise and universally accepted definition of drought adds to the confusion while declaring drought and its degree of severity. Adverse impacts of drought are non-structural and typically spread over a larger geographical region than are damages resulting from other natural hazards. It is critical that the stakeholders in the concerned geographic region understand their exposure to the drought hazard. Drought risks are the joint effects of both the region's exposure to the drought event and the vulnerability of its society to a drought at that point. However, there are various conventional ways to mitigate these adverse impacts, but modern technological advancement have shown a path to harness artificial intelligence methods towards facing drought challenges more efficiently with accurate prediction of its location-specific occurrence and duration estimation.

DOI: 10.4018/978-1-6684-3981-4.ch009

INTRODUCTION

Drought is an insidious natural hazard characterized by lower than expected or normal precipitation that, when a season or longer period extended over, is insufficient to meet the demands of human activities and the environment.

Drought is a temporary aberration, unlike aridity, which is a permanent feature of climate. Seasonal aridity, that is, a well-defined dry season, also needs to be distinguished from drought, as these terms often create confusion or used interchangeably. The differences need to be understood and properly incorporated in drought monitoring, early warning systems and preparedness plans. Having the capacity to monitor droughts in near–real time and providing accurate drought prediction from weeks to seasons in advance can greatly reduce the severity of social and economic damages caused by drought, a leading natural hazard for a country like India whose economy is greatly dependent on agriculture.

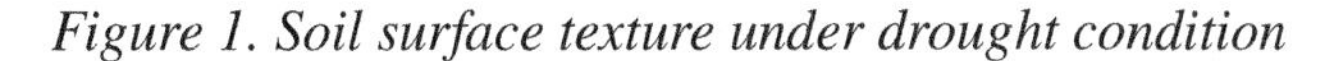

Figure 1. Soil surface texture under drought condition

Drought vs. Aridity

Aridity is defined, in meteorology and climatology as "the degree to which a climate lacks effective, life-promoting moisture" (Glossary of Meteorology, American Meteorological Society). Drought is "a period of abnormally dry weather sufficiently long enough to cause a serious hydrological imbalance". Aridity is permanent, while drought is temporary.

Aridity is measured by comparing long-term average water supply (precipitation) to long-term average water demand (evapotranspiration). If demand is greater than supply on average, then the climate is arid (Thompson, 1975). Drought refers to the moisture balance that happens on a month-to-month (or more frequent) basis. If the water supply is less than water demand for a given month, then that month is abnormally dry; if there is a serious hydrological impact, then a drought is occurring that month.

Classification of Droughts

Based on the implications of drought, it has been classified into four categories, mostly based on different parts of the hydrological cycle, as described below –

1. **Meteorological drought:** It is a measure of departure of precipitation from normal. Due to climatic differences, what might be considered a drought in one location of the country may not be a drought in another location.
   Droughts generally start with a lack of precipitation, possibly in combination with high evapotranspiration, resulting from the natural variability of the weather. Meteorological drought is based purely on rainfall. It is expressed solely based on the degree of dryness, in comparison to the normal or climatologically expected rainfall over a wide area on monthly, seasonal or annual time scale.
2. **Agricultural drought:** Agricultural drought prevails with a situation where the amount of moisture in the soil no longer meets the needs of a particular crop. The meteorological drought causes a lack of soil moisture, which is referred as soil moisture drought and which affects agricultural crops and/or the natural vegetation. The soil moisture drought is also frequently called agricultural drought.

Agricultural drought occurs when the soil moisture is depleted to the extent that crop yields are reduced to the extent of crop failure and crop production goes below the level of sustainable limit over space and time. Impact of agricultural drought depends on the duration and intensity of drought, nature of the crop, growth stage of the crop and on the soil characteristics.

3. **Hydrological drought:** It is indicated by the situation when surface and subsurface water supplies are below normal. Lack of precipitation may also cause low stream flows: the stream flow drought. The soil moisture drought causes a decrease in the amount of recharge, which in turn causes lower groundwater levels and decreasing groundwater discharge to the surface water system, which is referred as groundwater drought.

Both the stream flow drought and the groundwater drought are parts of the hydrological drought. Hydrological droughts are usually out of phase with or lag the occurrence of meteorological and agricultural droughts. It takes longer for precipitation deficiencies to show up in components of the hydrological system such as soil moisture, streamflow, groundwater and reservoir levels. As a result, these impacts are out of phase with impacts in other economic sectors. For example, a precipitation deficiency may result in a rapid depletion of soil moisture that is almost immediately discernible to agriculturalists, but the impact of this deficiency on reservoir levels may not affect hydroelectric power production or recreational uses for many months.

Moreover, water in hydrologic storage systems (*e.g.*, reservoirs, rivers) is often used for multiple and competing purposes (*e.g.*, flood control, irrigation, recreation, navigation, hydropower, wildlife habitat, *etc.*), further complicating the sequence and quantification of impacts. Competition for water in these storage systems escalates during drought and conflicts between water users increase significantly.

4. **Socio-economic drought:** The situation that occurs when physical water shortages begin to affect people. This condition is when supply of some goods and services such as energy, food and drinking water *etc.* are reduced or threatened by changes in meteorological, agricultural and hydrological conditions. The socio-economic drought expresses the deficit of water as an economic good and addresses the damages caused by all the different types of drought.

In most instances, the demand for economic goods is increasing as a result of increasing population and per capita consumption. Supply may also increase because of improved production efficiency, technology, or the construction of reservoirs that increase surface water storage capacity. If both supply and demand are increasing, the critical factor is the relative rate of change. Is demand increasing more rapidly than supply? If so, vulnerability and the incidence of drought may increase in the future as supply and demand trends converge.

Sequence of Drought Impacts

When drought (*i.e.* meteorological drought) begins, the agricultural sector is usually the first to be affected because of its heavy dependence on stored soil-water. Those who rely on surface water (*i.e,* reservoirs and lakes) and subsurface water (*i.e,* ground water) are usually the last to be affected.

A deficit of precipitation has different impacts on the ground water, reservoir storage, soil moisture, snowpack, and streamflow. Soil moisture conditions respond to precipitation anomalies on a relatively short scale, while ground water, streamflow, and reservoir storage reflect the longer-term precipitation anomalies.

Monitoring Drought: Physical, Biological and Social Indicators

- Physical indicators include rainfall, effective soil moisture, surface water availability, depth to groundwater, etc.
- Biological or agricultural indicators comprise of vegetation cover & composition, crop & fodder yield, condition of domestic animals, pest incidence, etc.
- Social indicators are mostly impact indicators and include food and feed availability, land use conditions, livelihood shifts, migration of population, etc.

In most of the cases only those indicators that measure the rainfall needs of following sectors are considered:

1. Agricultural needs,
2. Drinking water supply, and
3. Storage of reservoirs and ground water.

Drought Monitoring by India Meteorological Department

During 1965 and 1966, the major parts of India were under prolonged and severe drought conditions due to deficient monsoon rainfall. On the recommendations of the *Planning commission*, India Meteorological Department (IMD) has started 'Drought Research' and monitoring at Pune center in 1967.

IMD monitors drought by using three well-established drought indices. One of which is purely meteorological drought, the second one is agricultural drought and the third one started recently covers meteorological, agricultural as well as hydrological droughts.

Primary Drought Indices for Monitoring Meteorological Droughts

Different indices being used for drought monitoring are explained below:

- Percent of normal,
- Decile,
- Departure rain,
- Crop-moisture index,
- Palmer drought severity index,
- Standard precipitation index,
- Surface water supply index,
- Bhalme and Mooley drought index, and
- National rainfall index

1. **Percent of Normal:** The percent of normal precipitation is one of the simplest measurements of rainfall departure / deficiency for a location It is calculated by dividing actual precipitation by normal precipitation i.e. long period average (30 years or more) and multiplying by 100%. It is used in ranking the rainfall values in disciples dividing into hundred sectors grouped into five classes, whereas long duration data is required to compute normal rainfall. Analyses using the percent of normal are very effective when used for a single region or a single season. However, it is also easily misunderstood and gives different indications of conditions, depending on the location and season.

This can be calculated for a variety of time scales. Usually these time scales range from a single month to a group of months representing a particular season, to an annual or water year. Normal precipitation for a specific location is considered to be 100%.

One of the disadvantages of using the percent of normal precipitation is that the mean or average precipitation is often not the same as the median precipitation, which is the value exceeded by 50% of the precipitation occurrences in a long-term climate record. The reason for this is that precipitation on monthly or seasonal scales does not have a normal distribution. Use of the percent of normal comparison implies a normal distribution where the mean and median are considered being the same.

An example of the confusion this could create, illustrated by the long-term precipitation record in Melbourne, Australia for the month of January. The median January precipitation is 36.0 mm (1.4 in.), meaning that half the years recorded with precipitation less than 36.0 mm and half the years with more than 36.0 mm. The mean is 48.0 mm. However, a monthly January total of 36.0 mm would be only 75% of normal when compared to the mean, which is often considered to be quite dry.

Because of the variability in the precipitation records over time and location, there is no way to determine the frequency of the departures from normal or compare different locations. This makes it difficult to link a value of a departure with a specific impact occurring as a result of the departure, inhibiting attempts to mitigate the risks of drought based on the departures from normal and form a plan of response (Willeke *et al.,* 1994). Simplicity makes this index popular in India, but this has demerit also, that average precipitation is not always the same as median precipitation.

Criteria on meteorological drought: India Meteorological Department describes meteorological drought from rainfall departure from its long-term averages (normal) and declares meteorological drought on *SEASONAL* basis for a geographical region. Based on percentage departure of actual rainfall from normal, following scales are used to determine drought category as follows:

- 26 to - 49 ➔ Moderate drought
- 50 or more ➔ Severe drought

In India, a year is considered to be a *DROUGHT YEAR*, if the area affected by moderate and severe drought, either individually, or together is 20-40% of the total area of the country and seasonal rainfall deficiency during south-west monsoon season for the country as a whole is 10% or more. When the spatial coverage of drought is more than 40%, it is referred as *ALL INDIA SEVERE DROUGHT*.

2. **Deciles of Precipitation (DI):** Decile values are used to give an indication of the spread of the observations over the period of records, developed by Gibbs and Maher (1967). To determine deciles of a series of observations, they are first arranged in ascending order, and then divided into 5 equal groups. Decile 1 is the value at the top of the first grouping that, in 20% of the years on record, the monthly or yearly rainfall total did not exceed the decile 1 value (refer table 1).

Table 1. Deciles grouping and precipitation classification

Decile Classifications	
deciles 1-2: lowest 20%	much below normal
deciles 3-4: next lowest 20%	below normal
deciles 5-6: middle 20%	near normal
deciles 7-8: next highest 20%	above normal
deciles 9-10: highest 20%	much above normal

DI is simple to calculate, requires only precipitation data and fewer assumptions. It provides an accurate statistical measurement of precipitation. However, its demerit is that being too simplistic to inform about gravity of the problem in different sectors. Accurate calculations require a long climatic data record.

3. **Palmer Drought Severity Index (PDSI):** Palmer drought severity index measures abnormality in much of supply based on precipitation, temperature, and local available water content. Major abnormality of recent weather index is divided into nine groups – extreme drought to extreme wet (refer table 2).

Palmer (1965) developed soil moisture algorithm that uses precipitation, temperature data and available water content of the soil. This model relates regional soil moisture conditions to the normal using a water balance model. PDSI indicates standardized moisture conditions and allows comparisons to be made between locations and between months. PDSI values are normally calculated on a monthly basis.

This was the first comprehensive drought index developed in the United States. Although, its demerit is that Palmer values may lag emerging droughts by several months; less well suited for mountainous land or areas of frequent climatic extremes; complex – has an unspecified, built-in time scale that can be misleading.

The PDSI is a meteorological drought index, and it responds to weather conditions that have been abnormally dry or abnormally wet. When conditions change from dry to normal or wet, for example, the drought measured by the PDSI ends without taking into account streamflow, lake and reservoir levels, and other longer-term hydrologic impacts (Karl and Knight, 1985).

Table 2. Palmer Drought Severity Index based drought classification

PDSI Classifications	
4.0 or more	extremely wet
3.0 to 3.99	very wet
2.0 to 2.99	moderately wet
1.0 to 1.99	slightly wet
0.5 to 0.99	incipient wet spell
0.49 to -0.49	near normal
-0.5 to -0.99	incipient dry spell
-1.0 to -1.99	mild drought
-2.0 to -2.99	moderate drought
-3.0 to -3.99	severe drought
-4.0 or less	extreme drought

The PDSI is calculated based on precipitation and temperature data, as well as the local Available Water Content (AWC) of the soil. From the inputs, all the basic terms of the water balance equation can be determined, including evapotranspiration, soil recharge, runoff, and moisture loss from the surface layer. Human impacts on the water balance, such as irrigation, are not considered. Complete descriptions of the equations can be found in the original study by Palmer (1965) and in the more recent analysis by Alley (1984).

4. **Standardized Precipitation Index (SPI):** SPI is used for estimating wet or dry condition based on precipitation variable and first developed by McKee *et al.* (1993). This wet or dry condition can be monitored by the SPI on a variety of time scales from sub-seasonal to inter-annual scales. It is expressed as standard deviations that the observed precipitation would deviate from the long-term mean, for a normal distribution and fitted probability distribution for the actual precipitation record. Since precipitation is not normally distributed, a transformation is first applied, followed by fitting to a normal distribution. It is easy to calculate, but minimum 30 years rainfall data is required.

Computation of SPI is comprised of a transformation of one frequency distribution (*e.g.*, gamma) to another frequency distribution (normal, or Gaussian). First step to calculate SPI is to adequately choose a particular probability distribution (*e.g.*, gamma distribution, incomplete beta distribution (McKee *et al.*, 1993 & 1995), and Pearson III distribution (Guttman, 1998 & 1999) that reliably fits the long-term precipitation time series and conduct fitting to that distribution. Gamma distribution has been widely used due to its reliable fit to the precipitation distribution. The fitting can be performed through maximum likelihood estimation of the gamma distribution parameters. Percentile value from this probability distribution is then transformed to the corresponding value in the new probability distribution. As a result, the probability that the rainfall is less than or equal to any rainfall amount will be the same as the probability that the new variate is less than or equal to the corresponding value of that rainfall amount. The normal distribution is usually used for this another transformation so that the mean and standard deviation of the SPI for a certain station and long-term period is zero and one, respectively (Edwards & McKee, 1997).

Positive SPI values indicate wet condition greater than median precipitation, whereas negative values the dry condition less than median precipitation (refer table 3). More detailed description of the steps required to calculate the SPI is provided in Lloyd-Hughes & Saunders (2002).

Since the year 2013, Standardized Precipitation Index (SPI) for the cumulative four weeks period is being generated operationally in every week (Thursday / Friday) to identify the districts experiencing moisture stress situation for preparation of appropriate agri-met advisories. It may be noted that as per WMO guidelines, SPI is a unified index for meteorological drought and all most all the NMHS are using this index for monitoring drought.

SPI forecast based on IMD Global Forecasting system (GFS) model output was the first ever attempt to generate SPI Forecast that is very essential for drought monitoring and forecasting particularly for agricultural needs. In this forecast, actual district rainfall is used for the past three weeks and the district cumulative rainfall forecast based on IMD Global Forecasting system (GFS) model for the coming week.

Table 3. Cumulative probabilities for various SPI values and possible interpretation of wet (or dry) conditions

SPI	Cumulative Probability	Interpretation
-3.0	0.0014	extremely dry
-2.5	0.0062	extremely dry
-2.0	0.0228	extremely dry (SPI < -2.0)
-1.5	0.0668	severely dry (-2.0 < SPI < -1.5)
-1.0	0.1587	moderately dry (-1.5 < SPI < -1.0)
-0.5	0.3085	near normal
0.0	0.5000	near normal
0.5	0.6915	near normal
1.0	0.8413	moderately wet (1.0 < SPI < 1.5)
1.5	0.9332	very wet (1.5 < SPI < 2.0)
2.0	0.9772	extremely wet (2.0 < SPI)
2.5	0.9938	extremely wet
3.0	0.9986	extremely wet

IMD, Pune had started generating SPI forecast for the cumulative four weeks period since the 1st week of southwest monsoon-2014.

5. **Surface Water Supply Index (SWSI)**: SWSI is an indicator of surface water condition that incorporate both hydrological and climatological features. Input requires Snow packs, stream flow, precipitation, reservoir storage. It ranges between -4.2 to +4.2. It is not suitable for extreme events.
6. **Bhalme and Mooley Drought Index (BMDI):** Computation of BMDI needs only precipitation data. Basically, the calculation of Bhalme-Mooley Drought Index (BMDI) is based on the rainfall anomaly, being a recurrent index, which takes into account the influence of precipitations from the previous month and therefore can be considered as a simplified version of PDSI (Dunkel, 2009).
7. **Standardized Precipitation Evapotranspiration Index (SPEI)**: SPEI is a multi-scalar drought index based on climatic data. It can be used for determining the onset, duration and magnitude of drought conditions *w.r.t.* normal conditions in a variety of natural and managed systems such as crops, ecosystems, rivers, water resources, *etc.*
8. **Aridity Anomaly Index (AAI):** Aridity is the Thornthwaite's concept to describe water deficiency experienced by plants. Thornthwaite gave the following formula for computing aridity index (AI):

Table 4. Drought intensity distribution based on AAI values

Anomaly of Aridity Index	Agricultural Drought Intensity
1 – 25	Mild
26 – 50	Moderate
> 50	Severe

The Aridity Index is worked out on weekly / biweekly basis. The positive values of the anomalies have been classified into three different classes (refer table 4). PE (potential evapotranspiration) denotes the water need of the plants (Thornthwaite, 1948). AE denotes the actual evapotranspiration and (PE-AE) denotes the water deficit. PE is computed by Penman's equation. AE is obtained from the water balance procedure, which takes into account the water holding capacity of the soil at the place. Aridity Anomaly Map gives information about the moisture stress experienced by growing plant.

Adverse Effects of Drought

- Depletion in soil moisture.
- Lowering groundwater table.
- Decrease in electrical power generation.
- Decrease in industrial production.
- Possibilities of desertification.
- Sensation of surface run off.
- Decrease in agricultural production.
- Adverse effect on economy.
- Famine and Migration.

DROUGHT MONITORING AND DROUGHT PREPAREDNESS

Droughts mainly occur in arid regions because of the spatial variability of rainfall (Sen, 2008). Droughts have a slow initiation and they are usually only recognized when the drought is already well established. The slow initiation and undefined end of a drought make it very difficult to take defensive actions (Pereira *et al.,* 2002). There are various methods of drought monitoring and drought preparedness as follows:

- Preparation of land to absorb and hold maximum moisture.
- Agronomic practices to minimize run-off evaporation from soil and transpiration from crop.
- Selection of short duration and drought tolerant crops with low water requirement, *e.g.*, Castor sunflower Sorghum *etc.*
- Adjusting sewing date so to evade the growth and reproductive age from high probability period of drought.
- Select and grow suitable cropping pattern.
- Use of mulches application of anti-transparent on crop to reduce evapotranspiration.
- Preventing and recycling of excess runoff.
- Management of various inputs to suit the climate, to apply life-saving irrigation, cleaning up folliage to reduce evapotranspiration, to conserve soil moisture by agronomic practices, *etc.*

Physiological Changes in The Plants During Drought

- Decrease in photosynthesis,
- Change in enzyme activity,
- Change in form and content,
- Accumulation of compatible osmolytes, and
- Decrease in plant growth.

Pre-Mitigation Measures for Drought

- Ridges-furrows-ridges patterns are utilized for in-situ water harvesting.
- Deep tillage helps in soil pulverization – increased water infiltration and better root growth.
- Timeliness sowing of crops in dry land – onset of monsoon rains can improve the crop yield.
- Crop diversification - herbal, medicinal and aromatic plants can help to get higher profit in rainfed areas.
- Strip cropping in sandy soil, allay cropping, and rotation cropping.
- Horti or Agri-pasture system.
- Moisture economic cropping system – many long duration crops do not match period of water availability – replacing such crop with other appropriate crop.
- Nutrition management and organic recycling – apply small amount of nitrogen and phosphorus during sowing and nitrogen as a split application during later stage.

Drought Mitigation Plans

Various ways can help to mitigate drought, few of them are mentioned below:

- **Inventory of resources and constraints** – natural resources, human expertise, infrastructure, capital for financial crisis, and data collection cost.
- **Science and policy integration** – direct communication and understanding among stakeholders, and extensive contact.
- **Research recommendations from institution & plan evaluation** – evaluate and abduction of plans, operational evaluation, post drought evaluation.
- **Education** – understanding of water conservation, mitigation education and information.
- **Publicity –** implication presentation of media in public, and announcement a plan before drought sensitive seasons.
- **Drought mitigation procedures** – impact assessment and crisis mitigation.
- **Task force** – supervise and coordinate development plans, implication of plans during the time of drought.
- **Draft policy and plan** – purpose and object of drought plan.
- **Dissolve in water-users conflict** – discuss what use problems are, seek cooperative solutions.

Situations in Indian Subcontinent During the Drought

- High pressure in the Indian ocean,
- Low pressure in the Pacific ocean,
- Moving or decreasing weather systems in Indian ocean and southeast Asia,
- Unusual and low monsoon,
- Low no. of depressions during the drought season, and
- Negative Southern Oscillation Index.

Factors Important in Drought Studies

Information of drought indicator helps in planning, analyzing crop production, power production and social economic effects, few of them are as follows:

- Intensity of drought – degree of severity of drought and measured by the degree of rainfall anomaly.
- Duration of the drought – indicates prolonged deficient rainfall period.
- Spatial distribution – indicate expensive area coverage of drought.
- Temporal distribution of drought – important in crop production, *e.g.*, mid-season and late season drought, frequency of drought or its probability in specific period.

AI Can Help to Solve Drought Issues: Future Directives

- Artificial Intelligence can help in selection of most suitable drought index based on location, climate, cropping pattern, irrigation requirement, etc. E.g., during homogeneous drought analysis by Filho et al. (2020), the scale of the river basin was used to reduce random fluctuations when applying the point-based approach.

- It can be helpful in estimation of impact of drought. E.g., the applications of AI techniques in the domain of drought assessing, monitoring, forecasting, etc. shows a rapid growth and that the impact of these will be increasing in future (Kikon & Deka, 2021).
- Deep learning based models can be used for forecasting meteorological drought. E.g., Ibrahimi & Baali (2017) predict drought by using and comparing neuro-fuzzy adaptive inference systems (ANFIS), artificial neural network of multilayered perceptron (ANN-MLP) and the support vector model (SVR).
- Automated control mechanisms can help in precision irrigation. Dynamic irrigation scheduling is directed toward efficient water usage for each plant (Benyezza et al., 2018; Bigah et al., 2019; and De Oliveira, 2010).
- Xavier et al. (2020) used AI model for estimating severity of social and economic damage caused by drought.
- Estimation of drought prediction error with the use of machine learning, e.g., Genuer et al. (2010) found that AI tools could be used to obtain an unbiased estimate of the prediction error as well as an estimate of variable importance.
- Smart management of inventory of resources, e.g., Praveen et al. (2022) found that the overstock and stock-out of items are reduced as the stocks are ordered based on the demand using artificial neural networks outputs.
- Artificial Intelligence can help in solving practical problems, which can occur during the drought, e.g., Tyralis et al. (2019) found that versatility of artificial intelligence could solve practical problems, including those in the water sector.
- AI enabled Big data analytics can help in analyzing large amount of multi-sourced data, e.g., Villarin et al. (2019) considered machine learning techniques as an alternative approach to conventional statistical models.

CONCLUSION

In conclusion, increased emphasis should be enforced on drought mitigation policies and preparedness plans, as well as drought prediction and early warning capabilities, if society is to reduce the agricultural, economic and environmental losses occurring due to the drought events. This needs interdisciplinary research, cooperation and a collaborative effort by the stakeholders and policymakers at all levels. However, climate, weather and hazard events (including drought) prediction and estimation modelling capabilities are being improved with the use of artificial intelligence and deep learning methods, which can make our preparedness plans accurately effective to face challenges arising from drought events every year.

REFERENCES

Alley, W. M. (1984, July). The Palmer Drought Severity Index: Limitations and Assumptions. *Journal of Applied Meteorology*, *23*, 110–1109. doi:10.1175/1520-0450(1984)023<1100:TPDSIL>2.0.CO;2

Benyezza, H., Bouhedda, M., Djellout, K., & Saidi, A. (2018). Smart irrigation system based Thingspeak and Arduino. *IEEE International Conference on Applied Smart Systems (ICASS2018) Proceedings,* 1–4.

Bigah, Y., Rousseau, A. N., & Gumiere, S. J. (2019). Development of a steady-state model to predict daily water table depth and root zone soil matric potential of a cranberry field with a subirrigation system. *Agricultural Water Management, 213*, 1016–1027.

De Oliveira, L., & Talamini, E. (2010). Water resources management in the Brazilian agricultural irrigation. *Journal of Ecology and the Natural Environment, 2*, 123–133.

Dunkel, Z. (2009). Brief surveying and discussing of drought indices used in agricultural meteorology. *Quarterly Journal of the Hungarian Meteorological Service, 113*, 23–37.

Edwards, D. C., & McKee, T. B. (1997). *Characteristics of 20th century drought in the United States at multiple time scales.* Colorado State Univ., Climatology Report No. 97-2.

Genuer, R., Poggi, J. M., & Tuleau-Malot, C. (2010). Variable selection using random forests. *Pattern Recognition Letters, 31*(14), 2225–2236.

Guttman, N. B. (1998). Comparing the Palmer Drought Index and the Standardized Precipitation Index. *Journal of the American Water Resources Association, 34*(1), 113–121.

Guttman, N. B. (1999). Accepting the Standardized Precipitation Index: A calculation algorithm. *Journal of the American Water Resources Association, 35*(2), 311–322.

Ibrahimi, A. E., & Baali, A. (2018). Application of Several Artificial Intelligence Models for Forecasting Meteorological Drought Using the Standardized Precipitation Index in the Saïss Plain (Northern Morocco). *International Journal of Intelligent Engineering and Systems, 11*, 267–275.

Karl, T. R., & Knight, R. W. (1985). Atlas of Monthly Palmer Moisture Anomaly Indices (1931–1983) for the Contiguous United States. Historical Climatology Series, National Climatic Data Center, No. 3–9, 319.

Kikon, A., & Deka, P. C. (2021). Artificial intelligence application in drought assessment, monitoring and forecasting: a review. *Stoch Environ Res Risk Assess.* doi:10.1007/s00477-021-02129-3

Lloyd-Hughes, B., & Saunders, M. A. (2002). A drought climatology for Europe. *International Journal of Climatology*. Advance online publication. doi:10.1002/joc.846

McKee, T. B., Doesken, N. J., & Kleist, J. (1995). Drought monitoring with multiple time scales. *Ninth Conference on Applied Climatology*, 233-236.

McKee, T. B., Doesken, N. J., & Klesit, J. (1993). The relationship of drought frequency and duration to time scales. *Eighth Conference on Applied Climatology Proceedings*, 6(January), 17–22.

Palmer, W. C. (1965). *Meteorological drought.* United States Department of Commerce, Weather Bureau, Research Paper No. 45.

Peixoto Xavier, L. C., Oliveira da Silva, S. M., Carvalho, T. M. N., Pontes Filho, J. D., & Souza Filho, F. A. d. (2020). Use of Machine Learning in Evaluation of Drought Perception in Irrigated Agriculture: The Case of an Irrigated Perimeter in Brazil. *Water (Basel), 12*(1546), 1–20. doi:10.3390/w12061546

Pereira, L. S., Cordery, I., & Iaconides, I. (2002). *Coping with water scarcity.* UNESCO, International Hydrological Programme, IHP-VI, Technical Documents in Hydrology No. 58.

Pontes Filho, J. D., Souza Filho, F. A., Martins, E. S. P. R., & Studart, T. M. C. (2020). Copula-Based Multivariate Frequency Analysis of the 2012–2018 Drought in Northeast Brazil. *Water (Basel), 12*, 834.

Praveen, K. B., Pradyumna, K., Prateek, J., Pragathi, G., & Madhuri, J. (2022). Inventory Management using Machine Learning. *International Journal of Engineering Research & Technology (Ahmedabad), 9*(6), 866–868.

Sen, Z. (2008). *Wadi hydrology*. CRC Press.

Thompson, R. D. (1975). The climatology of arid world. University of Reading, Department of Geography, Paper No. 35, 39.

Tyralis, H., Papacharalampous, G., & Langousis, A. (2019). A brief review of random forests for water scientists and practitioners and their recent history in water resources. *Water (Basel), 11*, 910.

Villarin, M. C., & Rodriguez-Galiano, V. F. (2019). Machine Learning for Modeling Water Demand. *Journal of Water Resources Planning and Management, 145*, 1–15.

Chapter 10
Soil-Water Management With AI-Enabled Irrigation Methods

Ved Prakash Singh
Ministry of Earth Sciences, Bhopal, India

Shirish Khedikar
Ministry of Earth Sciences, Pune, India

Jimson Mathew
Indian Institute of Technology, Patna, India

Lucky Kulshrestha
Medi-caps University, Indore, India

ABSTRACT

In India, farming assumes many conventional and modern processes, having large impact on GDP. Modernizing the old methods diminishes the manual practices at fields. Irrigation is one of these processes, which involves several activities like soil parameters estimations, arrangement of watering instruments with substantial itemized and labour costs. It directly affects soil quality, water availability for the plants, and ultimately, affects the growth and yield of the crops. If irrigation is well managed with accurate soil-water availability estimation, minimizing water requirement, and intelligent techniques of applying the water in sites of interest, then significant reduction in overall farming cost can be achieved with retaining soil nutrients and reusable irrigation facilities in the fields.

INTRODUCTION

The artificial application of water to land is called as Irrigation. Some lands require irrigation before it is possible to use it for any agricultural production. In other places, irrigation is primarily a means to supplement rainfall and serves to increase production (Jones, 2022). Irrigation management primarily refers to the decision process and action of applying a chosen depth of irrigation water using a chosen

DOI: 10.4018/978-1-6684-3981-4.ch010

application method at a chosen time to achieve defined agronomic and economic objectives (Evett *et al.*, 2014).

Problems in Conventional Irrigation

1. First three days after irrigation
   a. Pores saturate with water
   b. Excess water suffocates roots
   c. Affect water and nutrient absorption, ultimately growth hampers
2. Middle three days due to Evapotranspiration
   a. Soil water reaches Field Capacity
   b. Optimum water, nutrient, air
3. Last two days
   a. Water moves below root zone
   b. Stress condition and growth restricts

Modern Irrigation Methods

1. Drip Irrigation/Trickle Irrigation or Micro Irrigation

An irrigation method, which saves water and fertilizer by allowing water to drip slowly to the roots of plants, either onto the soil surface or directly onto the root zone, through a network of valves, pipes, tubing, and emitters. Slower application of water to the root zone area. The drip irrigation system consist of Head, Main line and sub line, Lateral lines and Drip nozzles.

a. **Salient features**
   i. 50 to 65% water saving compared to control method,
   ii. Crop attains early maturity with increased crop quality and yield,
   iii. Requires least land leveling,
   iv. Poor quality water can be used,
   v. Overhead drip irrigation with good quality water can be used for crops like grapes.
b. **Advantages**
   i. The losses by drip irrigation and evaporation are minimized.
   ii. Precise amount of water is applied to replenish the depleted soil moisture at frequent intervals for optimum plant growth.
   iii. The system enables the application of water fertilizers at an optimum rate to the plant root system.
   iv. The amount of water supplied to the soil is almost equal to the daily consumptive use, thus maintaining a low moisture tension in soil.
c. **Disadvantages**
   i. The initial cost of the drip irrigation for large-scale irrigation is its main limitation.
   ii. The cost of the unit per hectare depends mainly on the spacing of the crop. For widely spaced crops like fruit trees, the system may be even more economical than sprinkler.

2. Partial Root-Zone Drying

One irrigation technique known as partial root-zone drying (or PRD) involves "tricking" the grapevine into thinking it is undergoing water stress when it is actually receiving sufficient water supply. In partial root-zone drying, half the roots are allowed to dehydrate which sends signals to the vine that is experiencing "water stress". Meanwhile the irrigated roots on the other side continue to provide sufficient amounts of water so that vital functions like photosynthesis does not cease. This is accomplished by alternating drip irrigation to where only one side of the grapevine receives water at a time.

The roots on the dry side of the vine produce abscisic acid that triggers some of the vine's physiological responses to water stress - reduced shoot growth, smaller berries size, *etc.* However, because the vine is still receiving water on the other side the stress does not become so severe to where vital functions such as photosynthesis is compromised. Partial root-zone drying has been shown to significantly increase a vine's water use efficiency. While PRD is shown to slightly reduce leaf area, this is generally not a problem as overall yield is unaffected.

3. Sprinkler or Overhead Irrigation

It is an application of water to soil in the form of spray, somewhat as rain, and mainly useful for sandy soils because they absorb water too fast. Soils that are too shallow, too steep or rolling can be irrigated efficiently with sprinklers. This method is suitable for areas with uneven topography and vulnerable to erosion hazards.

a. **Salient features**
   i. Ideally suitable for hilly terrains with undulating slope.
   ii. Suitable for estate crops like tea, coffee *etc.*
   iii. Suited to crops like groundnut, cotton and the crops, which are not susceptible to easy flower shedding.
   iv. 30-40% saving in cost than in conventional irrigation.

b. **Adaptations**
   i. A dependable supply of water
   ii. Uneven topography and shallow soils
   iii. Close growing crops

c. **Advantages**
   i. It ensures uniform distribution of water.
   ii. It is adaptable to most kinds of soil.
   iii. It offers no hindrance to the use of farm implements.
   iv. Fertilizer materials may be evenly applied through sprinklers by liquid fertilizer - time of application varies from 10 to 30 min.
   v. Water losses are reduced to a minimum extent and more land can be irrigated.
   vi. Costly land leveling operations are not necessary.
   vii. The amount of water can be controlled to meet the needs of young seedling or mature crops.

d. **Disadvantage**
   i. The initial cost is rather very high and suitable for high value crops.

ii. Wind interferes with the distribution pattern, reducing spread or increasing application rate near lateral pipe.
iii. There is often trouble from clogged nozzle or the failure of sprinklers to revolve.
iv. The cost of operations and maintenance is very high. Labour requirement for moving a pipe and related work approximately nearly one hour per irrigation.
v. It requires a dependable constant supply of water free from slit and suspended matter.

4. Surge Irrigation

It delivers irrigation flows into individual long furrows (25m to 200m) in an intermittent fashion of predetermined ON-OFF time cycles (5 min to 10 min) with the design duration of irrigation. During the ON time, waterfront advances into the furrow over a certain length and during the subsequent OFF time, the water applied partially saturates the soil and infiltration rate gets reduced on the advanced length. Water is delivered in the succeeding ON time, the water front advance gets accelerated due to the reduced intake rate and eventually it reaches the tail end of long furrow within 30 -50% of the design duration of irrigation.

Surge Irrigation is a variant of furrow irrigation where the water supply is pulsed on and off in planned periods (*e.g.*, on for ½ hour off for ½ hour). The wetting and drying cycles reduce infiltration rates, resulting in faster advance rates and higher uniformities than continuous flow. The reduction in infiltration is a result of surface consolidation, filling of cracks, micro pores, and the disintegration of soil particles during rapid wetting and consequent surface sealing during each drying phase. The effectiveness of surge irrigation is soil type dependent. Clay soils experience a rapid sealing behaviour under continuous flow therefore surge offers little benefit.

a. **Advantages**
   i. ON - OFF water supply and cutoff - highly minimized deep percolation and runoff losses (hardly exceeding 20%).
   ii. High uniformity of soil moisture distribution with in the effective root zone is achieved over the entire furrow length resulting in enhanced irrigation efficiencies of 85% to 95%.
b. **Limitations**
   i. Surge irrigation systems do not show marked differences in land and water saving in extremely clay or sandy soils.
   ii. Surge irrigation technology is still in the infant stage in India and requires popularization through extension methods.

5. Subsurface Drip Irrigation (SDI)

Subsurface irrigation or sub-irrigation may be natural or artificial. Natural sub surface irrigation is possible where an impervious layer exists below the root zone. Water is allowed into the series of ditches dug up to the impervious layer, which then moves laterally and wets root zone. In artificial sub surface irrigation, perforated or porous pipes are laid out underground below the root zone and water is led into the pipes by suitable means. It is very efficient in the use of water as evaporation is cut off almost completely. In either case, the idea is to raise the water by capillary movement. The method involves initial high cost, but maintenance is very cheap.

a. **Advantages**
   i. Evaporation losses are reduced or by weeds.
   ii. Period of irrigation is low (2-3 hours).
   iii. The growth of plants is homogeneous and ultimately, yield is improved.
   iv. Increasing the area of planting by saving the area of canals.
   v. Decreases the infestation with pests (insects, diseases and weeds).
   vi. Decreases the number of weeds.

6. Precision Irrigation

Irrigation aspires to be and should be a precision activity, involving both - accurate assessment of the crop water requirements and the precise application of this volume at the required time. Key steps in a precision irrigation System are Data acquisition, Interpretation, Control and Evaluation. Crop water stress mapping for site-specific irrigation is done by thermal imagery and artificial reference surfaces.

7. Variable Rate Irrigation (VRI)

Variable Rate Irrigation is an add-on to a centre-pivot or lateral-move irrigation system that allows different amounts of water to be applied along any part of the length of the irrigator at any one time. The VRI system provides precision control of all sprinklers on a centre-pivot or lateral-move irrigator. Individually pulsing sprinklers on and off, while also controlling the system speed to modify the application depth along the length of the irrigator.

Water savings can be made using variable rate irrigation because variable rate irrigation enables better use of stored soil water and aims to maintain maximum yield potential with no water stress to the plant. As global irrigation demands escalate, it is important to improve system efficiencies for the best use of freshwaters. VRI systems with real-time soil moisture monitoring of soil management zones provide the information required to maximize use of soil water storage and ensure adequate delivery of available soil water to the plant.

a. **Benefits of Variable Rate Irrigation (VRI):** Once a sprinkler system is modified for variable rate application, there are multiple benefits as given below:
   i. Keeping water off exclusion zones such as wet areas.
   ii. Increasing flexibility for mixed cropping.
   iii. Precision chemigation and fertigation.
   iv. Improving application accuracy at either end of the pivot.
   v. Maintaining optimum soil water levels for plant growth.

8. GPS End-Gun Controller

This small device can be installed on a center-pivot or linear-move irrigator to control up to three end-guns or other outputs depending on the position of the irrigator. Precision Irrigation's GPS end-gun controller is installed on the end-gun control line to turn the end-gun off or on in pre-programmed positions. It is simple and easy to install and ensures almost nil water wastage in the places where it is not needed.

9. Buried Wire Guidance System With Computer Designed GPS Layout for Corner-Arm and Z-Corner Pivots

While marking out corner-arms or z-corners, the calculation and manual measuring process for where to bury wire guidance is very slow and laborious. Given the field and centre-pivot specifications, Precision Irrigation can design the buried wire guidance positioning to be marked out with high-accuracy GPS. This is especially useful when several centre-pivots are working close to each other and you want to have your corner-arm or arms irrigate the maximum area possible while never being in risk of collision. We can design and simulate any brand of centre-pivot irrigator and provide the design as a set of GPS-coordinates to mark out yourself. Simulations can be done for standard corner-arms or z-corners, in either leading or lagging configuration.

10. Z and Corner Arm Programmable Control (The Precision Irrigation)

Corner-arm or z-corner control system is a cut down version of the full VRI system, which only controls valves on the corner or z sections of the irrigator. Every sprinkler is controlled individually to give maximum control. The amount of water applied under every part of the corner / z is fully programmable and updatable. A common problem with corner-arm and z-corners is that even though they sequence sets of valves to try and keep irrigation application as even as possible, there are still downfalls due to either the irrigator not cycling valves in the correct positions, or ground conditions that require more or less water than the corner-arm system applies.

IRRIGATION SITE MAPPING

For control of centre-pivot machines for site-specific irrigation (GSP), watering rates need to increase above 100%; the pivot will be slowed down. Similarly, if less than 100% is required in zones, the pivot will walk faster, saving energy and water. Modern ways of Available Water-Holding Capacity (AWC) mapping are discussed below -

EM Mapping

Electromagnetic (EM) mapping is a method for mapping the relative conductivity of soils in a field. Conductivity of soils is affected by soil properties including but not limited to soil moisture, clay content, compaction, salinity, cation exchange capacity and temperature. By creating an EM Map of a field being irrigated, we get a good insight into the variability of the field and this is a great starting point for creating a plan for Variable Rate Irrigation. Farm Mapping provides EM Mapping services in New Zealand and in association with Landcare Research, New Zealand Ltd, can create an Available Water-Holding Capacity (AWC) map to load into a VRI map or any other GIS software.

EM to AWC Map

EM survey, an accurate EM Map will show the spatial differences in soil conductivity. With local knowledge of field characteristics and by doing some ground truthing by digging up samples and lab testing, the

EM map can be interpreted and used as a basis for making decisions on how to best utilise VRI. EM Map created by Precision Irrigation, carry out a ground-truthing procedure on-site, and then extrapolate the ground truthing results using the EM Map to create an Available Water-Holding Capacity (AWC) map.

Combined with soil moisture monitoring, an AWC is a very useful tool for determining how much irrigation should be applied to different areas of the field. EM mapping can be used to map and quantify soil variability for site-specific management. AWC maps can be derived from an EM map, which can be adjusted on a daily basis using a water balance and/or daily soil moisture logging (*e.g.*, a wireless soil moisture sensor network), providing spatial information for accurate irrigation scheduling.

Types of Plant Based Sensing

1. Radiometric sensors - Normalized Difference Vegetation Index (NDVI)
2. Soil-Water Sensing
3. Thermal Sensing - crop canopy temperature – transpiration.

Scope of Advanced Irrigation

- Irrigation efficiency 90-95%,
- Uniformity in application,
- Optimum balance of nutrient, water, air in soil,
- Better growth and higher yield,
- Irrigation Efficiency.

Irrigation Efficiency

It is normally expressed in terms of the amount stored in the root zone as a percentage of the total water released at the project headworks. It is separated into three components: the conveyance efficiency (Ec), the field canal efficiency (Eb) and the field application efficiency (Ea) or project efficiency (Ep). Here, Ep = Ec x Eb x Ea. Main factors affecting irrigation efficiency are the size of the project, the number and the type of crops requiring adjustments in supply, the canal seepage, the size of the individual fields, the irrigation methods and practices, and managerial and technical facilities for water control. At the planning stage, irrigation efficiency is normally estimated on the basis of experience. The estimated values can be checked only after 5 to 10 years since construction of the project, when operators and farmers have become familiar with the water distribution and application.

Irrigation efficiency (Ep) is expressed in terms of water loss (m^3/m^3). For practical purposes, it is normally considered as constant over the growing season. However, in most cases efficiencies will vary over the season. For instance, under conditions of limited water supply to the field and when under-irrigation is practiced, most water applied to the fields is taken up by the crop and field application efficiency may be near 100 percent rather than the frequently assumed 50 to 70 percent. In addition, when water is in short supply, more attention is given to water distribution and scheduling, and project efficiency is consequently higher. Furthermore, the effect of limited water supply on crop yields varies over the growing season, and consequently the impact of water loss on yield varies over the growing season.

Besides expressing it in terms of water losses (m^3/m^3), irrigation efficiency can also be expressed in terms of yield losses by applying the relationships between water supply and crop yield. Under conditions of adequate supply, an increase in irrigation efficiency will save water. Thus, an extension of the irrigated area is possible and the total yield is subsequently increased. Under conditions of limited supply, an increase in irrigation efficiency will also save water, making a reduction in water deficits to the crop occurring during different growing periods possible, and yield per unit area is subsequently increased. Consideration of irrigation efficiency in relation to yield response to additional water supply assists in the evaluation of the need for improvement of irrigation efficiency, as for instance, the decision on canal lining.

Irrigation Water Management (IWM)

Irrigation water management (IWM) is the act of timing and regulating irrigation water application in a way that will satisfy the water requirement of the crop without wasting water, soil and plant nutrients, and degrading the soil resource. This involves applying water –

- according to crop needs,
- in amounts that can be held in the soil and be available to crops,
- at rates consistent with the intake characteristics of the soil and the erosion hazard of the site,
- so that water quality is maintained or improved.

A primary objective in the field of irrigation water management is to give irrigation decision makers an understanding of conservation irrigation principles, how they can judge the effectiveness of their own irrigation practices, make good water management decisions, recognize the need to make minor or major adjustments in existing systems, and/or to install new systems.

The net results of proper irrigation water management are typically:

- Prevent excessive use of water for irrigation purposes.
- Prevent excessive soil erosion.
- Reduce labor and minimize pumping costs.
- Maintain or improve quality of ground water and downstream surface water.
- Increase crop biomass yield and product quality.

Tools, aids, practices, and programs to assist the irrigation decision maker in applying proper irrigation water management include:

- Applying the use of water budgets, water balances, or both, to identify potential water application improvements.
- Applying the knowledge of soil characteristics for water release, allowable irrigation application rates, available water capacity, and water table depths.
- Applying the knowledge of crop characteristics for water use rates, growth characteristics, yield and quality, rooting depths, and allowable plant moisture stress levels.
- Irrigation scheduling techniques and water delivery schedule effects.
- Water flow measurement for on-field water management.
- Various methods of irrigation system evaluation.

Prerequisites for Integrated Water Management

- An institutional cooperative framework.
- Adequate monitoring system.
- A geo-referenced database and information system that includes relevant hydro-meteorological, water demand and other data and information.
- Shared-vision planning and water resources modelling, and management support tools and technical expertise.

Crop Simulation with DSSAT v4 or Higher

- The model integrates soil and weather during a growing season to predict yield of crops. Thus, they inherently account for the effects of amount and timing of rainfall during a season on crop growth and yield for specific soil conditions.
- Predict crop growth, yield, timing (outputs).
- Optimize management using climate predictions.
- Diagnose yield gaps (actual vs. potential).

Irrigation Management

- Climate-change effects on crop production, and Greenhouse climate-control.
- Quantify pest damage effects on production, and precision farming.

An irrigation water management system depends mainly on identifying the crop water requirements, the crop irrigation requirements and the crop irrigation schedule. The procedures reported by Doorenbos & Pruitt (1984) adopted to calculate the crop water requirements, the crop irrigation requirements, and the irrigation schedule for the development of the computerized irrigation-water management system. Calculation of Crop Water Requirements (ETcrop) considers:

- Effect of climate on crop water requirements.
- Effect of crop characteristics on crop water requirements.
- Effect of local conditions and agricultural practices on crop water requirements.

Effect of Climate on Crop Water Requirements

1. Reference crop evapotranspiration (ETo) describes the effect of climate on crop water requirements. The Penman–Monteith method was introduced by the American Society of Civil Engineers (ASCE) in 1990 and 1996 to predict ETo (Jensen *et al.*, 1990), further, selected for the calculation of the ETo for the CDIWMS.
2. The Penman–Monteith method combines thermodynamic and aerodynamic aspects, including resistance to sensible heat, and vapour transfer and surface resistance to vapour transfer.

AI / Neural Network Algorithm for Soil-Water Simulation

Neural Network Simulation (NNS) is a knowledge base of the DSS model. It can be defined as a computing system made up processing elements, which process information by their dynamic state response to external input. NNSs are typically organized in layers. Layers are made up of a number of interconnected 'nodes', which contain an activation function. Patterns are presented to the network via the input layers, which communicates to one or more 'hidden layers' where the actual processing is done via a system of weighted 'connection'. The hidden layers produces the answer as output. The scheme of the NNS, which is used in the DSS-model is described herein in succeeding texts.

The DSS model is composed of four units. The first unit a Database Management System (DBMS), used to manage all involved data for the model. A data base program is designed to manage the data in the program. The second unit is a forecasting simulation unit named FORSIM, used to simulate expected hydrology and climatic data for the input of SWAT unit.

Neural Network Simulation is employed in the FORSIM unit. The third unit is a soil water balance unit (SWAT), used to simulate water demand. The fourth unit is a Geographic Information Unit (GIS), used for the output presentation. This unit uses Active-X programming in Delphi. The spatial and temporal distribution of water demand can be presented in the screen through GIS technique in this program.

Functions of Soil Moisture (Water)

- Soil water serves as a solvent and carrier of food nutrients for growth of plant.
- Water acts as a nutrient itself, and is a principal constituent of plant.
- Water regulates soil temperature (keeps the soil norm from too cold or too hot).
- Water helps to create soil and differentiate it into horizons.
- Soil forming processes and weathering depends on water.
- Water leaches out plant nutrients.
- Controls the metabolic activities of microorganisms.
- Helps in chemical and biological activities of soil.
- Essential for photosynthesis.
- Carrier dissolved oxygen (O_2) in to the soil.
- Prevents entry of air into the soil.

Different Forms of Soil Moisture (Water)

- Wilting point or permanent wilting point (13.6 to 15.0 atmosphere): the soil moisture condition at which the case of release of water to plant roots is just barely too small to counter balance the transmission losses.
- Water of constitution/ interlayer water (> 7 pF).
- Hygroscopic water (pF 4.5-7.0): the hygroscopic coefficient of a soil is meant to be a condition at which absorption is completed, satisfied in an atmosphere completely saturated with water vapour.
- Best Tillage Range (pF 2.8-4.4): soil is to have enough consistence to maintain its small aggregates and to hold up the tillage implements.

- Field Capacity (1/3 bar of tension or pF of 2.5): when a thoroughly wetted soil is allowed to drain for few days, it reaches relatively to a stable moisture content called filed capacity. Normally, duration of 28-44 hours is sufficient to get the field capacity moisture.
- Capillary water (pF 2.5-4.5).
- Aeration porosity (PF 1.7 or tension or 1/20 atmosphere): aeration porosity of a soil is defined as that part of the pore space expressed in volume percent that is free of water by a tension of water column of a height of 50 cm.
- Gravitational water (pF 0-2.5).
- Ground Water-Zero (Saturation): when a soil whose pores are completely filled with water and then the soil is said to be saturated.

Here, pF= Energy with which water is held in soil as given by Schofield (1961).

Soil Moisture Measurement

Soil moisture plays a significant role in land-atmosphere interaction. Soil moisture is also a fundamental component in hydrology, climate and soil-vegetation interaction. Hence, soil moisture estimation helps in many natural resource applications such as hydrological modeling, stream flow, flood and drought mapping and monitoring. Several authors have revealed the importance of soil moisture estimation for drought monitoring (Mohanty & Skaggs, 2001). Popular methods to estimate soil moisture are discussed below –

1. **Point Measurement**
   a. **Thermo-Gravimetric Method:** It is most common and direct procedure for soil moisture estimation by removing a physical sample from the site in question, weighed and dried in an oven. This method is known as classified procedure as it is used for comparison of accuracy of other methods. Drying soil at 100 °C to 110 °C for 24 to 48 hours until reaching a constant weight and then re-weighed. (Reynolds, 1970a; Reynolds, 1970b; Gardner, 1986; and Evett, 2008). Following formulae is used to estimate moisture:

Moisture = ((Wt. of moist soil – Wt. of dry soil) / Wt. of dry soil) * 100

The water content is expressed in volumetric percent (Vol.-%), one has to consider the bulk density. As the amount of water lost by drying increases with increasing oven temperature for any inhomogeneous soil containing clay or organic matter, the temperature has to be controlled within a range between 100 and 110°C (Reynolds, 1970b). In gravimetric method of soil moisture estimation, about 10% or more variation can occur based on typical field samples (Holmes *et al.*, 1967). The gravimetric approach is rather unattractive in terms of a time efficient and repeatable ground data collection for remote sensing studies.

b. **Neutron Scattering Method:** Prob containing radio-active material such as Radium or Americium-241 and Beryllium used as a source of rapid moving neutrons. Fast Neutrons collides with Hydrogen ions and get slowed. The number of hydrogen atoms in soils changes because of the change in soil water content; therefore, the hydrogen content can be calibrated by counting neutrons on detection tube (Hignett & Evett, 2002).

Slower the neutron means greater water content of soil. Over estimation of soil moisture is possible due to presence of hydrogen in organic matter. Boron Tri-fluoride (BF_3) is used as detector. Prob having diameter 38 mm and length 750 mm is used, which can monitor soil moisture up to 4 meter depth. According to Evett (2008), a field-calibrated neutron moisture meter is the most accurate and precise indirect method for soil moisture measurement in the field.

c. **Gamma Rays Attenuation Method:** Gardner (1986) reports the use of gamma densiometry over the last fourty years. In the field, two access tubes are placed vertically into the soil and a gamma source (*e.g.*, Cesium-137) is lowered into one tube. A detector of gamma radiation is lowered into the other tube and the emitted radiation is recorded. In soil, the electron density of mineral particles is the same (Neiber *et al.*, 1991) and increased attenuation is due to the presence of water. If the combined attenuation is calibrated (via Beer's Law, 1852) and considered with the soil bulk density, the volumetric moisture content of water can be determined. It is doubtful that this method will become widely used in the field, as the operation is more complex than other techniques and due to safety problems handling the radioactive material.
d. **Electrical Resistance Block:** Measurement of the electrical resistance of a block embedded in porous media and in equilibrium with the media was one the first techniques used for determination of soil water status (Colman & Hendrix, 1949; and Cummings & Chandler, 1940). Electrodes are embedded in a porous block (including materials such as fiberglass, gypsum and nylon) and placed in the soil. Water in the soil will reach equilibrium with the water in the porous block and the electrical resistance is then determined and related to moisture content as a tension ($J\ kg^{-1}$). Measurement of the resistance is dependent on the soil temperature and bulk soil electrical conductivity. Considerations for operation of the porous blocks are detailed by Gardner (1986) and Campbell & Gee (1986).

To overcome the problem of changing soil electrical conductivity, gypsum is often used as the medium containing the electrodes. The gypsum dissolves, dominating the immediate soil solution with Ca^{2+} and SO_4^{2-} ions reducing the effect of surrounding soil electrical conductivity (Campbell & Gee, 1986). Gypsum blocks will dissolve over a period (with the rate of dissolution increasing in sodic soil) generally lasting for two to three seasons in good conditions (Gardner, 1986). Gypsum blocks are insensitive to moisture content changes at high potentials, 0 to -60 $J\ kg^{-1}$ (Gardner, 1986).

e. **Electro-Magnetic Induction:** As the generated EM wave travels along the extension cable and into the soil guided by the probes, proportions of the wave are reflected when a change in impedance occurs. The partial reflections reduce the energy in the EM wave and can interfere with the end-point determination of the waveform (White & Zegelin, 1995).
f. **Tensiometer Method:** Energy status of the water may be measured in-situ with the tensiometry technique as described in detail by Cassel & Klute (1986). Ritchards (1928) first identified the use of ceramic cups attached to monitoring equipment (following the work of Gardner *et al.,* 1922) for determination of the soil matric potential (Ψm). Tensiometers are applied to irrigation scheduling of crops in the late 1950's (Richards & Marsh, 1961). The use of tensiometers is expanding in

solute transport studies as detailed by Vanclooster *et al.* (1993); Yasuda *et al.* (1994); and Ward *et al.* (1995), amongst others.

g. **Hygrometric Method:** In unsaturated soil, the pressure potential (Ψp) is zero and the water potential (Ψw) is the sum of the matric potential (Ψm) and osmotic potential (Ψo) as shown as Ψw = Ψm + Ψs + Ψp. Thermocouple psychrometers (also termed hygrometers) are used extensively to determine the water potential as described by Rawlins & Campbell (1986); Campbell (1988); and Rasmussen & Rhodes (1995). Measurements are based on the theoretical relationship between the water potential and relative humidity shown as

$$\Psi w = \left(\frac{RT}{M_w}\right)\ln\left(\frac{p}{p_0}\right)$$

Brown & Oosterhuis (1992) describe some of the considerations required to obtain accurate measurements with psychrometers during in-situ operation and note the importance of temperature on reported near surface measurements.

h. **Soil Dielectric Method:** Dielectric measurement takes advantage of the differences in dielectric permittivity values between different soil phases (solid, liquid, and gas). Liquid water has a dielectric permittivity of »80 (depending on temperature, electrolyte solution, and frequency), air has a dielectric permittivity of »1, and the solid phase of 4 to 16 (Hallikainen *et al.,* 1985; and Wraith & Or, 1999). This contrast makes the dielectric permittivity of soil very sensitive to variation in soil moisture. The measurement of the bulk dielectric permittivity is then used to obtain the volumetric water content through calibration curves (Roth *et al.,* 1990; and Topp *et al.,* 1980).

Time Domain Reflectometry (TDR) prob

It determines soil moisture on volumetric basis. High-energy Electro-magnetic (EM) pulse is fed into soil between two metal rods. A part of pulse reelected back and the time interval between incident and reflected pulse is measured. Time interval is proportional to moisture content. TDR is based on principle of responding to changes in the relative dielectric constant. Employing hand held probes with data loggers is one of the most time efficient methods to measure distributed soil moisture patterns in the field.

Capacitance Prob

The frequency-domain technique is similar to that of TDR, in that the apparent dielectric (Ka) relationship to moisture content (θ) is exploited. Wyman (1930) identified the relationship between capacitance and soil moisture in the late 1920's. However, the use of capacitance based techniques was not possible due to the inability to select oscillating frequencies not influenced by the bulk soil electrical conductivity (σ). Malicki (1983), following the work of Thomas (1966), developed a working in-situ capacitance based sensor. Widespread use of frequency-domain (FD) sensors followed development of a down-hole portable instrument (Dean *et al.,* 1987).

Pulsed Nuclear Magnetic Resonance

The nuclear magnetic resonance (NMR) technique determines the soil water status due to the interaction between a static magnetic field and the nuclear magnetic dipole moments (Paetzold *et al.,* 1985). NMR has been successfully incorporated into an implement-pulled through the soil measuring moisture content (Paetzold *et al.,* 1985). Further, critical review of the technique has not been undertaken yet. An important aspect of this technique is the ability to distinguish between bound water and available water (Whalley & Stafford, 1992). The high cost of instrumentation is likely to limit in-situ technique application (Baker, 1990).

2. **Hydrological-Soil Models:** A model is a conceptual representation of a real world system and theoretically, it is possible to apply models to any hydrological problem. It is in hydrological modelling that modern hydrology has found its base. Different types of models are appropriate depending on the purpose, data availability, spatial scale, time scale, cost and computing resources. It can be empirical describing how the world behaves with little attempt to explain the underlying principles or concepts, based on limited representations of the processes occurring in the hydrological system, on a perceived system behavior. Alternatively, it can be physical model that represents all the relevant processes in the hydrological system under consideration in a physically meaningful way (Watts, 1997).
    a. Soil models are basically based on column mass balance, provide an alternative to directly or indirectly measuring soil moisture in the field based upon the conservation of mass, soil moisture in the system can be determined using the relationship given below:

$$SM_t = SM_{t-1} + P - R - L - E - T + C - Q$$

where,

SM_t is the soil moisture volume at time t,
SM_{t-1} is soil moisture volume at previous time,
P is the precipitation, R is surface runoff,
L is net lateral subsurface outflow,
E is the evaporation or condensation,
T is transpiration, Q is the percolation, and
C is capillary rise from lower levels.

Such model hence represents only a single column that is horizontally homogeneous at all levels. Contrary to this, actual systems are heterogeneous and those can be represented by spatial averages or by linked columns that account for the spatial variability. Most of the models based on soil-water relationship developed for practical application to agricultural activities like crop yield estimation, irrigation planning, runoff forecasting and use readily available meteorological data for inputs for one-day time step. The models can provide timely soil moisture information without the necessity of field visits. The main limitation of using such models is the error of their estimates.

In simulating soil moisture over the entire soil profile using a soil moisture model, large errors are unavoidable due to the highly dynamic nature of the near- surface zone. Therefore, when the measured soil moisture data are available, their use in place of the simulated data should improve the overall estimation of the soil moisture profile with the assumption that measurement errors are less than simulated errors (Arya *et al.,* 1983).

3. **Backscattering models:** The theoretical models used as a reference and the empirical & semi-empirical models tested are briefly described in this section. For the sake of simplicity, the complete formulation of the models is not included because they are well defined in the literature.
   a. **Theoretical/Physical Model:** Soil moisture content can be estimated from the mathematical inversion theoretical models. These models can be applied in the case of specific roughness conditions and assume that the rms height, the correlation length and the dielectric constant are known. So, it can be reversible. Some of the theoretical models
      i. Integral Equation Method (IEM)
      ii. Geometrical optics model (GOM)
      iii. Physical optics model (POM)
      iv. Small perturbation model

The Geometrical optics model (GOM) represents the stationary phase solution of the Kirchhoff models (KM). This solution is based on the assumption that the coherent backscattering term is much smaller than the non-coherent component and thus can be neglected (Ulaby *et al.,* 1986).

b. **Empirical model:** The statistical or empirical models are simplest and most widely used one. It requires huge amount of ground data, but this model is site specific.

SM = A + B * σþ

The empirical models are developed based on ground-based scatterometer observations acquired under a variety of roughness, moisture, and sensor configurations. These are derived from experimental measurements to establish useful empirical relationships for inversion of soil moisture from backscattering observations (Walker *et al.,* 2004).

c. **Semi-empirical models:** Practical application of empirical models can be problematic due to limited validity ranges within which such models can be used. Therefore, some researchers proposed backscattering models empirically fitted to simulations obtained by theoretical models (mainly Integral Equation Method), thus circumventing the site dependency of models based on observations. Those models provide simplified solutions to the theoretical algorithms that can be more easily applied and inverted.

FERTIGATION

Fertigation is the application of plant nutrients through the irrigation system. We take the chemical fertilizers and dissolve them into the irrigation system. The plant roots then receive WATER + FERTILIZERS at the same time and location. A brief history of fertigation is given below –

- 400 B.C. - Fertilization of gardens and orange groves in Athens by city sewage in canals.

- 1899 - First liquid fertilizer patent in USA.
- 1942 - Injection of anhydrous ammonia into soil.
- 1953 - First liquid fertilizer plant using ammonia, phosphoric acid and potash.
- 1958 - First reported application of commercial fertilizer through a sprinkler irrigation system.
- 1974 - 2800 fluid fertilizers plants in USA.
- 1980 - Fluids (inc. NH_3 & N solutions) = 32% of USA fertilizers
- 1994 - 90% of irrigated horticultural crops in Israel fertilized by fertigation.

Objectives of Fertigation

- Maximizes profit by applying the right amount of water and fertilizer.
- Minimizes adverse environmental effects by reducing leaching of fertilizers and other chemicals below the root zone.

Benefits of Water-Soluble Fertilizers (WSF) in Fertigation

- They are high quality imported fertilizer fully soluble in water.
- They contain one or more plant food nutrients along with micronutrients.
- WSF are pure and precipitate-free.
- Normally these fertilizers are acidic (pH 5.5-6.5), which helps in correction of soil pH and prevention of clogging of emitters.
- For acidic soils, liquid fertilizers are available in neutral or even higher pH.
- WSF ensures a regular flow of both water and nutrients, resulting increased growth rates and higher yields.
- WSF offers greater versatility in the timing of the nutrient application to meet specific crop demands.
- With WSF, three major plant nutrients are supplied in one solution as compared to single nutrient straight fertilizers.
- WSF are also chloride free (thus controlling negative effect of chlorides in plant metabolism), which are useful for high value crops and crops sensitive to chlorides, example-tobacco, citrus, grapes, arecanut and vegetables.
- WSF increases the availability and the uptake of nutrients.
- 100% soluble, hence wastage in soil (fixation) is minimal.
- High FUE as all the major micronutrients are fully available as preferable forms by the plants.
- WSF are easily movable in both soils and plants, hence higher uptake and associated higher FUE.
- Correction of micronutrient deficiencies is more convenient with EDTA / EDDHA forms through microirrigation system.
- Less quantum of fertilizer is required as Nutrient contents are relatively higher (ex: MKP 0-52-34) than conventional fertilizers.
- P nutrient availability is much greater with grades like MAP (12:61:0), MKP (0-52-34), Orthophosphoric acid (0-52-0) than conventional SSP and DAP (which are less soluble and fixed in soil).
- Saline water could be effectively managed by use of acid forming WSF like (injection of Phosphoric acid, MAP & MKP with pH of 2.6, 4.9, 5.5 respectively *etc.*).

- WSF are low in salt index, hence they do not leach or get fixed in soil or increase the salinity.
- Enforced with micronutrients and optional nutrients like Mg and Boron could be enforced.
- Special foliar grade WSF are available for correction of nutrient deficiencies and enhance yield and quality attributes.

Advantages of Fertigation

- Relatively uniform fertilizer applications,
- Flexibility in timing of applications,
- Less fertilizers used,
- Reduced costs.

Disadvantages of Fertigation

- Potential contamination hazard from equipment malfunctions.
- Backflow prevention devices are required.
- Careful handling of liquid fertilizers are required.

Which Fertilizers are Suitable for Fertigation

- High nutrient content in the solution
- Fully soluble at field temperature.
- Fast dissolution in irrigation water.
- Fine grade and flowable.
- No clogging of filters and emitters
    - Low content of insolubles,
    - Minimum content of conditioning agents.
- Compatible with other fertilizers.
- Minimal interaction with irrigation water.
- No drastic changes of water pH ($3.5 < pH < 9$).
- Low corrosivity for control head and system.

Micro Irrigation and Fertigation

- Method that is more efficient compared to conventional method.
- Scope for reduction of inorganic fertilizer without yield reduction.
- Yield and quality of crops increased.
- Increase in net income over conventional method.
- 30-40% water saving (drip) with 25-30% fertilizer saving.
- Minimizes water and soil pollution.

SUMMARY: FUTURE DIRECTIVES

Managing soil-water availability efficiently with natural or artificial (irrigation) methods is the key to ensure supply of water & nutrients to the plants, their maturity, optimal growth throughout the season and yield of crops with improved quality, thus becomes an important factor for enhanced income of the farmers. Nowadays, there are many water saving irrigation techniques are being adapted by the farming community, such as drip irrigation, sprinkler, variable rate irrigation, subsurface drip irrigation, precision irrigation, *etc.* to reduce the requirement of water in the fields and apply watering in desired amount only when it is needed by the plant. However, this cannot be achieved accurately, until economically suitable mechanism is followed to make dynamic schedule of watering with variable rate based on current and near-future estimations of soil-moisture (water) in the different sites of the fields. Artificial intelligence (AI) based simulation and prediction models can help precisely to achieve these goals at promising scale in irrigation-water management. Additionally, AI enabled automated irrigation-control solutions have the potential to reduce the labour costs involved in irrigation instrumentation on the fields. Moreover, implementing emerging ways of irrigation, such as dynamically scheduled micro-fertigation require customized modelling as per the crop and soil types that further needs broader research and technological adaptation among the farmers.

REFERENCES

Arya, L. M., Richter, J. C., & Paris, J. F. (1983). Estimating Profile Water Storage from Surface zone Soil Moisture Measurements Under Bare Field Conditions. *Water Resources Research*, *19*, 403–1.

Baker, J. M. (1990). Measurement of soil water content. *Remote Sensing Reviews*, *5*, 263–279.

Beer. (1852). Bestimmung der Absorption des rothen Lichts in farbigen Flüssigkeiten [Determination of the absorption of red light in colored liquids]. *Annalen der Physik und Chemie, 86*, 78–88.

Campbell, G. S. (1988). Soil water potential measurement: An overview. *Irrigation Science*, *9*, 265–273.

Campbell, G. S., & Gee, G. W. (1986). Water potential: miscellaneous methods. In A. Klute (Ed.), Methods of Soil Analysis, Part 1. Physical and Mineralogical Methods. Academic Press.

Cassel, D. K., & Klute, A. (1986). Water potential: Tensiometry. In A. Klute (Ed.), *Methods of soil analysis*. ASA and SSSA.

Colman, E. A., & Hendrix, T. M. (1949). The fibreglass electrical soil-moisture instrument. *Soil Science*, *67*, 425–438.

Cummings, R. W., & Chandler, R. F. (1940). A field comparison of the electrothermal and gypsum block electrical resistance methods with the tensiometer method for estimating soil moisture in situ. *Soil Science Society of America Proceedings*, *5*, 80–85.

Dean, T. J., Bell, J. P., & Baty, A. J. B. (1987). Soil Moisture Measurement by an Improved Capacitance Technique. Part I. Sensor Design and Performance. *Journal of Hydrology (Amsterdam)*, *93*, 67–78.

Doorenbos, J., & Pruitt, W. O. (1984). *Crop water requirements. FAO Irrigation and Drainage, Paper No. 24*. FAO.

Evett, S. R. (2008). Neutron moisture meters. In S. R. Evett, L. K. Heng, P. Moutonnet, & M. L. Nguyen (Eds.), *Field estimation of soil water content: A practical guide to methods, instrumentation, and sensor technology. IAEA-TCS-30* (pp. 39–54). International Atomic Energy Agency.

Evett, S. R., Colaizzi, P. D., O'Shaughnessy, S. A., Hunsaker, D. J., & Evans, R. G. (2014). Irrigation Management. In *Encyclopedia of Remote Sensing*. Springer. doi:10.1007/978-0-387-36699-9_73

Gardner, W., Israelsen, O. W., Edlefsen, N. E., & Clyde, D. (1922). The capillary potential function and its relation to irrigation practice. *Physical Review*, *20*, 196–204.

Gardner, W. H. (1986). Water content. In A. Klute (Ed.), Methods of Soil Analysis, Part 1. Physical and Mineralogical Methods. Academic Press.

Hallikainen, M. T., Ulaby, F. T., Dobson, M. C., El-Rayes, M. A., & Wu, L. K. (1985). Microwave dielectric behaviour of wet soil, Part 1. Empirical models and experimental observations. *IEEE Transactions on Geoscience and Remote Sensing*, *23*(1), 25–34.

Hignett, C., & Evett, S. R. (2002). Neutron thermalization. In J. H. Dane & G. C. Topp (Eds.), *Methods of soil analysis. Part 4. Physical methods* (pp. 501–521). Amer. Soc. Agron.

Holmes, J. W., Taylor, S. A., & Richards, S. J. (1967). In R. M. Hagan, H. R. Haise, & T. W. Edminster (Eds.), *Measurement of soil water in Irrigation of agricultural lands* (pp. 275–303). American Society of Agronomy.

Jensen, M. E., Burman, R. D., & Allen, R. G. (1990). Evapotranspiration and Irrigation Water Requirements. *ASCE Manuals and Reports on Engineering Practice,* 70.

Malicki, M. (1983). A capacity meter for the investigation of soil moisture dynamics. *Zesty Problemowe Postepow Nauk Rolniczych*, 201-214.

Paetzold, R. F., Matzkanin, G. A., & Santos, A. D. L. (1985). Surface soil water content measurement using pulsed nuclear magnetic resonance techniques. *Soil Science Society of America Journal*, *49*, 537–540.

Rasmussen, T. C., & Rhodes, S. C. (1995). Energy- related methods: psychrometers. In L. G. Wilson, L. G. Everett, & S. J. Cullen (Eds.), *Handbook of vadose zone characterization and monitoring* (pp. 329–341). CRC Press.

Rawlins, S. L., & Campbell, G. S. (1986). Water potential: thermocouple psychrometry. In A. Klute (Ed.), Methods of soil analysis. Part 1- Physical and mineralogical methods. Academic Press.

Reynolds, S. G. (1970a). The gravimetric method of soil moisture determination, Part 1. A study of equipment, and methodological problems. *Journal of Hydrology (Amsterdam)*, *11*(3), 258–273.

Reynolds, S. G. (1970b). The gravimetric method of soil moisture determination, Part 2. Typical required sample sizes and methods of reducing variability. *Journal of Hydrology (Amsterdam)*, *11*, 274–287.

Richards, S. K., & Marsh, A. W. (1961). Irrigation based on soil suction measurements. *Soil Science Society Proceedings*, 65-69.

Ritchards, L. A. (1928). The usefulness of capillary potential to soil moisture and plant investigators. *Journal of Agricultural Research*, *37*, 719–742.

Roth, K., Schulin, R., Fluhler, H., & Attinger, W. (1990). Calibration of time domain reflectometry for water content measurement using a composite dielectric approach. *Water Resources Research*, *26*(10), 2267–2273.

Schofield, R. K., & Taylor, A. W. (1961). A method for the measurement of the calcium deficit in saline soils. *Journal of Soil Science*, *12*(2), 269–275.

Thomas, A. M. (1966). In situ measurement of moisture in soil and similar substances by 'fringe' capacitance. *Journal of Scientific Instruments*, *43*, 21–27.

Topp, G. C., Annan, J. L., & Davis, A. P. (1980). Electromagnetic determination of soil water content: Measurements in co-axial transmission lines. *Water Resources Research*, *16*, 574–582.

Ulaby, F. T., Moore, R. K., & Fung, A. K. (1986). Microwave Remote Sensing Active and Passive. *From Theory to Applications,* 2136.

Vanclooster, M., Mallants, D., Diels, J., & Feyen, J. (1993). Determining local-scale solute transport parameters using time-domain reflectometry (TDR). *Journal of Hydrology (Amsterdam)*, *148*, 93–107.

Walker, J., Houser, P., & Willgoose, G. (2004). Active microwave remote sensing for soil moisture measurement: A field evaluation using ERS-2. *Hydrological Processes*, *18*, 1975–1997.

Ward, A. L., Kachanoski, R. G., Bertoldi, A. P., & Elrick, D. E. (1995). Field and undisturbed column measurements for predicting transport in unsaturated layered soil. *Soil Science Society of America Journal*, *59*, 52–59.

Watts, G. (1997). *Hydrological Modelling in Practice*. John Wiley and Sons Ltd.

Whalley, W. R., & Stafford, J. V. (1992). Real-time sensing of soil water content from mobile machinery: Options for sensor design. *Computers and Electronics in Agriculture*, *7*, 269–284.

White, I., & Zegelin, S. J. (1995). Electric and dielectric methods for monitoring soil-water content. In L. G. Wilson, L. G. Everett, & S. J. Cullen (Eds.), *Handbook of Vadose Zone Characterization and Monitoring* (pp. 343–385). CRC Press.

Wraith, J. M., & Or, D. (1999). Temperature effects on soil bulk dielectric permittiv- ity measured by time domain reflectometry: Experimental evidence and hypothesis development. *Water Resources Research*, *35*, 361–369.

Wyman, J. (1930). Measurements of the Dielectric Constants of Conducting Media. *Physical Review*, *35*, 623–634.

Yasuda, H., Berndtsson, R., Bahri, A., & Jinno, K. (1994). Plot-scale solute transport in a semiarid agricultural soil. *Soil Science Society of America Journal*, *58*, 1052–1060.

Chapter 11
Flood Assessment Using Hydrodynamic HEC-RAS Modelling

Vaishali I. Rana
Sardar Vallabhbhai National Institute of Technology, India

Dhruvesh P. Patel
https://orcid.org/0000-0002-2074-7158
Pandit Deendayal Energy University, India

Azazkhan I. Pathan
Sardar Vallabhbhai National Institute of Technology, India

Prasit G. Agnihotri
Sardar Vallabhbhai National Institute of Technology, India

Samir B. Patel
https://orcid.org/0000-0002-4280-6446
Pandit Deendayal Energy University, India

ABSTRACT

River flooding causes several human and financial casualties, and hence, it is necessary to perform research studies and implement subsequent actions consistent with the nature of the river. To minimize flood damage, floodplain zoning is a prominent non-structural measure in planning the areas surrounding the river. The present study uses HEC-RAS Version 5 to develop a flood model for the Chandan River, situated in the southern part of Balaghat district. The Digital Elevation Model (DEM) used for this analysis is 30m open source CartoDEM V-3 R1. The peak floods of 1990, 2002, and 2006 are taken into consideration. The river reach is divided into 48 cross sections, and a one-dimensional steady flow analysis is performed on HEC-RAS to assess the flood. The depths observed in the floods of 1990, 2005, and 2006 are 5.99 m, 3.2 m, and 3.49 m, respectively. The coefficient of correlation (R2) is obtained as 0.954 which shows the consistency and accuracy of the model. This study can help governing bodies to plan the city and attenuate the losses caused by floods in the Chandan River.

DOI: 10.4018/978-1-6684-3981-4.ch011

INTRODUCTION

A Flood is a natural calamity that occurs due to excessive rainfall. When the discharge rises to the point that the banks can no longer hold the water, it overtops its banks and floods the surrounding areas, impacting the lives of the people residing in the vicinity of the river. This process in which the water overtops the banks of the river causes riverine flooding (Eccles & Hamilton, 2019). It is the most common, hazardous and widespread natural calamity that disrupts the normal life of mankind. Some of its hazardous effects include damage of crops, loss of livestock, and worsening of health and hygiene conditions owing to waterborne ailments. It also hinders the economic growth and development of a country as a huge amount of funding goes into post-relief operations. According to a survey conducted by UNICEF (United Nations Children's Fund), India is the second most flood affected country in the world after Bangladesh. It is not possible to prevent such events, but with the help of engineering techniques, their effect can be minimized.

Due to the unpredictable behaviour of the rivers during floods, computer modelling is the most efficient and economical tool to simulate the flows. It involves the minimum cost to study and replicate the behaviour of the channels and natural streams. Flood mapping is usually done by hydraulic mathematical models. These models can be very useful in determining the water levels and flood prone areas. Hydraulic modelling is a powerful tool in planning infrastructure development as it gives a complete conceptual representation of the actual problem area (Monte et al., 2016). Hydrodynamic models with the integration of GIS is the most suited way to develop flood model and assessment of flooding in the local areas (Ullah et al., 2016). HEC-RAS (Hydrologic Engineering Centre River Analysis System) is a powerful tool for dealing with such issues. It is well developed and the models created by the software are quite realistic. It is very easy to use and has found wide acceptance in the hydraulic community (Ahmad et al., 2016).

HEC-RAS was developed by the U.S. Army Corps of Engineers to manage rivers, canals, and other public works under water resource engineering. It was developed in 1995 and since then, it had a wide acceptance and popularity. It is easy to use very handy software with geospatial capabilities. HEC-RAS has a broad range of data entry capabilities. HEC-RAS Version 5 now offers geospatial functionality for obtaining river geometry data, which was previously done with the HEC-GeoRAS module in ARC-GIS, but now HEC-RAS can perform these functions very easily. The hydraulic modelling approach has become more practical as a result of the introduction of geospatial approaches, and it saves a lot of time on pre-processing operations.

This aim of this research work is to indicate the applicability of HEC-RAS in one dimensional flood modelling in balaghat district. The steady flow condition has been utilised to simulate the model on the Chandan River. The output of the model was obtained in context of water depth. The results were also validated by performing error analysis to acquire the accuracy and precision of the HEC-RAS model. The study can be helpful for the local government to plan flood control and management plan for the study area.

BACKGROUND

The tactic of flood modelling by HEC-RAS is elaborated by Husain, (2017) and the principles on which HEC-RAS works are also described in detail. He had also done a review of assorted literature based upon

HEC-RAS, which is incredibly useful in the study of the software. Sun et al. (2017) have found applications of HEC-RAS for flood forecasting in perched river situated in China. Stimulation forecasting of river flood banks is becoming a key interest of researchers now a days. They have derived the strategy of working out the height of the levee to safeguard the banks. One-dimensional steady flow analysis using HEC-RAS on the Jhelum River in the Kashmir valley was performed by Ahmad et al., (2016). They recorded the maximum flow and used it as input in the HEC-RAS model to search for the expected flood levels in the coming years. The results of this study could be further useful in policy making and flood mitigation to attenuate the losses caused by the disaster. Azaz and Agnihotri (2020) performed 1D hydrodynamic food modelling on the Purna River with special reference through geospatial techniques. Ongdas et al. (2020) have prepared 2D Flood Hazard Maps for the Yesil River in Kazakhstan. They have used HEC-RAS 2D to simulate flood scenarios and have also compared the model performance by using different mesh sizes of 25, 50, and 75m. They found a remarkable difference in simulation time. Flood plain zoning simulation using HEC-RAS and CCHE2D software was performed by ShahiriParsa et al. (2016). The flood level and spatial extent of the inundation in the SungaiMaka river flood plain were calculated using this software. This can eventually help the government agencies and developers to create a correct plan for the future expansion of the town. The flood modelling of the Vistula Basin was performed by Paper and Kowalczuk(2018). They have introduced necessary modifications like adding cross sections and alterating in flow rates, which might be critical while predicting the flood. Steady flow analysis in the Anambe River utilizing HEC-RAS done by Traore. A case study on the Kayanga River Basin, Senegal using HEC-RAS was also performed in which a HEC-RAS model for hydraulic simulation of the river was prepared by V. B. Traore et al. (2015).1D Unsteady Flow Analysis using HEC-RAS Modeling Approach for Flood in Navsari City was performed by Patel and Yadav, (2019). For that data was taken of 2004 and 2013 flood. River reach of 32 km was taken for the analysis purpose, from Pinsad Village to Borsi village. River reach was divided into 128 cross sections to perform analysis. Critical cross sections were identified using HEC-RAS. The objective of this study was to find the sufficiency of river reach to hold the flood and to identify the critical cross sections of the river. Conclusions on the basis of results were derived. Construction of several structures such as retaining wall, sluice gate, stone pitches, embankments etc were proposed.

STUDY AREA

Balaghat district is located in the south eastern part of Madhya Pradesh. It's latitude and longitude range are 21°19' to 22°24'N and 79°31' to 81°3'E respectively. It shares its border with Seoni, Mandla and Dindori districts of Madhya Pradesh, Gondia and Bhandara districts of Maharashtra, and Rajnandgaon district of Chhattisgarh. The total geographical area covered by Balaghat is 9245 square km. The study area map of the district is shown in figure 1. Wainganga River and its tributaries are the major sources of water in the district. It is a part of the Godavari basin. The normal annual rainfall in the district is 129.4 cm. Balaghat faced severe floods in the years 1990, 2002, and 2006. The river reach selected for this study is from Jhaliwada village to Atri village, which is approximately 15 km.

Figure 1. Location map of study area

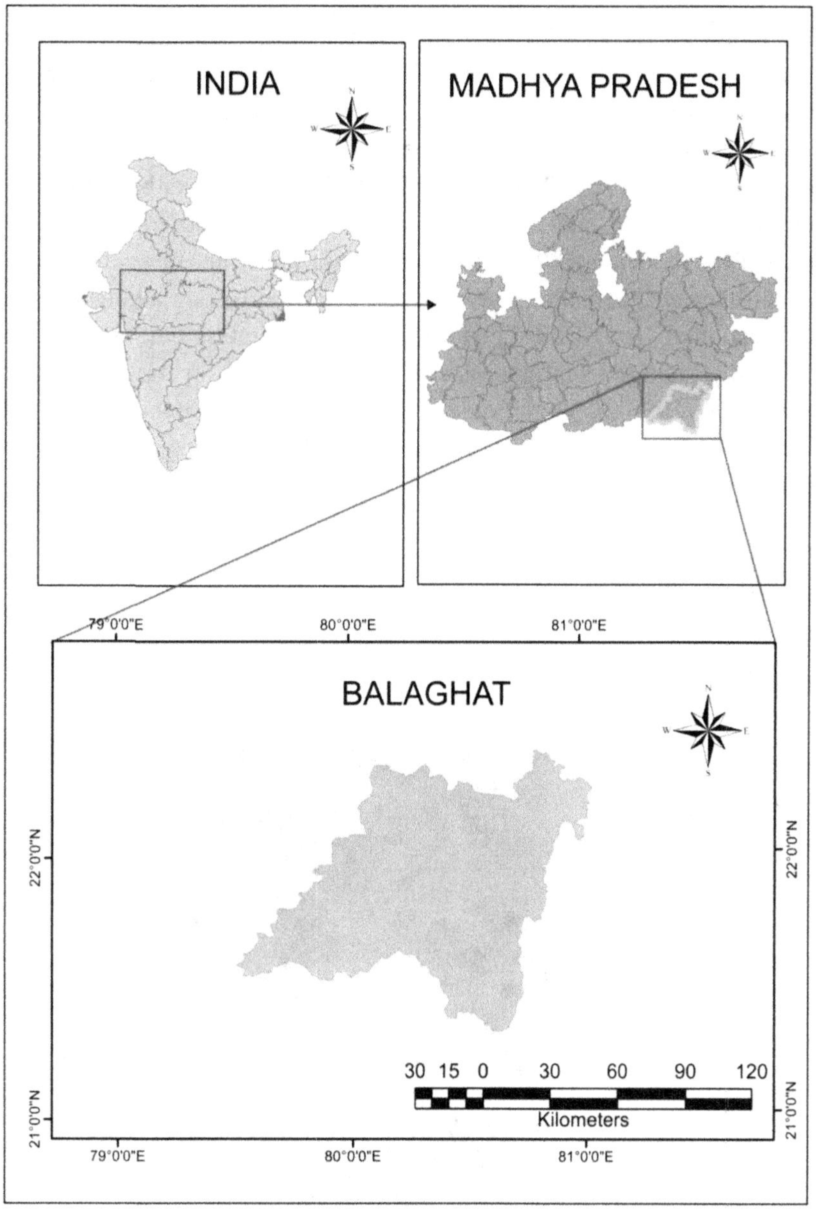

Data Required for the Study

Daily discharge data for the period of 1988-2016 is obtained from MPWRD. A gauging station named Rampayli is used for data collection. Cartosat -I Digital Elevation Model (DEM) having a 30m resolution can be obtained from (https://bhuvan-app3.nrsc.gov.in/data/download/index.php). Projection file of the area is downloaded from (https://spatialreference.org/). The projection used for this study is EPSG 32644-WGS 84/UTM zone 44N.

RESEARCH METHODOLOGY

To start flood modelling using HEC-RAS, the data required are the Digital elevation model, Projection file, shape file, and Manning's roughness coefficient. The present study uses 30m resolution Cartosat-I DEM for generating the terrain profile and converting the DEM data to DTM. Initially, the projection file (UTM zone 44N) is assigned to the RAS Mapper and after that DEM is uploaded.

Now, by geometries feature of RAS Mapper delineate centre line, river bank lines, flow paths which are shown by the colours blue, red and green respectively. The Geometric Data Window of HEC-RAS is shown in figure 2. Peak discharge at the Rampaily gauge site from the years 1988 to 2014 is given in graphical form in figure 3. Thereafter, draw the cross-section lines, making sure that they are perpendicular to the direction of flow. It must be taken into consideration that these five lines must never intersect each other. Add Manning's roughness coefficient as 0.035 for the main channel and 0.04 for flood plains (Chow et al., 1998). Contraction and Expansion coefficients as per the HEC-RAS user manual are taken as 0.1 and 0.3 respectively. After creating this geometry in the RAS Mapper, the next step is to perform steady flow analysis.

Figure 2. Geometric data window

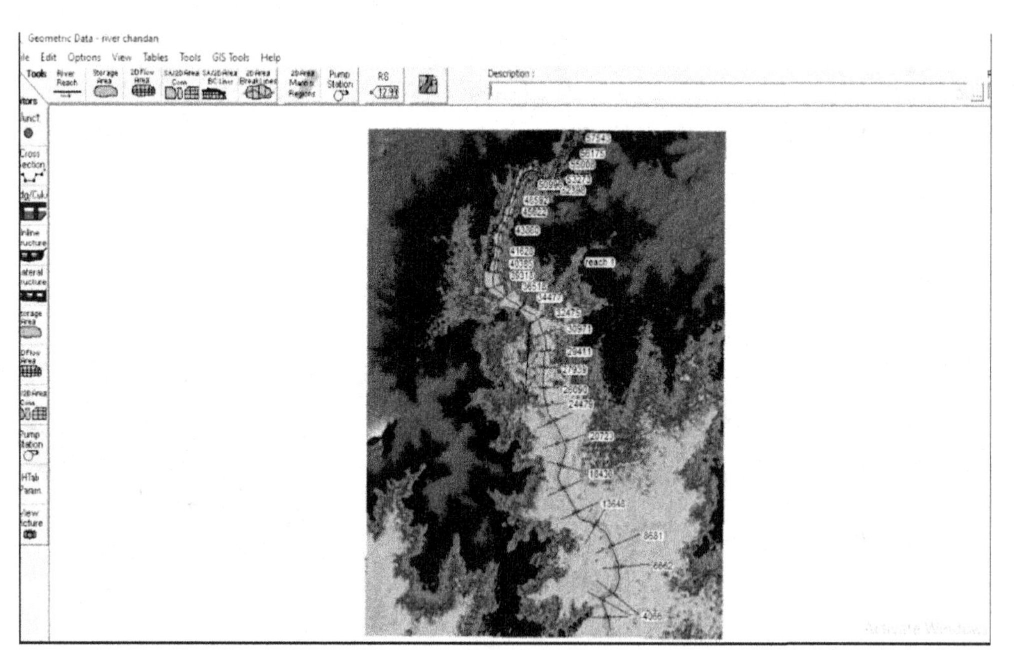

For this purpose, peak flows from 1990, 2005, and 2006 are used as steady flow discharges (2400.8 m^3/s, 721.68 m^3/s, and 895.68 m^3/s). The upstream boundary condition for Chandan River is these discharges, and the downstream boundary condition is a normal slope of 0.001. (calculated by plotting the terrain profile). The next step is to run the steady flow simulation for the given peak discharges. Now compare the simulated water surface elevation with the actual data obtained from MPWRD. Complete flow chart of the proposed work is shown in figure 4.

Cross-section 1 obtained after performing the steady flow analysis is shown in the figure 5.

Figure 3. Peak discharge at Rampaily Gauge site

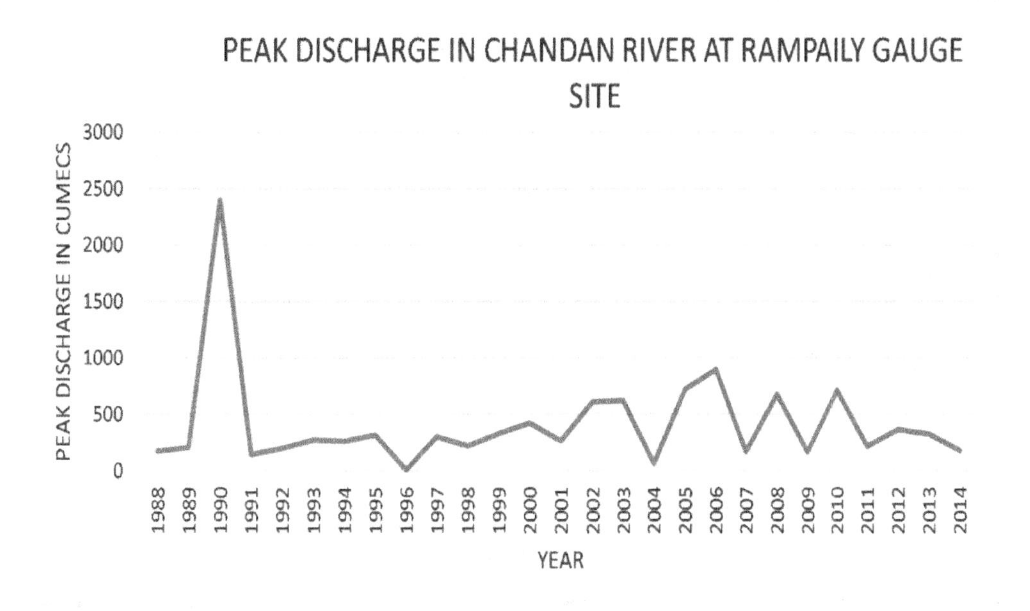

Figure 4. Flow chart of methodology

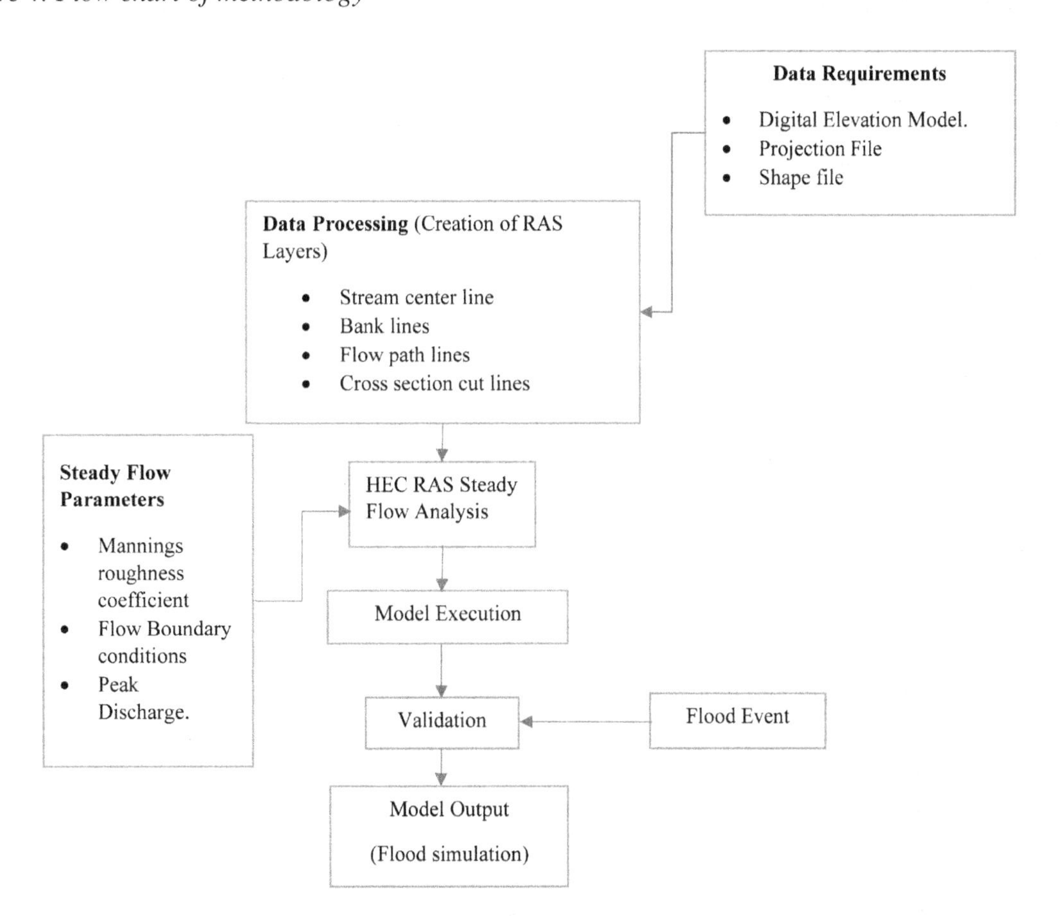

Figure 5. Cross-sectional view of the river at CS-1

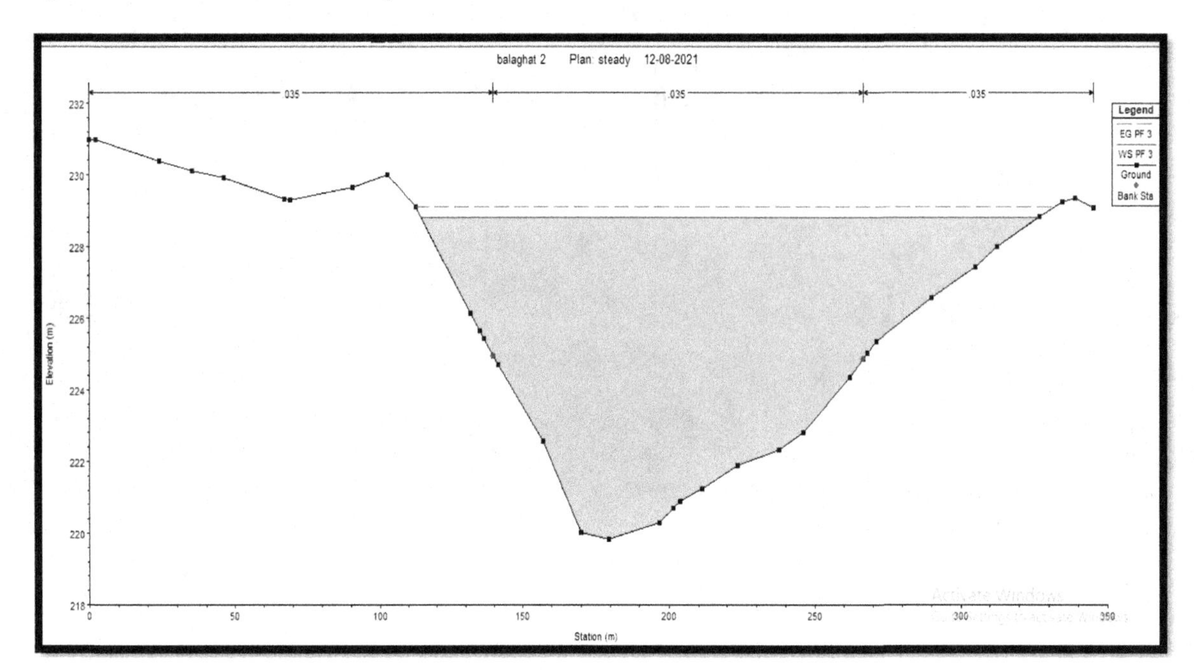

Principles and Equations Used in HEC-RAS

A one-dimensional energy equation is used for the basic computations of HEC-RAS (Equation 1). Water surface profiles are computed by solving the energy equation called the standard step method, which follows an iterative procedure.

$$Z_1 + Y_1 + \frac{V_1^2}{2g} = Z_2 + Y_2 + \frac{V_2^2}{2g} + h_e \tag{1}$$

Z_1, Z_2 = elevation of main channel inverts
Y_1, Y_2 = depth of water at cross section
V_1, V_2 = average velocities at section 1 and 2 respectively
g= acceleration due to gravity
h_e = energy head loss

Energy losses are evaluated which are caused by friction, contraction, and expansion (Equation 2). If the water surface profiles are rapidly varied, as in the case of hydraulic jumps, hydraulics of bridges, profiles at river confluences etc, momentum equation may be used.

$$h_e = LS_f + C_L \left| \left(\frac{\pm_2 V_2^2}{2g} \right) - \left(\frac{\pm_1 V_2^2}{2g} \right) \right| \tag{2}$$

SOLUTIONS AND RECOMMENDATIONS

The simulated values of water surface elevation are approximately similar to the actual data obtained from MPWRD at gauge station Rampaily (Table 1). Below is the graph (figure 6) drawn between the simulated water depth and the actual water depth at the gauging site. The results obtained are quite promising. The output of the simulated result indicates that cross-sections of the lower region of the river were more critical during the floods of 1990 and 2006. Simulated flood inundation of the area around the river for the events is shown in figure 7, 8, and 9 for the discharge of 610.05 m^3/s, 895.68 m^3/s, and 2400.8 m^3/s respectively.

Figure 6. Regression relationship between simulated and observed water depth at Rampaily gauging site

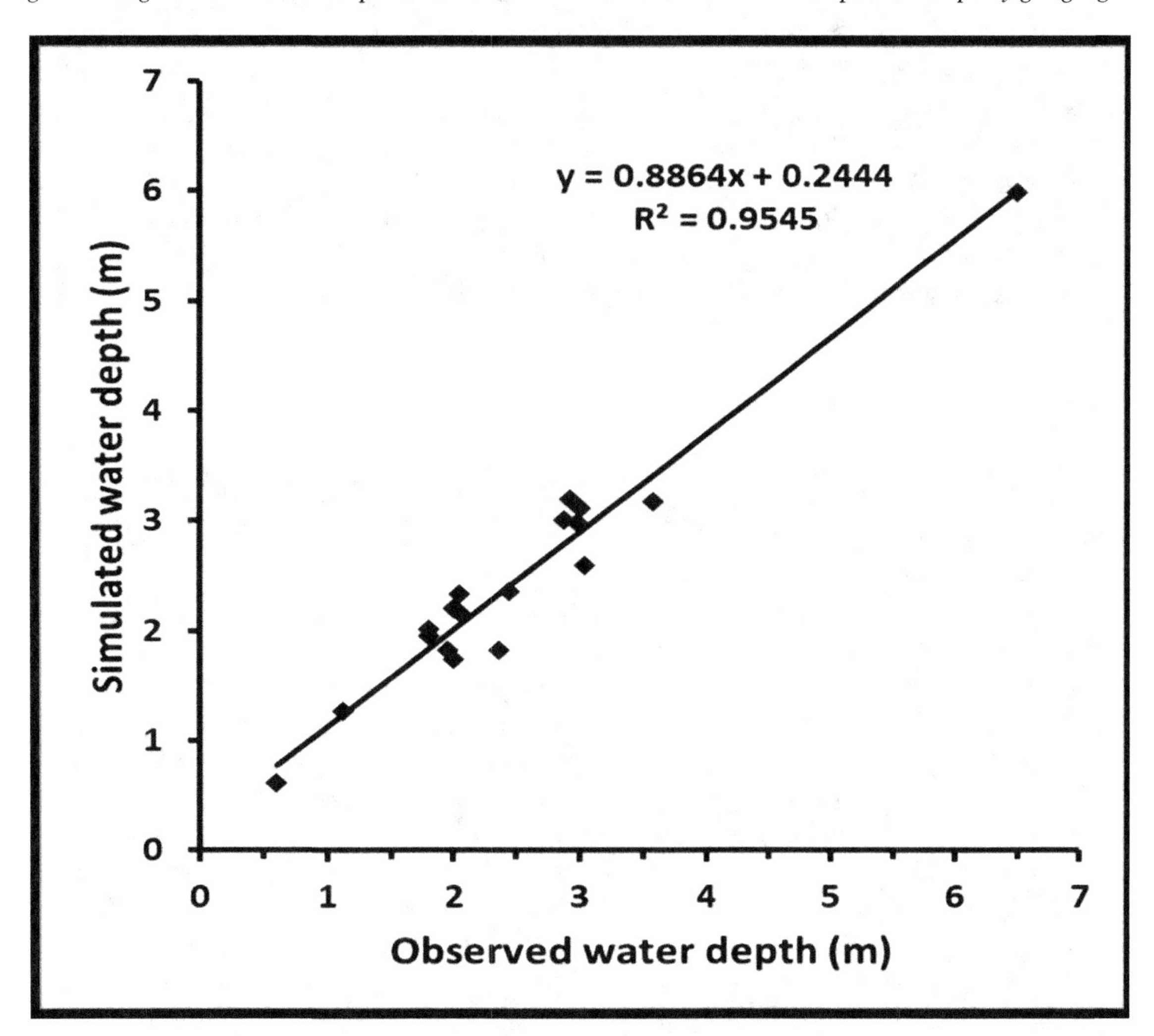

Figure 7. Simulated flood inundation at a discharge of 721.68 m^3/s (2005)

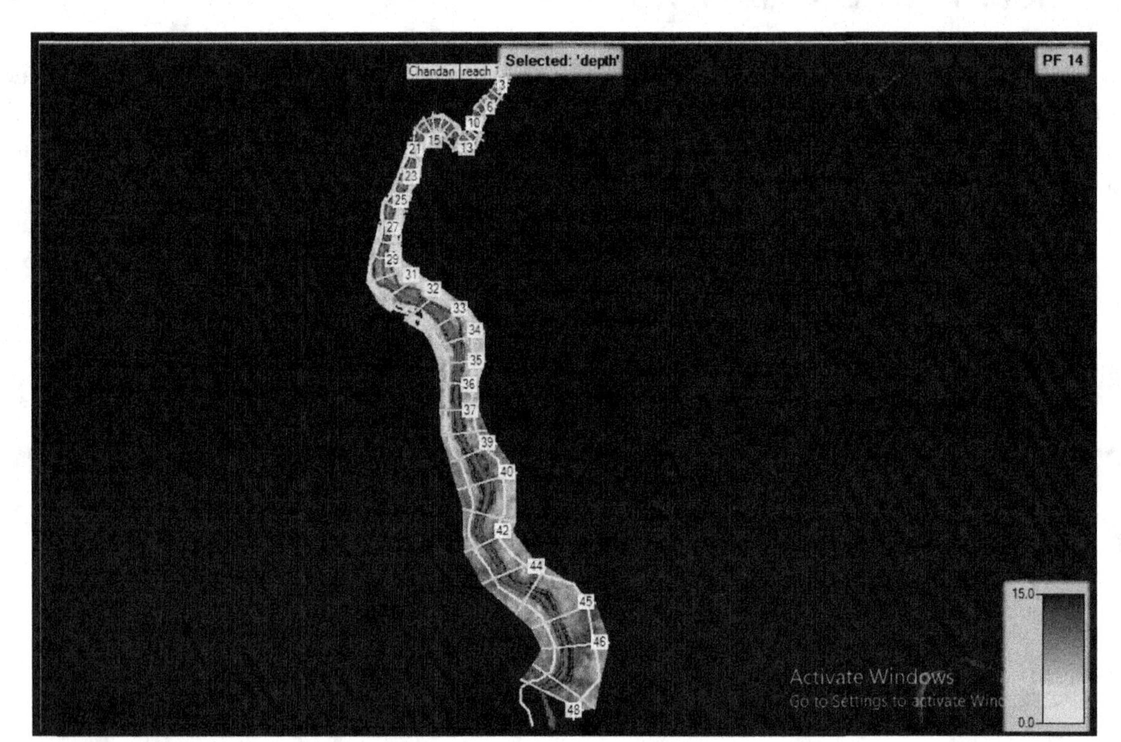

Figure 8. Simulated flood inundation at a discharge of 895.68 m^3/s (2006)

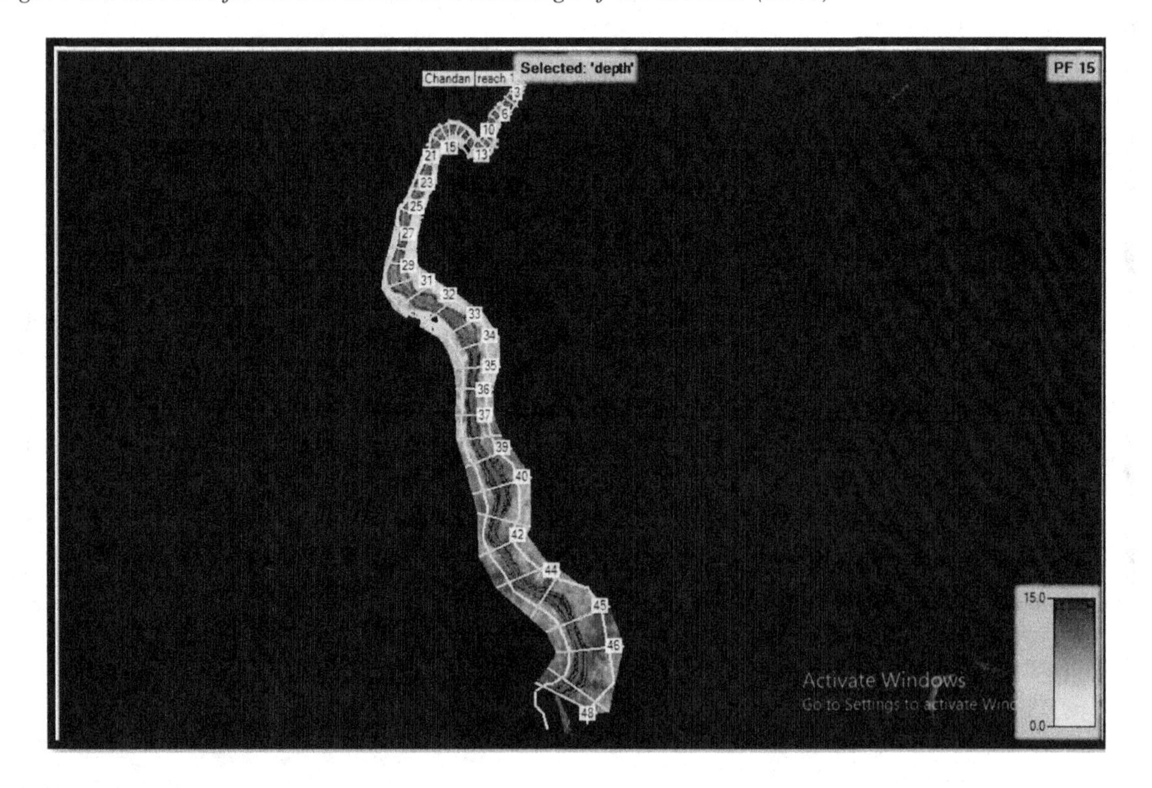

Figure 9. Simulated flood inundation at a discharge of 2400.8 m3/s (1990)

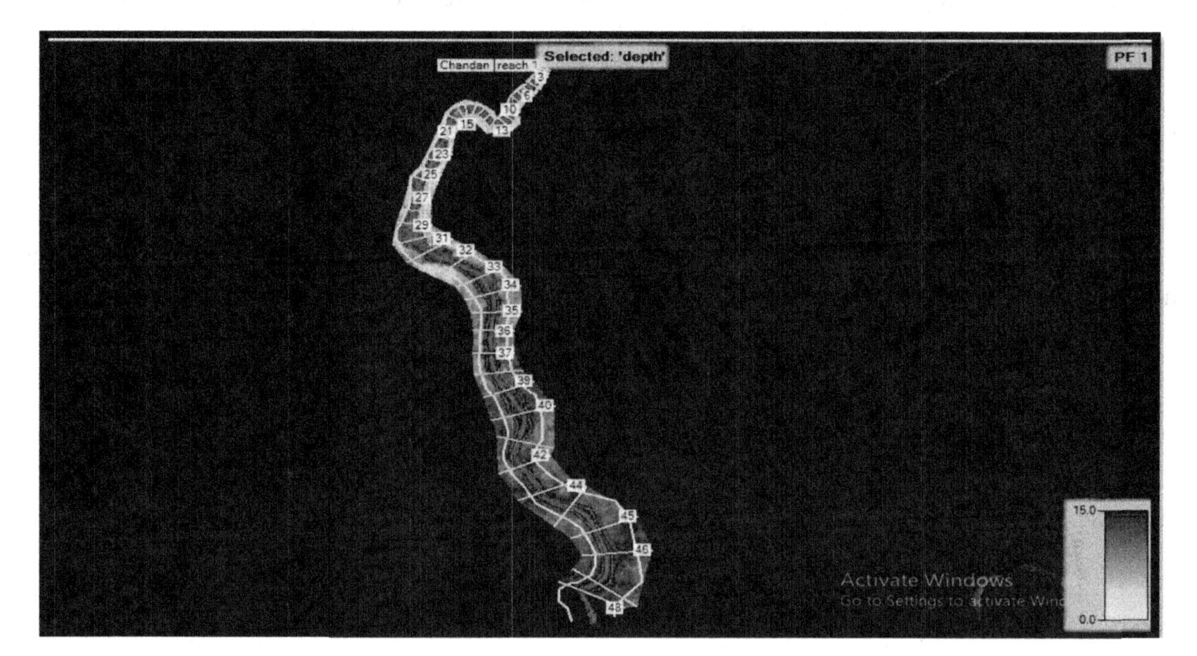

Furthermore, the RMSE value is calculated as 0.381 and the coefficient of correlation is obtained as 0.954. From the analysis of the study area; it is observed that Cross Sections from 32 to 48 are at high risk because the depth observed here is relatively higher than the other cross sections. To reduce the risk of inundation in that area; embankments should be raised to a certain extent to prevent submergence. This study also indicates that the right bank of the Chandan River is more susceptible to flood as compared to the left bank, hence more protection should be provided to the right side of the river.

No further encroachment should be allowed in the nearby river areas to prevent the loss of lives and properties during the time of high floods. Also, no new structures should be constructed in the flood-plain zone of the river.

Moreover, several limitations were observed in this study which can be stated as:

1. This study has been performed in one dimensional modelling; results obtained by two-dimensional modelling would be much more accurate.
2. Since there was only one gauging station for monitoring discharge on the Chandan River, the lack of data may have resulted in lower accuracy.

FUTURE RESEARCH DIRECTIONS

The present research is performed by utilizing the 1D HEC-RAS hydraulic modelling to prevent future flooding and reduce the risk of flood damage in low lying areas, and can thus provide valuable support in the creation of flood mitigation methods for any similar case around the world. The study will also give officials advice on how to operate major dams and expand levees in the future. Further in this study,

2D hydraulic modelling can be performed to obtain more accurate results. Also, validation can be done using 2 or more gauging sites. This will improve the overall accuracy of the model.

CONCLUSION

The application of hydrodynamic modelling in HEC-RAS provides the capability to simulate natural conditions like floods to a great extent. The output of the HEC-RAS model can be useful in determining the extent of overtopping of barrages/dams in the study reach which are subjected to a given magnitude of flood. From this analysis, the depths observed in the floods of 1990, 2005, and 2006 are 5.99 m, 3.2 m, and 3.49 m, respectively. The coefficient of correlation (R^2) is obtained as 0.954, which shows the consistency and accuracy of the model. From the present analysis, it can be concluded that further encroachment near the river will increase the risk of inundation, so it should be prevented. Therefore, it can be concluded that human and financial causalities can be prevented with the help of prior analysis and study.

ABBREVIATION

DEM – Digital Elevation Model
1D – One Dimensional
2D – Two Dimensional
UTM- Universal Transverse Mercator
CS- Cross Section

ACKNOWLEDGMENT

The first author would like to express her sincere gratitude to Dr. P.G. Agnihotri, Professor, Sardar Vallabhbhai National Institute of Technology, Surat, for providing her the opportunity to take up this study. The author also would like to express her gratitude thank Azazkhan I. Pathan for providing all the necessary support in technical assessment during the entire study. Special thanks to Dr. Dhruvesh Patel, Department of Civil Engineering, School of Technology, Pandit Deendayal Energy University, Gujarat, for giving this golden opportunity and his invaluable guidance. Also, thanks to MPWRD for providing the necessary data required for this study.

REFERENCES

Ahmad, H. F., Alam, A., Bhat, M. S., & Ahmad, S. (2016). One Dimensional Steady Flow Analysis Using HECRAS – A case of River Jhelum, Jammu and Kashmir. *European Scientific Journal, ESJ*, *12*(32), 340. doi:10.19044/esj.2016.v12n32p340

Chow, V. T., Maidment, D. R., & Mays, L. W. (1998). *Applied Hydrology Chow 1988*. http://ponce.sdsu.edu/Applied_Hydrology_Chow_1988.pdf

Eccles, R., Zhang, H, & Hamilton, D. P. (2019). *A review of the effects of climate change on riverine flooding in subtropical and tropical regions*. doi:10.2166/wcc.2019.175

Husain, A. (2017). Flood Modelling by using HEC-RAS. *International Journal of Engineering Trends and Technology*, *50*(1), 1–7. doi:10.14445/22315381/IJETT-V50P201

K B. P., & S.M, Y. (2019). One Dimensional Unsteady Flow Analysis Using HEC-RAS Modelling Appoach for Flood in Navsari City. *Proceedings of Recent Advances in Interdisciplinary Trends in Engineering & Applications (RAITEA)*. doi:10.2139/ssrn.3351780

Kowalczuk, Z., Swiergal, M., & Wróblewski, M. (2017, September). River Flow Simulation Based on the HEC-RAS System. In *International Conference on Diagnostics of Processes and Systems* (pp. 253-266). Springer. 10.1007/978-3-319-64474-5

Monte, B. E. O., Costa, D. D., Chaves, M. B., Magalhães, L. D. O., & Uvo, C. B. (2016). Hydrological and hydraulic modelling applied to the mapping of flood-prone areas. *RBRH*, *21*(1), 152–167. doi:10.21168/rbrh.v21n1.p152-167

Ongdas, N., Akiyanova, F., Karakulov, Y., Muratbayeva, A., & Zinabdin, N. (2020). Application of HEC-RAS (2D) for Flood Hazard Maps generation for Yesil(Ishim) river in Kazakhstan. *Water (Basel)*, *12*(10), 2672. doi:10.3390/w12102672

Pathan, A. I., & Agnihotri, P. G. (2021). Application of new HEC-RAS version 5 for 1D hydrodynamic flood modeling with special reference through geospatial techniques: A case of River Purna at Navsari, Gujarat, India. *Modeling Earth Systems and Environment*, *7*(2), 1133–1144. doi:10.100740808-020-00961-0

ShahiriParsa, A., Noori, M., Heydari, M., & Rashidi, M. (2016). Floodplain zoning simulation by using HEC-RAS and CCHE2D models in the Sungai Maka river. *Air, Soil and Water Research*, *9*, 55–62. doi:10.4137/ASWR.S36089

Sun, P., Wang, S., Gan, H., Liu, B., & Jia, L. (2017, April). Application of HEC-RAS for flood forecasting in perched river-A case study of hilly region, China. *IOP Conference Series: Earth and Environmental Science* (Vol. 61, No. 1, p. 012067). IOP Publishing. doi:10.1088/1755-1315/61/1/012067

Traore, V. B., Bop, M., Faye, M., Malomar, G., Gueye, E. H. O., Sambou, H., ... Beye, A. C. (2015). Using of Hec-ras model for hydraulic analysis of a river with agricultural vocation: A case study of the Kayanga river basin, Senegal. *American Journal of Water Resources*, *3*(5), 147–154. doi:10.12691/ajwr-3-5-2

Ullah, S., Farooq, M., Sarwar, T., Tareen, M. J., & Wahid, M. A. (2016). Flood modeling and simulations using hydrodynamic model and ASTER DEM—A case study of Kalpani River. *Arabian Journal of Geosciences*, *9*(6), 439. doi:10.100712517-016-2457-z

APPENDIX

Table 1. Difference between observed and simulated water depth

Year	Discharge (Cumecs)	Observed Depth (M)	Simulated Depth (M)	Difference
1990	2400.81	6.5	5.99	0.51
1991	145.08	2	1.74	0.26
1992	197.56	1.8	1.95	-0.15
1993	273.62	2	2.2	-0.2
1995	316.9	2.05	2.33	-0.28
1996	8.56	0.6	0.61	-0.01
1998	217.2	1.8	2.02	-0.22
1999	329.69	2.45	2.36	0.09
2000	420.97	3.04	2.59	0.45
2001	262.97	2.05	2.17	-0.12
2002	610.05	2.98	2.99	-0.01
2003	621.81	2.88	3.01	-0.13
2004	64.35	1.13	1.27	-0.14
2005	721.68	2.925	3.2	-0.275
2006	895.68	3.375	3.49	-0.115
2007	167.11	1.955	1.83	0.125
2008	676.83	3.005	3.12	-0.115
2009	164.68	2.365	1.82	0.545
2010	712.25	3.575	3.18	0.395

Chapter 12
Water Quality Time-Series Modeling and Forecasting Techniques

Rashmiranjan Nayak
National Institute of Technology, Rourkela, India

Mogarala Tejoyadav
National Institute of Technology, Rourkela, India

Prajnyajit Mohanty
National Institute of Technology, Rourkela, India

Umesh Chandra Pati
https://orcid.org/0000-0001-9805-2543
National Institute of Technology, Rourkela, India

ABSTRACT

Water pollution is a global problem. In developing countries like India, water pollution is growing exponentially due to faster unsustainable industrial developments and poor waste-water management. Hence, it is essential to predict the future levels of pollutants from the historical water quality data of the reservoir with the help of appropriate water quality modeling and forecasting. Subsequently, these forecasting results can be utilized to plan and execute the water quality management steps in advance. This chapter presents a comprehensive review of time series forecasting of the water quality parameters using classical statistical and artificial intelligence-based techniques. Here, important methods used to calculate the water quality index are discussed briefly. Further, a problem formulation for the modeling of water quality parameters, the performance metrics suitable for evaluating the time-series methods, comparative analysis, and important research challenges of the water quality time-series modeling and forecasting are presented.

DOI: 10.4018/978-1-6684-3981-4.ch012

INTRODUCTION

Water is an essential natural resource that plays a significant role not only for human beings but also for all living organisms for life's survival purpose (Budiarti et al., 2019). Water quality represents the physical, chemical, and biological characteristics of the water depending on the standards for various applications (Swenson, 1965), (Johnson et al., 1997). Hence, the quality of the water plays a critical role in different aquatic systems such as oceans, reservoirs, ponds, lakes, etc., for the sustainability of aquatic bodies and living organisms that depend on the particular water system (Y. Chen et al., 2020). Although the water is abundantly available, all of it is not suitable for drinking purposes. Hence, the limited freshwater should be conserved and used properly for the continuation of the living things. Most of the natural resources are getting damaged due to the unsustainable use of several innovations for maintaining an easy lifestyle. Subsequently, many diseases like typhoid, dysentery, cholera, skin infections, eyes, and gastroenteritis, hepatitis is spreading among people due to due to usage of poor quality water (U. Ahmed et al., 2019). Particularly, in India, water bodies have been destroyed due to unsustainable urbanization and mechanization-driven growth. According to the estimations of expertise, nearly 70% of surface water in India is not suitable for human use (Chawla et al., 2015). Each day, water bodies and water sources are getting mixed with 40 million liters of sewage on an average with only a small portion of appropriate treatments (Priyank Hirani, 2019). The fast-developing of industrialization and a greater increase in farming growth combined with the new improvement as well as agricultural compost have increased the water pollution levels to a great degree (Geetha & Gouthami, 2016). Low water quality results have been known as the significant element of the rise in intolerable diseases (U. Ahmed et al., 2019). The major rivers like the Ganga, Yamuna, Godavari, Kaveri, Narmada of India are getting polluted due to the involvement of large industrial discharge, domestic products, and pesticides used in the farming field, which requires special treatment and attention (Aktar et al., 2010). The quality of water depends on the chemical, physical, and biological parameters. The major physical parameters include turbidity, water temperature, electrical conductivity, total dissolved solids, taste, and odor. The major chemical and biological parameters are the potentials of Hydrogen (pH), nitrates, Biochemical Oxygen Demand (BOD), Dissolved Oxygen (DO), Chloride, bacteria, and algae, respectively. Accessing all the water pollutants from water sources is complex and time-consuming. Further, the huge set of data acquired after exploring all the samples and comparing it with the predefined rules is hard to handle (Iqbal et al., 2019). So, the Water Quality Index (WQI) measurement method has been considered, which can represent the total water quality in a single index (Iqbal et al., 2019). The WQI will give a comprehensive idea related to the water quality level. The traditional water quality monitoring technique is costly, slow-moving, tedious, and frequently produces manual errors. Moreover, it only permits examining a finite number of samples owing to a need for infrastructure and resources (Priyank Hirani, 2019). Hence, it is essential to develop advanced methodologies for observing and analyzing water quality. The main motto of this research work is to study different statistical and Artificial Intelligence (AI)-based models for efficient prediction and future forecasting of the WQI.

Motivation

Many natural resources, including water, have been severely affected due to unsustainable industrialization, increase in population, and urbanization developments. Continuous water quality monitoring is the best way to deal with the unhygienic and poor quality of water. The manual time consuming and risky water quality analysis are not effective to get the proper results. Hence, some advanced data collection, processing, and modeling approaches using trending fields of Machine Learning (ML) and Deep Learning (DL) techniques must be explored for proper water quality forecasting. DL techniques have a strong ability of decision-making and automatic feature learning from their experience. Hence, DL-based methods can learn most powerful learning features and efficiently perform the time series forecasting task. These are the major motivating factors for carry out the research in this topic.

Contributions

The major contributions of the research work are outlined as follows.

- A comprehensive review of the time-series water quality forecasting techniques using both classical statistical and artificial intelligence-based techniques is presented.
- Important methods used for the calculation of the WQI are discussed briefly.
- A problem formulation for the modeling of water quality parameters is presented.
- A brief analysis of the performance metrics suitable for the evaluation of the time-series methods is presented.
- A brief comparative analysis of the time series forecasting techniques for the water quality prediction is presented.
- Important research challenges involved in Water Quality time-series modeling and forecasting are discussed.

Organization

The rest of the article is organized as follows. A comprehensive literature survey on various WQI calculation methods is presented in Section 2. Problem formulation for the modeling of the water quality parameters is briefly presented in Section 3. Different modeling approaches of time series data are briefly discussed in Section 4. Basic definitions and formulas of different evaluation metrics are discussed in Section 5. Similarly, various key features of the time series modeling techniques with major research challenges and the brief comparative analysis of the water quality forecasting techniques article are addressed in Sections 6 and 7, respectively. Finally, the chapter is concluded in Section 8.

WATER QUALITY INDEX

The quality of water mainly depends on its physio-chemical biological parameters and some heavy metals such as arsenic, zinc, copper, etc., but a single parameter cannot sufficiently determine the overall water quality (Kachroud et al., 2019). WQI is the beneficial and unique quantity to represent the overall water status in a single index (Shah & Joshi, 2017; Tyagi et al., 2013). The common steps involved in all approaches for calculating WQI are as follows (Kachroud et al., 2019): (1) parameter selection, (2) transformation of original parameters into sub-index values to reflect them on a common range of scale, and (3) aggregation of sub-indices to compute the final WQI.

Selection of Water Pollutants

The selection of the water parameters from the four major classes, namely oxygen concentration, total dissolved solids, biochemical components, and health aspects, which have a remarkable impact on water condition, is recommended [10]. The common parameters involved in most studies (Chawla et al., 2015; Kachroud et al., 2019; Tyagi et al., 2013) are pH, DO, BOD, turbidity, temperature, etc. The standard pH range of water is between 7.0 - 7.5. Water with a pH of 11 or greater is considered strongly alkaline (basic), which can annoy the eyeballs, skin, and mucous membrane (Geetha & Gouthami, 2016). Acidic water with a pH of 4 or less than that can also result in skin and eye diseases (Geetha & Gouthami, 2016). The value of DO provides the total amount of oxygen that is dissolved in water in molecular form (Shakhari et al., 2019). All water-living creatures are sensitized to high temperatures as the solubility of oxygen is lower in water with high temperatures (Ysi, 2019). The BOD is nothing but the total dissolved oxygen concentration required for all water-living creatures. Table 1 represents the standard range and corresponding weight factor of each parameter.

Table 1. Water parameters with its standard range and NSFWQI weight (U. Ahmed et al., 2019; Chawla et al., 2015; Kachroud et al., 2019)

Parameters	Standard Range	NSFWQI Weight Factor
pH	7 to 8.5	0.11
Fecal Coliform (CFU)	1 to 10	0.16
DO (mg/L)	>7	0.17
BOD (mg/L)	0 to 3	0.11
Nitrate (mg/L)	0 to 20	0.10

Methodologies for the Development of WQI

The different types of WQIs that have been developed till now are discussed clearly in (Tyagi et al., 2013). Similarly, in (Kachroud et al., 2019), the summary of clear structure, aggregation formula, number of variables, and location of each study used in further research to develop various water quality indices are discussed. Two widely used WQI calculation methods are described as follows.

National Sanitation Foundation Water Quality Index (NSFWQI)

Initially, the NSFWQI approach was first developed by Brown et al. in 1970 (Tyagi et al., 2013), but later it was observed that this method is not performing accurately. Hence, in (Kumar & Alappat, 2009), an improved method of NSFWQI calculation is discussed using nine common water parameters. The mathematical equation of NSFWQI is given in Eq. (1),

$$NSFWQI = \sum_{i=1}^{N} Q_i \times W_i \tag{1}$$

where, N= total water quality parameter count, Q_i = sub-index of ith water pollutant, and W_i = weight factor of ith water pollutant.

Weighted Arithmetic Water Quality Index (WAWQI)

Various researchers have widely used the WAWQI method (Chawla et al., 2015; Iqbal et al., 2019; Kachroud et al., 2019). This technique classifies the water quality in the range of pureness by using the most common and widely used water parameters. The WAWQI calculation formula is expressed in Eq. (2) (Chawla et al., 2015).

$$WAWQI = \left| \frac{1}{100} \left(\frac{\sum_{i=1}^{N} Q_i W_i}{\sum_{i=1}^{N} W_i} \right) \right| \tag{2}$$

The quality rating (Q_i) and weight factor (W_i) for all water parameters can be calculated as follows:

$$Q_i = 100\left(A_i - S_i\right) \tag{3}$$

where, A_i and S_i corresponds to the actual and standard value of parameter, respectively.

$$W_i = \frac{k}{P_i} \tag{4}$$

where, k is proportionality constant, P_i represents the higher allowable value. The advantages and disadvantages of the above-mentioned water quality index development methods have been briefly described in (Kachroud et al., 2019; Tyagi et al., 2013).

PROBLEM FORMULATION FOR THE MODELING OF WATER QUALITY PARAMETERS

The primary objective of water quality modeling is to forecast the water quality parameters by analyzing the historical time-series data of the water quality parameters corresponding to various water resources such as oceans, rivers, lakes, etc. Recently, data-driven methods are preferred over rule-based methods to model the hidden trends of the historical water quality data due to the availability of the large datasets and high-end computational platforms. Various water quality parameters such as DO, BOD, TDS, Turbidity, pH, temperature, etc., are time and location dependent. Hence, water quality parameters can be treated as time series data. Subsequently, water quality modeling for forecasting the water quality parameters can be treated as time series modeling. Broadly, the problem formulation for the time-series forecasting of the water quality parameters and the associated water quality index corresponding to any water bodies can be outlined as follows.

- Assumptions: The time-series data of the water quality parameters correspond to the same water bodies for which model training and testing will be carried out. Sufficiently, a good and big dataset is available for the learning of the model.
- Inputs: Benchmarked water quality datasets covering different possible stretches of the water body are readily available for training. The training data must contain at least one time-dependent water quality parameter for training and validation. Otherwise, the training dataset may be comprised of multiple time-dependent multiple water quality parameters for training and validation. Depending on the number of water quality parameters, either univariate time series modeling (for one time-dependent water quality parameter) or multivariate time series modeling (for more than one time-dependent water quality parameter) approach may be followed. The training dataset can be inputted for the development of the model that can be used for water quality forecasting.
- Objectives: The main task is to develop an efficient and robust time series model for water quality forecasting using either the statistical or Artificial Intelligence-based modeling approaches. Subsequently, an efficient water quality forecasting method can be finalized corresponding to the application for controlling and managing the water pollution levels with the help of proper decisions.
- Methods: One critical step is selecting appropriate modeling methods (either statistical or Artificial Intelligence-based methods) to forecast the water quality parameters and their associated WQI from the historical time series water quality data. A systematic review of various widely used time series modeling methods will be discussed in Section 4.
- Field trials: Once the efficient and robust model for the time series forecasting of the water quality parameters corresponding to a specific water resource is developed, it can be validated practically before the real deployment. The offline mode of validation uses the test samples that are the stored water quality parameters, whereas the online mode of validation requires test samples that are streaming from the live sensors. Upon successful field trials with multiple iterations of fine-tuning, the model can be deployed for real-time applications.

TIME SERIES MODELING FOR WATER QUALITY FORECASTING

Time series data are the recorded values that vary irregularly over time, corresponding to any phenomenon (Kitagawa, 2020). Here, the water quality parameter and the associated WQI index vary irregularly over time due to the biochemical process caused by water pollution. Hence, water quality data and the associated WQI corresponding to a particular water source can be treated as time-series data. Generally, time-series data are chronologically ordered data that are collected and ordered with even intervals of time. The frequency of time series data is defined as the time intervals at which data are collected. The models which are used to work on time series data are called time series models. Further, forecasting the water quality parameters by an appropriate mathematical model from the historical data is known as time series water quality modeling for water quality forecasting. Based on the number of time-dependent input variables, time series modeling can be of two types as follows:

- Univariate modeling: When the time series modeling involves only one time-dependent variable, then it is known as univariate modeling. For example, forecasting of DO by modeling the historical time series data comprises only dissolved oxygen values. In the case of univariate modeling, it is assumed that a single input variable or parameter is self-sufficient to explain the process. In other words, the univariate modeling does not consider the relationship among the input variables while modeling. Further, most of the natural phenomena like water pollution, air pollution, etc., are always affected by multiple input parameters. Hence, univariate models are less efficient than the multivariate models for the input data where more than one inter-dependent variable is present. However, univariate models are computationally efficient as compared to multivariate models.
- Multivariate modeling: When the time series modeling involves more than one time-dependent variable, it is known as multivariate modeling. For example, forecasting of DO by modeling the historical time series data comprises values of dissolved oxygen, temperature, turbidity, etc. In the case of multivariate modeling, it is assumed that multiple interdependent input variables or parameters are required to explain the process entirely. In other words, multivariate modeling considers the interdependent relationship among the input variables while modeling. Hence, multivariate models are efficient as compared to the univariate models for the input data where more than one inter-dependent variable is present. However, multivariate models require a little bit higher computational resources as they involve many parameters.

In addition to this, broadly, there are two categories of time series modeling for water quality forecasting depending on the used techniques, as follows.

Classical Statistical Techniques for Time-Series Forecasting

Time series analysis is nothing but a statistical approach that works on the variables with seasonal effects or trends regarding time. For the past few years, classical time series models like Auto Regression (AR) (Pu & Bai, 2014), Auto-Regressive–Moving-Average (ARMA) (Halim & Bisono, 2008), Auto-Regressive Integrated Moving Average (ARIMA) (Chang, 2019), Seasonal Auto-Regressive Integrated Moving Average (SARIMA) (Gocheva-Ilieva et al., 2014), Seasonal Auto-Regressive Integrated Moving Average with eXogenous factors (SARIMAX) (Vagropoulos et al., 2016), Fb-Prophet, etc. have been widely used in many studies. A general flowchart for the development of the classical statistical technique-based models for time-series forecasting of the water quality parameters may be outlined as shown in Fig. 1. The input time-series water quality parameters data comprises of BOD, DO, temperature, turbidity, etc., have been collected. In the preprocessing steps, outliers and missing values are removed. Generally, missing values can be filled by using the back-fill methods. Outliers can be removed by using appropriate time series anomaly detection techniques. The time-series data having no trend or seasonal effect over time is known as stationary time-series data. Stationary data is very much effective for time series modeling. Hence, the stationarity of the time-series water quality data has been checked using appropriate techniques such as the augmented dicky-fuller test. Here, the augmented dicky-puller test may be performed separately for the individual water quality parameters or pollutants for checking the stationarity. The nonstationary time-series data should be converted to stationary time-series data by using the difference method. Subsequently, Auto-Correlation Function (ACF) and Partial Auto-Correlation Function (PACF) are plotted to investigate the interdependency among the water quality parameters. Further, this analysis helps to find the optimal parameters that can be used in statistical modeling. Now, the stationary data is split into two subsets, namely, training data and testing data. The training data is used to build the appropriate time-series statistical models such as ARIMA (Chang, 2019), Vector Autoregressive Integrated Moving Average (VARIMA) (Rusyana et al., 2020), etc. The testing data is used to test the efficacy of the developed statistical model. When the performance metrics such as MAE, MSE, RMSE, etc., are not satisfactory, the built model is improved by using model parameter tuning and improved training strategies. Once the model's performance is found to be satisfactory during the testing phase, it can be deployed in the field trial or production phase. Here, the database may contain online (real-time streaming water quality data from the sensor modules) or offline (previously collected water quality data from the sensor modules). When the data are fed to the final model for water quality forecasting, the appropriate results are disseminated to make necessary management decisions to maintain or improve the reservoir's water quality. Finally, these forecasted values can be compared with actual values as and when that are available to check the forecasting efficacy. Subsequently, these data can be used to fine-tune the model for getting better performance. Further, there are two separate sets of classical statistical techniques for univariate and multivariate forecasting as follows.

Univariate Time-Series Forecasting

In Univariate time series models, the dependent variable is of only a single time series, i.e., these models will summarize only one variable at a time. There are many models used to forecast univariate time series data. The classic models like AR (Pu & Bai, 2014), ARMA (Halim & Bisono, 2008), ARIMA (Chang, 2019), SARIMA (Gocheva-Ilieva et al., 2014), SARIMAX (Vagropoulos et al., 2016), etc. are the few important univariate time-series forecasting techniques. The time series consists of a trend that describes the variation along with the time interval, and seasonal effects represent seasonal fluctuations, cyclical variations are similar to seasonal effects. But, it responds to periodical variations along with the interval, and irregular variations show the other nonrandom sources of fluctuations in times series. ARIMA model is able to overcome the deficiency of deterministic factorization (An & Zhao, 2017). Before applying any time-series data to the ARIMA model, it is essential to check for the stationarity of the series. The time-varying series can be transformed to stationary using log, square root, cube root transformation, and using the differencing method. Further, SARIMA provides more accurate results when seasonal and cyclical variations are involved in input time-series data (Gocheva-Ilieva et al., 2014). Many research studies have proved that the SARIMA model's performance is higher compared to that of the ARIMA method. In order to build the SARIMA model following few steps are required: (i) Preprocessing of data to check for Gaussian, stationary, and stochastic process [19]; (ii) Model-identification; (iii) Parameter estimation and diagnosis; (iv) Defining the model with an appropriate order. Facebook designed Fb-Prophet in 2017, which is a popular open-source, designed to execute time series prediction and analysis by an additive model approach (Battineni et al., 2020). Fb-Prophet consists of trend parameters such as linear and logistic trend function, which can be set with yearly, weekly, daily seasonality, and holiday effects. Fb-Prophet model is very robust to missing values and works efficiently on nonlinear data. Recently, various univariate models such as ARIMA, SARIMA, and Fb-Prophet models have been implemented to forecast the DO and BOD levels and their associated WQI for the river Ganga (Kogekar et al., 2021c). Here, Fb-Prophet has outperformed the other two techniques while predicting the WQIs.

Multivariate Time-Series Forecasting

Multivariate models are designed to compare more than two variables at a time that are interdependent. For example, the water quality parameters such as pH, temperature, BOD, DO, etc., are interdependent. Hence, multivariate time-series forecasting is more effective for water quality modeling. The multivariate time series models like SARIMAX (Vagropoulos et al., 2016), VAR (Keng et al., 2017), VARIMA (Rusyana et al., 2020), and Vector Error Correction Model (VECM) (Xiong & Wu, 2008) are widely used. SARIMAX is the extension of SARIMA with exogenous variables (Vagropoulos et al., 2016). It is the multivariate form of the SARIMA model, which can have multiple external variables that are correlated to dependent features. The Vector Autoregressive (VAR) model (Keng et al., 2017) is one of the famous statistical models that is best suitable for multivariate time series data and can easily grasp the nonlinear trend and seasonality of any time series data.

Figure 1. Generalized steps involved in Classical Time Series technique

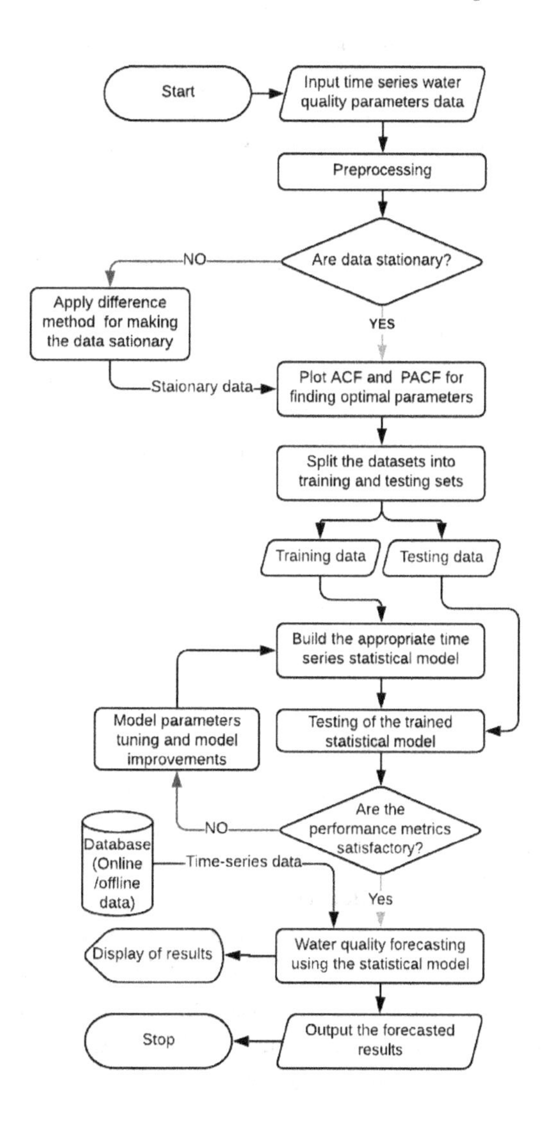

Artificial Intelligence-Based Time-Series Forecasting

AI-based time-series forecasting has been widely used for water quality forecasting. Depending on the availability of data, targeted problem, and feature engineering, either machine learning or deep learning may be used for water quality forecasting.

Machine Learning-based time-series forecasting

ML is a subgroup of AI that allows computers or any mainframe to learn automatically from previous information without human intervention. The time-series data can be used as an input sequence to the machine learning algorithms. The primary ML regression models include linear regression (Jalal & Ezzedine, 2019), Support Vector Regression (SVR) (U. Ahmed et al., 2019). A generalized block diagram of machine learning-based time-series forecasting of the water quality parameters is represented in Fig. 2. Similar to the statistical modeling, the input time-series water quality parameters data is segregated into two subsets, namely, training data and testing data. The training data is preprocessed to remove the outliers and missing values. Subsequently, appropriate features are extracted and used for building the suitable machine learning models based on SVR, Random Forest (RF), decision tree, etc. The testing data is used to test the efficacy of the trained machine learning model. When the performance or quality metrics such as MAE, MSE, RMSE, etc., are not satisfactory, the built model is improved by using the hyper-parameter tuning and improved training strategies. Once the model's performance is found to be satisfactory during the testing phase, it can be deployed in the field trial or production phase. When the data are fed to the final model for water quality forecasting, the appropriate results are displaced and intimated to the concerned authorities so that necessary management decisions can be taken to maintain or improve the reservoir's water quality. Finally, these forecasted values can be compared with actual values as and when it is available to check the forecasting efficacy. Subsequently, these data can be used to fine-tune the model for getting better performance. Generally, ML models are found to be more effective when a good amount of qualitative data is available and effective features are extracted as well as used in the suitable model-building process.

Few important machine learning approaches widely used for water quality forecasting are discussed as follows.

The Support Vector Machines (SVM) can be utilized for regression as well as classification problems. SVR implementation concept is the same as SVM, but quite different from other regression models because it takes kernel functions while defining the SVR model (Samal et al., 2020). The different types of kernel functions used in SVM are the sigmoid kernel, Radial Basis Function (RBF), linear, and polynomial (Haghiabi et al., 2018). The Least Squares Support Vector Machine (LS-SVM) is an improved version of general SVM (Cao & Wang, 2018), which is based on the adaptive Particle Swarm Optimization algorithm and can forecast the water quality time series more effectively than other ML methods. Random Forest (RF) (Li et al., 2016) model can also be used for both regression and classification problems. The random forest model uses a number of supportive models on a subgroup of the input data and makes decisions based on all the models (U. Ahmed et al., 2019). The base model of the random forest algorithm is a decision tree. The most widely used classification models include SVM, Random forest classifier, Decision tree, and Gradient boosting classifier due to their high accuracy performance (U. Ahmed et al., 2019). One of the most optimized ML algorithms, namely Extreme Gradient Boosting (XGBoost), comes with a faster computational process and more accurate features

(Joslyn, 2018). In (Ragi et al., 2019), a brief method to predict the unknown parameters like alkalinity, Chloride, and sulfate values using known parameters such as pH, Conductivity, TDS, etc., using the Levenberg Marquadart algorithm is proposed. But, this method is a laboratory process in which samples are taken, and laboratory testing is performed. An Extreme Learning Machine (ELM) method uses a neural network with a single hidden layer that randomly selects input weights and hidden biases. The output weights are determined analytically by using Moore-Penrose generalized inverse to forecast the sales in fashion retailing (Z.-L. Sun et al., 2008). ELM has the advantage over many problems faced by the Gradient-based algorithms like learning rate, stopping criteria, learning epochs, over tuning, and local minima problems (Z.-L. Sun et al., 2008). In this, the most affecting factors of the sales are taken as inputs to the method. In (Lv & others, 2020), Improved Extreme Learning Machine Algorithm based on Gravitational Search algorithm (IGSAELM) is proposed. The IGSA is used to optimize the hidden thresholds and input weights which are randomly selected by ELM for prediction. It uses Chinese city's traffic time-series data to verify this IGSA model. It showed highly accurate results with MSE 0.33 and RMSE as 0.109. Further, a new strategy of machine double layer learning was implemented (G. Chen & Hou, 2007). It combines the advantages of ANN/SVM and that of the Genetic Algorithm (GA), where ANN/SVM is implemented to obtain the model's inner parameters and GA is implemented to obtain model outer parameters. When compared to common machine learning algorithms, this method possesses a stronger self-adaptive ability.

Deep Learning-Based Time-Series Forecasting

DL is a subtype in machine learning that easily works on unstructured and unlabeled data. Figure 3 represents the generalized form of a DNN with its layers. Many traditional approaches, i.e., ARIMA and basic linear regression model, are available to predict future water quality. The main disadvantage of the ARIMA model is that it needs a stationary time series to process the data. This model cannot capture the non-stationarity and seasonality effect involved with the series (Y. Chen et al., 2020). On the other hand, Multiple Linear Regression (MLR) can't detect the nonlinearity of the input sequence. A generalized block diagram of deep learning-based time-series forecasting of the water quality parameters is represented in Figure 4. The DL-based time series forecasting techniques are developed similar to the machine learning-based time-series forecasting techniques, except that there is no exclusive manual feature extraction step involved in the deep learning approaches. DL-based approaches are completely data-driven, and hence, they are more suitable as robust features are automatically extracted by the Deep Neural Network (DNN). It is observed that deep learning-based time series forecasting techniques are the best techniques provided that good and big time series water quality datasets, as well as good computing facilities, are available. However, deep learning-based models are a little bit more computationally expensive as compared to the other two approaches. However, recently appropriate techniques such as pruning and quantization are used to reduce the computational complexity without significantly decreasing the model's performance.

Few important deep learning approaches widely used for water quality forecasting are discussed as follows.

The DNN architecture consists of three layers in its memory cell, i.e., the input layer, hidden layers, and output layer. Hence, this will automatically grab and manipulate complex input features structures as well as correlation in dependency [36]. The stand-alone DNNs are Artificial Neural Network (ANN) (Y. Chen et al., 2020), Recurrent neural network (RNN), Convolutional neural network (CNN) (Hoseinzade

& Haratizadeh, 2019), Gated Recurrent Unit (GRU), Long-Short Term Memory (LSTM) (Wang et al., 2017), and Bidirectional GRU (Yan et al., 2021), Simple Recurrent Unit (SRU) (Liu et al., 2018), etc. The ANN model's performance is better as compared to the SVM model. But, this doesn't mean that the ANN is the best method. They are individually best suitable for each different problem (Yu et al., 2018). In (Yu et al., 2018), ANN and SVM are used to predict Total Nitrogen and Total Phosphorus. SVM parameters are optimized using a Genetic algorithm to achieve the lowest prediction errors. Generally, the CNN layer is used to extract spatial input features, whereas LSTM and Bidirectional LSTM networks can automatically capture long and short-term dependencies. GRU and Bidirectional GRU are the updated version of LSTM networks. The Bidirectional Recurrent Neural Network (Bi-RNN) can process the input features from both forward and reverse directions. In (Liu et al., 2018), a Simple Recurrent Unit (SRU) is implemented to predict the key quality parameters of water like temperature and pH for Mariculture. Further, the results of the SRU method are compared with that of the RNN method. Before implementing the model, the quality parameters are preprocessed, and the correlation between them is obtained using the Pearson correlation method. SRU is just an improved version of the RNN network structure, and it has the merits like fast convergence, simple structure, and good stability. In similar conditions, RNN will consume less time than that of SRU, but SRU will give 98.91 percent of prediction accuracy (Liu et al., 2018).

Figure 2. Generalized block diagram of ML-based time-series forecasting technique

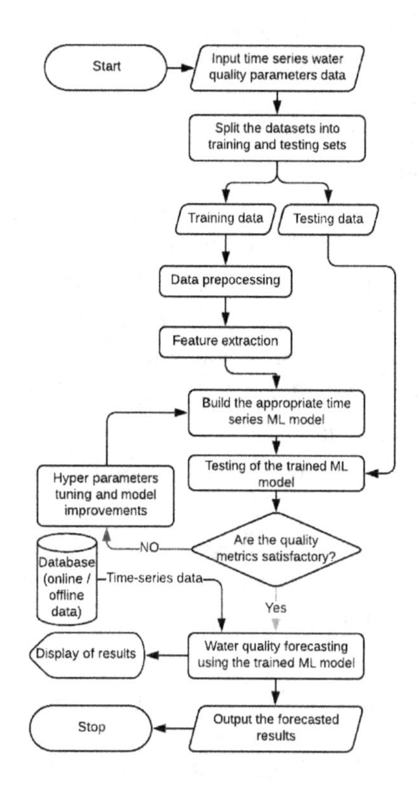

Figure 3. Generalized architecture of Deep Neural Network

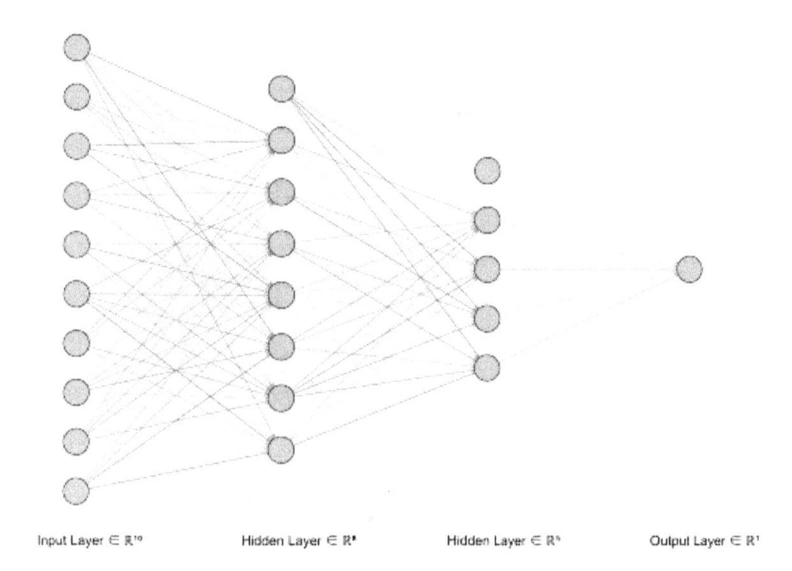

Multi-Layer Perceptron - Artificial Neural Network (MLP-ANN) (A. N. Ahmed et al., 2019) model is implemented to estimate the parameters of water quality like Ammonia Nitrates (AN), Suspended Solids (SS), and pH. However, this MLP-ANN model faced delayed convergence during the training as its hidden layer contains more neurons. It also fails to acquire the values outside the scope of values in the calibration data. Alternatively, RBF- ANN (A. N. Ahmed et al., 2019) has equal abilities of MLP-ANN in solving problems related to function estimations. Shorter network architecture and estimating the best solution without managing the local minima are two main benefits of RBF over the MLP. It also used the Adaptive Neuro-Fuzzy Inference System (ANFIS), a multi-layer feed-forward that employs both neural networks and fuzzy reasoning. Various methods of LSTM are used to predict the time series data and achieve significant results. But, most of the methods ignored the water quality changes in space-time correlation (H. Zhang & Jin, 2020). In (H. Zhang & Jin, 2020), the water quality features are selected by using an Automatic Encoder (AE). An unsupervised neural network like self-encoding dimensional reduction is used to encode low dimensional data from high dimensional data. The data which is preprocessed will be given as encoder's input to learn the features automatically, and this AE consists of one encoder and one decoder. This Automatic Encoder code is input to the LSTM model to establish a model that predicts spatiotemporal-based features. This method is comparatively robust and accurate compared to the LSTM time series and LSTM spatiotemporal correlation models (H. Zhang & Jin, 2020).

Hybrid Modeling for Time-Series Forecasting

The previous research work (Yan et al., 2021) observed that efficient monitoring and forecasting of time series data is complicated. The commonly used independent deep learning and statistical models such as CNN, LSTM, GRU, BiGRU, BiLSTM, and ARIMA are insufficient to make accurate predictions due to the trend and seasonal effects of time series data. However, time series forecasting can be done proficiently using the suitable integration of these individual methods. The different reported hybrid models from the past research studies are discussed in this section.

Figure 4. Generalized block diagram of DL-based time-series forecasting technique

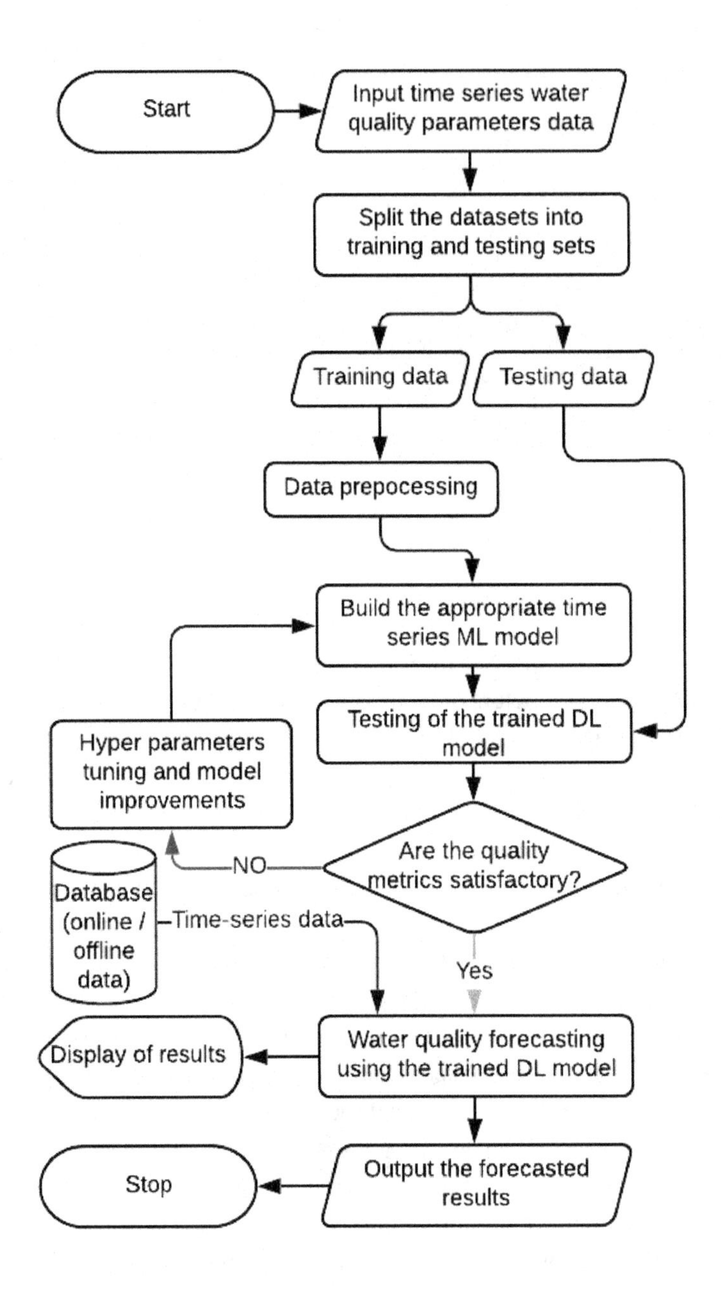

The ANN - ARIMA (Faruk, 2010) integrates two models, i.e., ANN and ARIMA. The individual ANN model cannot grasp the seasonality associated with time series, whereas the ARIMA model is not able to solve complex nonlinear problems. A water quality time series is a combination of the linear auto-correlation function and nonlinear variables. A hybrid version of both these models can easily capture linear and nonlinear components of given water quality time series. In this model, error deviation can be modeled by using ANN to encounter nonlinearity. Similarly, SARIMA - LSTM is another hybrid model in which SARIMA and LSTM are combined to get multi-feature predictions (Xu et al., 2019). Further, the CNN layer is utilized to extract spatial features from the source data in the Convolutional Neural networks - Long Short Term Memory (CNN-LSTM) (Lu et al., 2020) model. For defining the CNN-LSTM model, the LSTM layer should be added after the CNN layer. The main feature of the LSTM layer is it provides the solution to longstanding problems of gradient explosion in RNN (Lu et al., 2020). The ARIMA-CNN-LSTM (Ji et al., 2019) is the integration of the ARIMA, CNN, and LSTM models. In the research paper (Ji et al., 2019), other three deep learning models, such as CNN, LSTM, and ARIMA, are implemented as a baseline to the ARIMA-CNN-LSTM model. This deep learning hybrid model can be effectively used for forecasting water pollutants. Another efficient hybrid model is an attention-based CNN-LSTM-BiLSTM model (Wu et al., 2021). It consists of an attention mechanism technique, which can be used in visual observation modeling to remove high-rated information and drop the irrelevant data. An attention mechanism focuses on dominant input features for model prediction by providing the separate weights to feature. After employing the attention mechanism, the CNN-LSTM-BiLSTM approach performs better for long-term series (Wu et al., 2021). The Bidirectional-Gated Recurrent Unit (BiGRU) is one of the types of LSTM neural networks. The general unidirectional GRU model can process the past information only in a single portion. On the other hand, the BiGRU model can work in backward-proportion as well as forward-proportion in order to pass the past information. One-dimensional residual convolutional neural network - Bidirectional gated recurrent unit (1DRCNN-BiGRU) (Yan et al., 2021) is constructed using a hybrid neural network to forecast water quality. In this, the 1DRCNN model is particularly used to extract nonlinear input features of river Luan, China. Here, the BiGRU layer is used for capturing the long-term and short-term dependencies.

Another hybrid model is Convolutional Neural Network - Gated Recurrent Neural Network (CNN - GRU) (Jichang et al., 2019). In the construction of CNN - GRU architecture, the GRU's accuracy in prediction and CNN's efficiency in prediction both are in good balance. The CNN model collects the relationship among continuous water quality data, and it forms multiple feature vectors. Then, these vectors are given as input to the GRU after constructing them in time series. Then the GRU network grabs the logical relationship between these vectors. In (Jichang et al., 2019) also compares the results of CNN-GRU with that of SVR, RNN, and GRU. This model takes less time for training and improves the accuracy of predicted results. In (Deng et al., 2014), a hybrid network of ARIMA and Radial Basis Function Neural Network (RBF-NN) model is developed. ARIMA model is employed to examine the linear components in the problem. ACF and PACF are used in the identification stage as basic tools to train the data in order to recognize the order of ARIMA. The RBF-NN network will be trained from the outputs of the ARIMA model. RBF, which is a three-layered feed-forward network, is used one step in front of forecasting, constructed to predict the water quality parameters. In (Deng et al., 2014), this model is employed in the prediction of DO and NH3-N.

Recently, two hybrid models, namely CNN-BiLSTM-SVR (Kogekar et al., 2021a) and CNN-GRU-SVR (Kogekar et al., 2021b), have been proposed to forecast the water quality parameters such as DO and BOD with their associated WQI of the river Ganga using the historical data. The reported comparative analysis showed that the proposed models provide better prediction accuracy in terms of RMSE and MAE as compared to individual models such as LSTM, BiLSTM, GRU, BiGRU as well as other hybrid models such as CNN-LSTM, CNN-BiLSTM, CNN-GRU, CNN-BiGRU, and CNN-BiGRU-SVR. Another hybrid method to forecast the multivariate time series of stock prices is implemented in (Lusia & Ambarwati, 2018), which is the combination of VARIMA and Feed Forward Neural Network. The hybrid method gives comparatively accurate results than individual methods (Lusia & Ambarwati, 2018).

PERFORMANCE METRICS

In order to evaluate the performance of different models, there is a lot of variation in choosing the performance metrics from many research studies. The most commonly used error metrics are Root Mean Square Error (RMSE), Mean Absolute Error (MAE), Mean Square Error (MSE), Mean Absolute Percentage Error (MAPE), and R^2 scores.

MAE

MAE is defined as the mean of the absolute value of prediction error as given in Eq. (5).

$$MAE = \frac{1}{n}\sum_{i=1}^{n}\left|e_i\right| \tag{5}$$

Here, e_i is the deviation between predicted and actual result, $i = 1, 2, 3, \ldots .n.$ The n value denotes the total number of samples in the data set. Value of MAE gives the amount of inaccuracy that we estimate in forecast values on an average. Usually, the performance of a model is evaluated through the coefficient of determination, while MSE is employed to correct the fitness between the model output and the actual output (A. N. Ahmed et al., 2019). MAE is useful in the case of absolute error measurement, but it is not handy for extreme values.

MSE and RMSE

MSE is defined as the average of the squared values of the difference between predicted and actual values, whereas RMSE is simply the square root of MSE (Samal et al., 2020), as given in Eq. (6).

$$RMSE = \sqrt{MSE} = \sqrt{\frac{1}{n}\sum_{i=1}^{n}e_i^2} \tag{6}$$

Lower values of MSE, RMSE, and MAE denote that the defined model is performing better. The values of RMSE and MAE are compared to check for abnormal inaccuracies in the forecast. If the values of RMSE and MAE differ by a lot more is the error size with low data cases.

R^2 score

The coefficient of determination (R^2) is the proportion of variance in the dependent feature, which is predictable from the independent features. It is defined as in Eq. (7).

$$R^2 = 1 - \frac{SS_{res}}{SS_{tot}} \tag{7}$$

A high value of R^2 indicates that the variance of the model is almost the same as true values and a low value of the same indicates that the two values are not much related. R^2 score should be higher and closer to 1 in the range of 0 to 1.

COMPARATIVE ANALYSIS OF THE TIME-SERIES TECHNIQUES

A comparative analysis of the important approaches used for the time-series forecasting that are reported for the water quality forecasting is presented in Table 2.

RESEARCH CHALLENGES INVOLVED IN WATER QUALITY TIME-SERIES MODELING AND FORECASTING

There is enough research works carried out till date to forecast the water quality parameters. However, still, there is a wide scope of investigation and improvement in the water quality time series modeling methods. A few of the important research challenges involved in time series modeling and forecasting is listed as follows.

Raw Data Collection Issue

Water quality data collection is a high-priced and slow process due to the involvement of expensive sensors and other electronic gadgets. In order to get the future water quality, it is necessary to have past information or data. Nowadays, government data programs are making the data available on open-source platforms. Still, it is tough to get complete real-time data.

Unavailability of the Raw Data in Suitable Forms

According to many studies, there is no such systematic way to determine how many samples or volume of data is needed for a particular problem. The requirement of data points is varied as per the requirements of the different approaches. Some deep learning methods, such as RNN, require more sample points, whereas GRNN can easily handle a small number of data points.

Involvement of Outliers

The dataset can consist of outliers, which means it can have some values on a different range of scales. There are requirements of the historical data that need to be consistent, and outlier detection is required in many studies to achieve higher accuracy and lower errors.

Table 2. Comparative analysis of the time-series techniques for water quality forecasting

Approach	Models	Basic principle	Advantages	Disadvantages
Classical statistical technique-based	ARIMA (An & Zhao, 2017; Kogekar et al., 2021c), SARIMA (Gocheva-Ilieva et al., 2014; Kogekar et al., 2021c), Fb-Prophet (Kogekar et al., 2021c; Kumar Jha & Pande, 2021), VAR (Chien et al., 2007; Keng et al., 2017; Mingyuan & Shiying, 2009), VARIMA (Rusyana et al., 2020), SARIMAX (Vagropoulos et al., 2016)	Classical statistical techniques use the trend and seasonality of the time-series data for modeling and forecasting.	• Efficient for small-sized structured time-series data. • An advanced model like Fb-Prophet is able to handle the nonlinearity issue.	• Poor response for long term forecast • Efficient for stationary data only. • Performance is not guaranteed for nonstationary data. Hence, nonstationary to stationary data conversion is performed in the preprocessing step.
Machine learning-based	SVM (B.-J. Chen et al., 2004; Mutavhatsindi et al., 2020), random forests (Li et al., 2016), XGBoost (Joslyn, 2018), Decision tree, and Gradient boosting classifier (U. Ahmed et al., 2019)	ML-based techniques use feature engineering to find efficient descriptors used for time-series modeling.	• Efficient for moderate-sized structured time-series data. • Models are computationally less expensive as compared to that of DL and hybrid modeling-based techniques. • More model transparency is there.	• Efficiency decreases when the dimensionality of the data increases. • Less accurate for the large-sized qualitative data as compared to that of the DL and hybrid modeling-based techniques.
Deep learning-based	ANN (Y. Chen et al., 2020), RNN and CNN (Hoseinzade & Haratizadeh, 2019), GRU, LSTM (Wang et al., 2017), and Bidirectional GRU (Yan et al., 2021), SRU (Liu et al., 2018), LSTM (H. Zhang & Jin, 2020).	DL-based techniques use hierarchical learning with completely automatic feature representation to find more efficient and robust descriptors used for the time-series modeling.	• It uses complete automatic feature extraction and representation. • Efficient for large-sized structured time-series data. • More robust as models are capable of handling inherent data issues. • It provides better accuracy with more forecasting durations. • It is able to model both linear and nonlinear data efficiently.	• No exclusive manual feature extraction step is involved. • Less model transparency is there due to the black box nature of the DNNs. • Models are computationally more expensive as compared to that of ML and statistical-based techniques.
Hybrid modeling-based	Modified genetic algorithm combined with random forest (L. Zhang et al., 2018), SARIMA-LSTM (Q. Sun et al., 2020; Xu et al., 2019), ARIMA-RBF-NN (Deng et al., 2014), CNN-LSTM (Hu et al., 2019; Lu et al., 2020), CNN-LSTM-BILSTM (Wu et al., 2021), CNN-GRU (Jichang et al., 2019; Nana et al., 2019), ARIMA-CNN-LSTM (Ji et al., 2019), CNN-BiLSTM-SVR (Kogekar et al., 2021a), CNN-GRU-SVR (Kogekar et al., 2021b)	Hybrid modeling-based techniques use the best of the various base models to efficiently model the time-series data for better forecasting results. Generally, DL models are used for extracting robust and efficient features in an end-to-end pipeline. Subsequently, ML-based models are used for classification and forecasting. Finally, the overall outputs outperform that of the statistical models, ML, and DL-based models.	• It uses complete automatic feature extraction and representation. • Efficient for moderate and large-sized structured time-series data. • Most robust as models are capable of handling inherent data issues. • It provides better accuracy with more forecasting duration as compared to other approaches.	• Models are computationally expensive. However, computational complexity can be reduced further with the help of proper pruning and quantization techniques. • Modeling is difficult. Hence, it requires more expertise both in coding and domain knowledge.

Input Feature Selection Issue

The output of many time series problems depends on a large number of input variables. Selecting input feature vectors to get the specific result is not an easy task. It is very important to select the input features wisely, but from many studies, it is observed that researchers chose input based on their interest or in a random fashion which can ultimately affect the output result.

Data Splitting Issue

The data splitting method is not a constant or pre-specified method. Researchers split the input data with their own choice; some use 80% of training data, 10% of validation data, and the remaining 10% of testing data while defining the machine learning models. The most common data splitting percentages of training, testing, and validation are 70%, 15%, 15%, as well as 50%, 25%, 25%, respectively. In order to get higher accuracy, rather than choosing train test split, some literature performed different types of cross-validation methodology.

Modeling Issue

In the ARIMA modeling technique, it is important to check for the stationarity of input time series. Again if the series is nonstationary, the application of the differencing method can make this series stationary, which is a different method to use while defining the ARIMA model. Also, this model cannot handle nonlinear data. Deep learning models also have a few limitations. A separate CNN model can only use for classification purposes. Similarly, ANN models are not able to capture the trend or seasonal effect of time series data. More hybrid model development is necessary to get accurate prediction results.

CONCLUSION

The Internet of Things (IoT) infrastructure has started to improve. Due to advancements in the IoT and machine learning fields, it is now possible to get complete real-time data. This real-time data can be used as input to predict future values using different machine learning modeling approaches. From the analysis and experience, new approaches have been come into the picture to solve the research problems associated with water quality time series models. Firstly, researchers worked only on individual methods, and slowly, they moved to the implementation of hybrid models. Many classification-based models such as SVM, RF, Gradient boosting are also used for prediction problems. Integration of deep learning algorithms with statistical and some machine learning models proved its prediction capability. Recently, hybrid models became the most popular approach in the field of time series forecasting due to their long-term forecasting abilities and fast computational speed.

REFERENCES

Ahmed, A. N., Othman, F. B., Afan, H. A., Ibrahim, R. K., Fai, C. M., Hossain, M. S., Ehteram, M., & Elshafie, A. (2019). Machine learning methods for better water quality prediction. *Journal of Hydrology (Amsterdam), 578*, 124084. doi:10.1016/j.jhydrol.2019.124084

Ahmed, U., Mumtaz, R., Anwar, H., Shah, A. A., Irfan, R., & García-Nieto, J. (2019). Efficient water quality prediction using supervised Machine Learning. *Water (Basel), 11*(11), 1–14. doi:10.3390/w11112210 PMID:32021704

Aktar, M. W., Paramasivam, M., Ganguly, M., Purkait, S., & Sengupta, D. (2010). Assessment and occurrence of various heavy metals in surface water of Ganga river around Kolkata: A study for toxicity and ecological impact. *Environmental Monitoring and Assessment, 160*(1), 207–213. doi:10.100710661-008-0688-5 PMID:19101812

An, Q., & Zhao, M. (2017). Time Series Analysis in the Prediction of Water Quality. *Proc. 7th Int. Conf. on Education, Management, Information and Mechanical Engineering (EMIM 2017).* 10.2991/emim-17.2017.11

Battineni, G., Chintalapudi, N., & Amenta, F. (2020). Forecasting of COVID-19 epidemic size in four high hitting nations (USA, Brazil, India and Russia) by Fb-Prophet machine learning model. *Applied Computing and Informatics.*

Budiarti, R. P. N., Tjahjono, A., Hariadi, M., & Purnomo, M. H. (2019). Development of IoT for Automated Water Quality Monitoring System. *Proc. IEEE Int. Conf. on Computer Science, Information Technology, and Electrical Engineering (ICOMITEE)*, 211–216. 10.1109/ICOMITEE.2019.8920900

Cao, S., & Wang, S. (2018). Design of River Water Quality Assessment and Prediction Algorithm. *2018 Eighth International Conference on Instrumentation \& Measurement, Computer, Communication and Control (IMCCC)*, 1625–1631. 10.1109/IMCCC.2018.00335

Chang, T.-C. (2019). The Performance of Grey Model and Auto-Regressive Integrated Moving Average for Human Resources Prediction in China. *2019 IEEE International Conference on Computation, Communication and Engineering (ICCCE)*, 245–248. 10.1109/ICCCE48422.2019.9010801

Chawla, P., Kumar, P., Singh, M., Hasteer, N., & Ghanshyam, C. (2015). Prediction of pollution potential of Indian rivers using empirical equation consisting of water quality parameters. *Proc. IEEE Technological Innovation in ICT for Agriculture and Rural Development (TIAR)*, 214–219. 10.1109/TIAR.2015.7358560

Chen, B.-J., Chang, M.-W., & Lin, C.-J. (2004). Load forecasting using support vector Machines: A study on EUNITE competition 2001. *IEEE Transactions on Power Systems, 19*(4), 1821–1830. doi:10.1109/TPWRS.2004.835679

Chen, G., & Hou, R. (2007). A new machine double-layer learning method and its application in nonlinear time series forecasting. *2007 International Conference on Mechatronics and Automation*, 795–799. 10.1109/ICMA.2007.4303646

Chen, Y., Song, L., Liu, Y., Yang, L., & Li, D. (2020). A review of the artificial neural network models for water quality prediction. *Applied Sciences (Basel, Switzerland), 10*(17), 5776. doi:10.3390/app10175776

Chien, H.-F., Lee, S.-H., Lee, W., & Tsai, Y. (2007). Forecasting Monthly Sales of Cell-phone Companies - the Use of VAR Model. *Second International Conference on Innovative Computing, Informatio and Control (ICICIC 2007)*, 459. 10.1109/ICICIC.2007.314

Deng, W., Wang, G., Zhang, X., Guo, Y., & Li, G. (2014). Water quality prediction based on a novel hybrid model of ARIMA and RBF neural network. *2014 IEEE 3rd International Conference on Cloud Computing and Intelligence Systems*, 33–40.

Faruk, D. Ö. (2010). A hybrid neural network and ARIMA model for water quality time series prediction. *Engineering Applications of Artificial Intelligence*, *23*(4), 586–594. doi:10.1016/j.engappai.2009.09.015

Geetha, S., & Gouthami, S. (2016). Internet of things enabled real time water quality monitoring system. *Smart Water*, *2*(1), 1–19. doi:10.118640713-017-0005-y

Gocheva-Ilieva, S. G., Ivanov, A. V., Voynikova, D. S., & Boyadzhiev, D. T. (2014). Time series analysis and forecasting for air pollution in small urban area: An SARIMA and factor analysis approach. *Stochastic Environmental Research and Risk Assessment*, *28*(4), 1045–1060. doi:10.100700477-013-0800-4

Haghiabi, A. H., Nasrolahi, A. H., & Parsaie, A. (2018). Water quality prediction using machine learning methods. *Water Quality Research Journal*, *53*(1), 3–13. doi:10.2166/wqrj.2018.025

Halim, S., & Bisono, I. N. (2008). Automatic seasonal auto regressive moving average models and unit root test detection. *International Journal of Management Science and Engineering Management*, *3*(4), 266–274. doi:10.1080/17509653.2008.10671053

Hoseinzade, E., & Haratizadeh, S. (2019). CNNpred: CNN-based stock market prediction using a diverse set of variables. *Expert Systems with Applications*, *129*, 273–285. doi:10.1016/j.eswa.2019.03.029

Hu, P., Tong, J., Wang, J., Yang, Y., & de Oliveira Turci, L. (2019). A hybrid model based on CNN and Bi-LSTM for urban water demand prediction. *2019 IEEE Congress on Evolutionary Computation (CEC)*, 1088–1094. 10.1109/CEC.2019.8790060

Iqbal, K., Ahmad, S., & Dutta, V. (2019). Pollution mapping in the urban segment of a tropical river: Is water quality index (WQI) enough for a nutrient-polluted river? *Applied Water Science*, *9*(8), 197–213. doi:10.100713201-019-1083-9

Jalal, D., & Ezzedine, T. (2019). Performance analysis of machine learning algorithms for water quality monitoring system. *Proc. IEEE Int. Conf. on Internet of Things, Embedded Systems and Communications (IINTEC)*, 86–89. 10.1109/IINTEC48298.2019.9112096

Ji, L., Zou, Y., He, K., & Zhu, B. (2019). Carbon futures price forecasting based with ARIMA-CNN-LSTM model. *Procedia Computer Science*, *162*, 33–38. doi:10.1016/j.procs.2019.11.254

Jichang, T. U., Xueqin, Y., Chaobo, C., Song, G. A. O., Jingcheng, W., & Cheng, S. U. N. (2019). Water Quality Prediction Model Based on GRU hybrid network. *2019 Chinese Automation Congress (CAC)*, 1893–1898.

Johnson, D. L., Ambrose, S. H., Bassett, T. J., Bowen, M. L., Crummey, D. E., Isaacson, J. S., Johnson, D. N., Lamb, P., Saul, M., & Winter-Nelson, A. E. (1997). Meanings of environmental terms. *Journal of Environmental Quality*, *26*(3), 581–589. doi:10.2134/jeq1997.00472425002600030002x

Joslyn, K. (2018). *Water quality factor prediction using supervised machine learning*. Academic Press.

Kachroud, M., Trolard, F., Kefi, M., Jebari, S., & Bourrié, G. (2019). Water quality indices: Challenges and application limits in the literature. *Water (Basel), 11*(2), 361–387. doi:10.3390/w11020361

Keng, C. Y., Shan, F. P., Shimizu, K., Imoto, T., Lateh, H., & Peng, K. S. (2017). Application of vector autoregressive model for rainfall and groundwater level analysis. *AIP Conference Proceedings, 1870*(1), 60013. doi:10.1063/1.4995940

Kitagawa, G. (2020). *Introduction to Time Series Modeling with Applications in R*. CRC Press. doi:10.1201/9780429197963

Kogekar, A. P., Nayak, R., & Pati, U. C. (2021a). A CNN-BiLSTM-SVR based Deep Hybrid Model for Water Quality Forecasting of the River Ganga. *2021 IEEE 18th India Council International Conference (INDICON)*, 1–6. 10.1109/INDICON52576.2021.9691532

Kogekar, A. P., Nayak, R., & Pati, U. C. (2021b). A CNN-GRU-SVR based Deep Hybrid Model for Water Quality Forecasting of the River Ganga. *2021 International Conference on Artificial Intelligence and Machine Vision (AIMV)*, 1–6. 10.1109/AIMV53313.2021.9670916

Kogekar, A. P., Nayak, R., & Pati, U. C. (2021c). Forecasting of Water Quality for the River Ganga using Univariate Time-series Models. *2021 8th International Conference on Smart Computing and Communications (ICSCC)*, 52–57. 10.1109/ICSCC51209.2021.9528216

Kumar, D., & Alappat, B. J. (2009). NSF-water quality index: Does it represent the experts' opinion? *Practice Periodical of Hazardous, Toxic, and Radioactive Waste Management, 13*(1), 75–79. doi:10.1061/(ASCE)1090-025X(2009)13:1(75)

Kumar Jha, B., & Pande, S. (2021). Time Series Forecasting Model for Supermarket Sales using FB-Prophet. *2021 5th International Conference on Computing Methodologies and Communication (ICCMC)*, 547–554. 10.1109/ICCMC51019.2021.9418033

Li, B., Yang, G., Wan, R., Dai, X., & Zhang, Y. (2016). Comparison of random forests and other statistical methods for the prediction of lake water level: A case study of the Poyang Lake in China. *Hydrology Research, 47*(S1), 69–83. doi:10.2166/nh.2016.264

Liu, J., Yu, C., Hu, Z., Zhao, Y., Xia, X., Tu, Z., & Li, R. (2018). Automatic and accurate prediction of key water quality parameters based on SRU deep learning in Mariculture. *2018 IEEE International Conference on Advanced Manufacturing (ICAM)*, 437–440. 10.1109/AMCON.2018.8615048

Lu, W., Li, J., Li, Y., Sun, A., & Wang, J. (2020). A CNN-LSTM-Based Model to Forecast Stock Prices. *Complexity, 2020*, 2020. doi:10.1155/2020/6622927

Lusia, D. A., & Ambarwati, A. (2018). Multivariate Forecasting Using Hybrid VARIMA Neural Network in JCI Case. *2018 International Symposium on Advanced Intelligent Informatics (SAIN)*, 11–14. 10.1109/SAIN.2018.8673351

Lv, C., & ... (2020). The Time Series Prediction Algorithm Based on Improved GSA-ELM. *2020 IEEE International Conference on Artificial Intelligence and Computer Applications (ICAICA)*, 91–94. 10.1109/ICAICA50127.2020.9181944

Mingyuan, G., & Shiying, Z. (2009). Study on VaR Forecasts Based on Realized Range-Based Volatility. *2009 International Conference on Business Intelligence and Financial Engineering*, 860–862. 10.1109/BIFE.2009.197

Mutavhatsindi, T., Sigauke, C., & Mbuvha, R. (2020). Forecasting Hourly Global Horizontal Solar Irradiance in South Africa Using Machine Learning Models. *IEEE Access: Practical Innovations, Open Solutions*, *8*, 198872–198885. doi:10.1109/ACCESS.2020.3034690

Nana, H., Lei, D., Lijie, W., Ying, H., Zhongjian, D., & Bo, W. (2019). Short-term Wind Speed Prediction Based on CNN_GRU Model. *2019 Chinese Control And Decision Conference (CCDC)*, 2243–2247. 10.1109/CCDC.2019.8833472

Priyank Hirani, V. D. (2019). *Water parameter information*. Academic Press.

Pu, T., & Bai, J. (2014). An auto regression compression method for industrial real time data. *The 26th Chinese Control and Decision Conference (2014 CCDC)*, 5129–5132.

Ragi, N. M., Holla, R., & Manju, G. (2019). Predicting Water Quality Parameters Using Machine Learning. *2019 4th International Conference on Recent Trends on Electronics, Information, Communication \& Technology (RTEICT)*, 1109–1112.

Rusyana, A., Tatsara, N., Balqis, R., & Rahmi, S. (2020). Application of Clustering and VARIMA for Rainfall Prediction. *IOP Conference Series. Materials Science and Engineering*, *796*(1), 12063. doi:10.1088/1757-899X/796/1/012063

Samal, K. K. R., Babu, K. S., Acharya, A., & Das, S. K. (2020). Long term forecasting of ambient air quality using deep learning approach. *2020 IEEE 17th India Council International Conference (INDICON)*, 1–6.

Shah, K. A., & Joshi, G. S. (2017). Evaluation of water quality index for River Sabarmati, Gujarat, India. *Applied Water Science*, *7*(3), 1349–1358. doi:10.100713201-015-0318-7

Shakhari, S., Verma, A. K., & Banerjee, I. (2019). Remote Location Water Quality Prediction of the Indian River Ganga: Regression and Error Analysis. *Proc. IEEE 17th International Conference on ICT and Knowledge Engineering (ICT\&KE)*, 1–5. 10.1109/ICTKE47035.2019.8966796

Sun, Q., Wan, J., & Liu, S. (2020). Estimation of Sea Level Variability in the China Sea and Its Vicinity Using the SARIMA and LSTM Models. *IEEE Journal of Selected Topics in Applied Earth Observations and Remote Sensing*, *13*, 3317–3326. doi:10.1109/JSTARS.2020.2997817

Sun, Z.-L., Choi, T.-M., Au, K.-F., & Yu, Y. (2008). Sales forecasting using extreme learning machine with applications in fashion retailing. *Decision Support Systems*, *46*(1), 411–419. doi:10.1016/j.dss.2008.07.009

Swenson, H. A. (1965). A primer on water quality. US Department of the Interior, Geological Survey. doi:10.3133/7000057

Tyagi, S., Sharma, B., Singh, P., & Dobhal, R. (2013). Water quality assessment in terms of water quality index. *American Journal of Water Resources*, *1*(3), 34–38. doi:10.12691/ajwr-1-3-3

Vagropoulos, S. I., Chouliaras, G. I., Kardakos, E. G., Simoglou, C. K., & Bakirtzis, A. G. (2016). Comparison of SARIMAX, SARIMA, modified SARIMA and ANN-based models for short-term PV generation forecasting. *2016 IEEE International Energy Conference (ENERGYCON)*, 1–6. 10.1109/ENERGYCON.2016.7514029

Wang, Y., Zhou, J., Chen, K., Wang, Y., & Liu, L. (2017). Water quality prediction method based on LSTM neural network. *2017 12th International Conference on Intelligent Systems and Knowledge Engineering (ISKE)*, 1–5.

Wu, K., Wu, J., Feng, L., Yang, B., Liang, R., Yang, S., & Zhao, R. (2021). An attention-based CNN-LSTM-BiLSTM model for short-term electric load forecasting in integrated energy system. *International Transactions on Electrical Energy Systems*, *31*(1), e12637. doi:10.1002/2050-7038.12637

Xiong, J., & Wu, P. (2008). An analysis of forecasting model of crude oil demand based on cointegration and vector error correction model (VEC). *2008 International Seminar on Business and Information Management, 1*, 485–488.

Xu, R., Xiong, Q., Yi, H., Wu, C., & Ye, J. (2019). Research on Water Quality Prediction Based on SARIMA-LSTM: A Case Study of Beilun Estuary. *2019 IEEE 21st International Conference on High Performance Computing and Communications; IEEE 17th International Conference on Smart City; IEEE 5th International Conference on Data Science and Systems (HPCC/SmartCity/DSS)*, 2183–2188. 10.1109/HPCC/SmartCity/DSS.2019.00302

Yan, J., Liu, J., Yu, Y., & Xu, H. (2021). Water Quality Prediction in the Luan River Based on 1-DRCNN and BiGRU Hybrid Neural Network Model. *Water (Basel)*, *13*(9), 1273. doi:10.3390/w13091273

Ysi. (2019). *Water Quality Parameters*. Academic Press.

Yu, T., Yang, S., Bai, Y., Gao, X., & Li, C. (2018). Inlet water quality forecasting of wastewater treatment based on kernel principal component analysis and an extreme learning machine. *Water (Basel)*, *10*(7), 873. doi:10.3390/w10070873

Zhang, H., & Jin, K. (2020). Research on water quality prediction method based on AE-LSTM. *2020 5th International Conference on Automation, Control and Robotics Engineering (CACRE)*, 602–606.

Zhang, L., Alharbe, N. R., Luo, G., Yao, Z., & Li, Y. (2018). A hybrid forecasting framework based on support vector regression with a modified genetic algorithm and a random forest for traffic flow prediction. *Tsinghua Science and Technology*, *23*(4), 479–492. doi:10.26599/TST.2018.9010045

Section 3

Chapter 13
Intelligent Agrometeorological Advisory System

Shirish Khedikar
Ministry of Earth Sciences, Pune, India

Ved Prakash Singh
https://orcid.org/0000-0002-2281-5687
Ministry of Earth Sciences, Bhopal, India & Indian Institute of Technology, Patna, India

Jimson Mathew
Indian Institute of Technology, Patna, India

Vaibhavi Bandi
Medi-Caps University, Indore, India

ABSTRACT

Agrometeorological inputs to the agriculture can play a significant role, mainly in the countries like India, where agriculture and allied sectors are the key pillars of its economy. To facilitate substantial growth in the sector and to improve the socio-economic status of farmers, it becomes inevitable to advise the agri-user community on how best they can avail the advantages of the meteorological parameters and to minimize the damage to agriculture, livestock, caused due to hazardous weather elements. Operationally useful forecast of meteorological variables that are important to current farming operations together with agriculture interpretations are essential to achieve the goal, and ultimately to deliver the customized agrometeorological advisory service. Moreover, inclusion of intelligent technological developments such as artificial intelligence and natural language processing can enrich these services at its best for the wellness of farmer's community, such as providing farm-specific advice in the farmer's local language.

DOI: 10.4018/978-1-6684-3981-4.ch013

INTRODUCTION

Successful farming is about forgetting the weather. However, it has been found that while farmers are eagerly awaiting the rains coming from the southwest monsoon, many are curious about it, everyone has the name of the meteorological department on their face; but once it starts raining, everyone ignores the weather. This should not be the case; the use of weather forecasting is not limited to sowing but is used in every activity in agriculture, from plowing the fields to reaching the market. Let's learn more about weather and climate-based farming.

CLIMATE ABNORMALITIES THAT CAUSE CROP DAMAGE

Primary climate abnormalities, which cause crop damage, are described below:

1. **Drought**

Figure 1. Cracked soil texture due to the rought

Drought has been identified as the leading cause of agricultural production losses. Drought is responsible for about 34% of crop and livestock production losses in Least Developed Countries (LDCs) and Low-to-Middle-Income Country (LMIC), costing the sector USD 37 billion in total. Drought mostly affects agriculture; it accounts for 82 percent of all drought effects, compared to 18 percent in all other industries.

2. **Heat wave and cold wave:** Extreme heat has an impact on a variety of agricultural areas. Heat waves have a significant influence on livestock. During heat waves, millions of birds have died. During heat waves, milk production and cattle reproduction can suffer. Extreme heat has a negative influence on pigs as well. We do know that high temperatures at the wrong time inhibits a crop yield. Wheat, rice, maize, potato, and soybean crop yields can all be significantly reduced by extreme high temperatures at key development stages. Primary effects of high temperature are:
    a. Membrane stability affected
    b. Imbalance between photosynthesis and respiration

Figure 2. (a) Heat stress during summer, and (b) Fog and cold waves during winters

c. Declining of photosynthesis than respiration
d. PSII affected more than PS1
e. RuBisCO, RuBisCO activase, PEP carboxylase
f. C3 plants affected more than C4 plants
g. Increased PUFA
h. Increased Hormones – ABA & Ethylene

Figure 3. (a) Affected plant leaves due to excessive sunlight, and (b) due to low sunlight

3. **Excessive sunlight or low sunlight**

Category wise, some of the direct effects of UV-B radiation on plants are:

a. Changes in leaf secondary metabolites
b. Alterations in leaf anatomy & morphology
c. Reductions in photosynthesis
d. Changes in biomass allocation and growth
e. Altered gene activity
f. Non-specific damage to DNA

g. Alterations in plant hormones or nucleic acids

Some of the indirect effects of UV-B radiation on plants are:

a. Affect Nutrient mobilization.
b. Susceptibility of plants to insects and pathogens.
c. Changes result in either a decrease or increase in susceptibility
d. Susceptibility to abiotic stress

Low light has been found to have a significant impact on agronomic qualities of plant and hinder physiological metabolic activities such as photosynthesis, antioxidant properties, and carbon and nitrogen fixation. It causes delayed growth, a reduction in leaf weight, and a reduction in the number of flower buds. In addition, this environmental condition lowers the sugar and starch content as well as changing the coloring and lengthening the maturity time.

Figure 4. Effect of shockwaves during thunderstorms on crops on field

4. **Thunderstorms**: The intense heat and shock waves generated by the current may inflict the most serious damage to plants; however, additional detrimental effects are likely to occur. Plant cells and tissues, including roots, stems, branches, and fruits, become blackened, necrotic, and burned, resulting in wilting and blackened, necrotic, charred tissues.
5. **Duststorms and sandstorms:** Dust has a number of detrimental effects on agriculture, including burying seedlings, causing plant tissue loss, lowering photosynthetic activity, and increasing soil erosion.
6. **Storms and hurricanes:** It may cause plant damage such as fractures, bends, or other sorts of injuries that result in productivity loss. Rainfall that is too heavy can drown plants and create soil erosion. Plants are frequently carried down by mudslides, ripping away their roots and killing the plant.
7. **High winds:** The force of the wind can shred or remove leaves from the plant. Intermittent or continuous rubbing can cause abrasion to dense plant canopies. Plant tissue can be damaged by soil particles that have been raised into suspension by the wind. The intensity of the wind can literally blow plants over, making harvesting crops harder.

Figure 5. Storms caused damages to Banana crop

Figure 6. Effect of high winds with rain on crops on fields

Figure 7. Damages to crops due to Forest fire

8. **Forest fire:** Forest fire not only affect the crop productivity (quality and quantity) in areas which are directly affected by forest fire but affecting to orchards, crops, livestock, and farm infrastructure in those locations which are far away from the fire's borders and that is mainly due to fire pollution.

Figure 8. Damages to crops due to snowfall

9. **Snowfall:** It is well-known certainty that melting snow enhances soil moisture that favoring plant growth but there are negative effects of snowfall these as follows:
    a. It shortens growing-season length.
    b. It is impeding seedling establishment and growth.
    c. Snowfall puts added weight and cause damage to plant structures.
    d. Snowfall in water-logged areas can make soil soggy and damages root systems.
    e. If freezing and thawing occurred repeatedly then it can cause plants to heave out of the soil.
    f. Snowfall can also damage susceptible flower buds.
    g. Ice layer can blocks exchange of gases between the soil and air trapping these toxic gases beneath the soil surface causing plants to suffocate and eventually die off.
    h. Lifting of crown or the growing point of the plant above soil surface making it more susceptible to winter kill.
10. **Frozen dew point**

Figure 9. Damages to crops due to frost

When temperature goes near or below freeing point (0 to -5 °C), it causes freezing of liquid water that results in frost, it is usually called as frozen dew. Temperature equivalent to a specific level of damage is usually called as "critical temperature" or "critical damage temperature" and frost injury in crop escalated as the surrounding temperature falls below this point. When ice form inside the plant tissue, which draws water out that make cell dehydrated that causes injury to the plant cells.

CROP MANAGEMENT BASED ON WEATHER

1. **Preparing the sowing schedule**
    a. Sowing is preferred when the rainfall is more than or equal to half of the potential evaporation.
    b. Management of agricultural labor
2. **Intercultural operations**
    a. Complementary conditions of the disease can be successfully predicted.
    b. Harvesting, weeding, mulching are done in dry climate.
    c. To identify maturity, *e.g.*, the temperature unit is calculated from the weather data that helps in determining the date of harvest.
    d. Manages frozen dewdrops
    e. Management of inter-cultivation work
3. **When to use fertilizers and apply spray / irrigation**
    a. Use fertilizer if low rainfall, adequate soil moisture (less than 90%) and wind speed less than 25 kmph.
    b. Irrigation scheduling, *e.g.,* irrigate if soil moisture is less than 50% along with determining when and how much water to be applied.
    c. Completion of irrigation or cutting of water (in cash)
    d. When spraying
4. **Accurate weather forecasts can minimize potential crop damage such as:**
    a. Crop selection
    b. Short-term adjustments in daily agricultural activities
    c. Improving the quality of agricultural products.
    d. Guidance on seasonal work
    e. Low cost involved
    f. Substantial loss of total crop yields can be avoided.

Utility of Weather Forecast In Agriculture

There are various ways to use the weather forecast such as:

- Sowing according to climate, *e.g.*, using the annual and seasonal weather forecast
- Climate change, *e.g.*, stopping before hail, reducing cold, creating artificial rain.
- Subtle (around tree) climate change, *e.g.*, Farming in a glass house.
- Changes in temperature, humidity and time of sunlight.
- Changes in crop response to climate, *e.g.*, artificial sunlight, cooling of seeds before sowing.

Usefulness of seasonal weather forecasting

- Manages water resources,
- Timely planning of pests and diseases,
- Selection of varieties according to climate,
- Fixed area for crop or sowing,
- Determining crop yield,
- Crop fixing.

Usefulness of long-term weather forecasting (valid for 30 days)

- Fodder management,
- To determine irrigation frequency,
- Short term storage of harvested crop,
- Stop chemical spraying,
- For soil moisture management.

Usefulness of medium-term weather forecast (valid for 3 to 5 days)

- How deep to sow the seeds,
- Whether to sow,
- Irrigation management based on incoming rainfall,
- Effective use of sprays,
- Protection of crop from frozen dew point,
- Management of agricultural labor,
- Management of farm implements,
- Management of fodder for cattle,
- Whether to harvest.

Usefulness of short-term weather forecasting (valid for 3 hours to 48 hours)

- To prepare irrigation schedule,
- Field work hours,
- Effective use of chemicals,
- Reducing the effects of pests and diseases,
- Soil efficiency,
- Protection from frozen dewdrops.

OBJECTIVES OF OPERATIONAL AGRICULTURAL METEOROLOGICAL SERVICE

1. To advise the user community on how best they can avail the advantages of the meteorological parameters.
2. To minimize the damage to agriculture, livestock, caused due to hazardous weather elements.

Operationally useful forecast of meteorological variables that are important to current farming operations, together with agriculture interpretations are essential to achieve the goal, and ultimately to formulate the agrometeorological advisory. Primary steps involved in Agrometeorological Advisory Service (AAS) are mentioned below -

- Realized weather
- Weather forecast
- Current crop condition
- Stages of crops
- Weather based pest and disease forewarning
- Fertilizer application
- Irrigation scheduling
- Farm operations

Agronomic Data Required / Used During Preparation of Agri-Met Advisory

- Pattern of cropping in a certain districts or blocks.
- The dominant verities in a certain district or block.
- The area of a district or block that is irrigated and rainfed.
- Sowing area in various crop-over districts / blocks.
- Normal date of sowing of different crops and present season sowing dates of particular district or block.
- Stage and state of major crops over district / blocks.
- Soil moisture data of particular district or block.
- Weather & pest relationship for particular crop

Weather Data Required / Used During Preparation of Agri-Met Advisory

- Historical rainfall data for at least 30 years of particular district or block, to study climatology of particular district or block.
- Information of different products on IMD or related websites, *e.g.*, NDVI (Normalized Vegetation Difference Index) and SPI (Standard precipitation Index) maps, Satellite images *etc.*
- Past weather over the crop season.
- Weather forecast and warnings in different time scale (Short, Medium, Extended and Long-range weather forecast of particular district or block).

COMPONENTS OF AGROMETEOROLOCIAL ADVISORY SERVICES (AAS)

Weather Information

Weather summary for the past three or four days including the highest maximum temperature (HMT) during summer, the lowest minimum temperature (LMT) during winter, both HMT and LMT during transitional months (*i.e.* October, March), chief amount of rainfall, heat wave condition / cold wave condition *etc.* Forecast valid for five days with warning, if any.

Crop Information

Major standing crops, their growth stages (from sowing to harvest) and state (*i.e.* crop condition in qualitative terms *i.e.* mostly satisfactory, satisfactory/ normal, partially satisfactory or unsatisfactory. In addition, pests and diseases infestation, information on water stagnation or drought like situation are also incorporated).

Weather Based Crop Advisories

Agricultural advisories during the period including control measures against pests and diseases and agri-meteorological advisories depending on forecast and ongoing agricultural operations.

Role of Agri-Meteorologist In Preparation Of AAS Bulletin

In finalizing weather-based crop advisories based on crop information (depending on stage and state of crop), agri-meteorologist has a very important role. Interpretations of forecast depending upon ongoing agricultural operations and their dissemination are also important activities of the agri-meteorologist.

Methodology of Preparation of Advisory Bulletin

- Consider the stage and state of major crop in district / block including major biotic and abiotic stress (infestation of pest and pest disease deficiency of access rainfall, weather/nutrient stress).
- Whether the crop is irrigated or non-irrigated and source of irrigation.
- Past weather information in the area including its adverse impacts on the different crop.
- Probability of infestation of pest/disease based on crop-weather crop-pest relationship in the area.
- Weather forecast and warning.
- Considering all above information, the panel of experts (including Agri-meteorologist, agronomist, plant protection specialist and extension specialist) should prepare crop specific advisory of the respective district / block. Advisory committee should be formulated consisting of Subject Matter Specialists from different disciplines.

Format of Agrometeorological Advisory Bulletin

Agromet Advisory Bulletin will comprise of following info-points -

- Past Weather: Observed weather information of last five days in tabular format
- Quantitative medium range weather forecast for next five days based on forecast.
- Values weather summary for next 5 days also be kept in the bulletin.
- Based on weather forecast, abiotic weather-related stress for crops livestock, and poultry of the district / block need to be formulated. Crop specific advisory should take care of the phenological

phases of the crop and their relation with the weather variables based on which weather forecast based agromet advisories are formulated.

- Category rainfall forecast for the outlook of succeeding week (*i.e.* 6th to 12th days) to be included in bulletin. Categories are Above normal (≥ 20%), Normal (-19% to +19%) and below normal (≤ -20%) applicable at Met Sub-division scale.

CLASSIFICATION OF AGRI-STRESSES

Abiotic Stresses

- Floods
- Droughts
- Extreme temperatures
- Cyclones
- Frost
- High wind
- Dust storms

Biotic Stresses

- Pests infestation
- Plant diseases
- Animals
- Birds
- Nematodes
- Fungal infection
- Weeds
- Parasites

AGRI-MET DATA AND PRODUCTS USED FOR PREPARATION OF ADVISORIES

Satellite Products

- NDVI Composite (Country and State wise)
- Temperature Condition Index (TCI)
- Vegetation Condition Index (VCI)
- Insolation
- Reference Evapotranspiration

Weather Parameters

- Daily Spatial Distribution
- Weekly Spatial Distribution

- Fortnightly Spatial Distribution
- Monthly Spatial Distribution
- Seasonal Spatial Distribution

Soil Temperature

- Daily Soil Temperature
- Weekly Soil Temperature
- Soil Moisture and its forecast
- Standard Precipitation Index
- Weekly rainfall distribution
- District wise rainfall maps / seasonal district wise rainfall maps

Types of Agrometeorological Advisories Bulletin

- National AAS Bulletin
- National AAS Bulletin based on ERFS State Composite AAS Bulletin
- District AAS Bulletin
- Block Level AAS Bulletin

Broad Spectrum of Agro-Met Advisories

- Kharif crop sowing / transplanting based on monsoon start; while rabi crop sowing depending on residual soil moisture.
- Fertilizer application based on wind conditions; while delay in fertilizer application based on rain intensity.
- Prediction of pest and disease occurrence dependent on weather and proactive actions to eradicate pests and diseases at the appropriate time.
- Weeding and thinning at regular intervals for improved crop growth and development.
- Irrigation during a vital stage of the crop's development and quantity and timing of irrigation based on meteorological thresholds.
- Recommendations for crop harvesting in a timely manner.

ADVANCEMENTS IN THE AGRO-MET ADVISORY SECTOR: FUTURE DIRECTIONS

Advisories Do Not Always Usefully Combine Weather and Agriculture Data

Agrometeorological Advisory should be based on agrimet products, mere weather and agriculture data will not suffice, to solve the purpose, machine learning can help. ML models can give appropriate suggestion regarding which agrimet products and tools should be used based and condition of crop. *E.g.*, Forecast of crop yield which is accurate and timely can not only help the government of any country in taking various strategical decisions planning import / export, formulating future policies like cost /

selling price of crops and timely gauging the future threats. It can also be a great help to a farmer whose livelihood is totally based on the expected yield of the crop (Lipper *et al.,* 2014).

Micro-Scale Advisories

Advisories are prepared generally for district scale; hence, many times advices given to the farmers does not matches with farmers requirements. To make advisories more relevant, Machine Learning and Artificial Intelligence can add desired targeted value.

Trained Agri-Meteorologists

The trained agri-meteorologist can use modern tools during preparation of advisories. They should be able to use forecasting and do predictive analytics to reduce adverse effect of weather on plant.

Regular and Reliable Advisories

Currently, agrometeorological advisories are prepared by Technical officers / Subject matter specialties manually, their work is time bound and hence, issuing advisories based on weather condition is not possible every time. However, application of Machine vision and Artificial intelligence can help to monitor the state of plants and weather which help to issue regular and reliable advisories automatically.

Decision Support Tools for Translating Weather Forecast into Advisories

Big data and Artificial intelligence platforms can accommodate and process significantly more data with faster rate helps during decision-making. *E.g.*, Image processing helps to detect fungal or pest activities in their early stage (Sharvane *et al.,* 2021).

Application of Crop Weather Calendar

Real-time monitoring of crop growth with respect of weather helps to enhance the crop productivity and reduces the adverse effect of harmful weather but for this purpose, it is necessary to know whether crop is following weather calendar. Dynamic machine learning models can be developed to assess the status of crop during particular stage. *E.g.,* machine learning techniques can help in crop selection to maximize crop yield based on weather and soil parameters (Jain & Ramesh, 2020).

Application of Dynamic Crop Simulation Models

Dynamic crop simulation models can predict effect of various biotic and abiotic parameters on crop growth and development. This data is huge and for the application of this information, intelligent big data analytics tools can be very useful.

Diagnose Weather Related Stress

Weather sensitivity of crops varies from plant to plant, while microclimate plays a major role. Management practices can reduce the adverse effect and help in encasing benefits of favorable weather, though, for this purpose proper infusion of application intelligence is required.

Crop Yield Forecasting Techniques

With the help of AI enabled geo-spatial remote sensing, now, it is possible to combine ground sensor data with crop stage / data, which help to get real-time information which was never had access before. *E.g.*, Machine learning can be used for yield prediction and disease spots (Murlidharan *et al.,* 2021; and Singh *et al.*, 2019).

Microclimatic Study

Effect of various weather parameter near the plant canopy (microclimate) is not fully explored yet; Machine learning can help to study microclimate in deep and better way.

Water Use Management

Machine learning and Artificial intelligence can able to do automated irrigation and this can enhance the productivity and reduce the workload on the farmers.

Use of Remote Sensing techniques

Image recognition and perception can be done by using intelligent machine vision tools, which can help to take more appropriate and timely decisions. *E.g.*, Remote sensing can provide massive amounts of data about crop condition and health via plant and fruit characteristics (Zheng *et al.,* 2021)

Feedback from End Users

Natural Language Processing with Deep learning techniques can help to enhance advisories by implementing suggestions received from the stakeholders and to prepare tailor made advisories that are more suitable.

CONCLUSION

Agrometeorological advisors need the knowledge about farm-level inputs and weather inputs to make famers capable of taking feasible decisions on various activities and technology usage to gain maximum possible crop outcomes from their fields and minimizing the losses. To achieve these objectives, meteorologist's knowledge needs to be merged with smart farming data for hybrid knowledge systems with the help of various AI and ML tools available or with customized prediction models. However, the large heterogeneous datasets are being captured through remote sensing and in-situ observations every day, but these data has to be analyzed *w.r.t.* particular crop and field conditions in an automated way, so that it can be utilized as decisive input to agrometeorological bulletins. Further, scientific versions of these bulletins can be transformed in easily interpretable form for common stakeholders (farmers and agri-business workers) with the help of deep learning based natural language processing models. Moreover, use of AI/ ML based decision support systems may help farming community to suggest various critical actions dependent upon irrigation and drought conditions, weather and yield forecasts, micro-climate stress, disease and pests, *etc.* Therefore, agri-meteorologists are encouraged to get involved in aligned R&D activities and ultimately, to play an important role in farming capacity enhancement.

REFERENCES

Jain, S., & Ramesh, D. (2020). Machine learning convergence for weather based crop selection. *IEEE International Students Conference on Electrical, Electronics and Computer Science (SCEECS-2020)*, 1-6, 10.1109/SCEECS48394.2020.75

Lipper, L., Thornton, P., Campbell, B. M., Baedeker, T., Braimoh, A., Bwalya, M., Caron, P., Cattaneo, A., Garrity, D., Henry, K., Hottle, R., Jackson, L., Jarvis, A., Kossam, F., Mann, W., McCarthy, N., Meybeck, A., Neufeldt, H., Remington, T., ... Torquebiau, E. F. (2014). Climate-smart agriculture for food security. *Nature Climate Change*, *4*(12), 1068–1072. doi:10.1038/nclimate2437

Sharvane, M., Shukla, V. K., & Chaubey, A. (2021). Application of Machine Learning in Precision Agriculture using IoT. *2nd International Conference on Intelligent Engineering and Management (ICIEM)*, 34-39. 10.1109/ICIEM51511.2021.9445312

Singh, V. P., Khedikar, S., & Verma, I. J. (2021). Improved yield estimation technique for rice and wheat in Uttar Pradesh, Madhya Pradesh and Maharashtra States in India. *Mausam (New Delhi)*, *70*(3), 541–550. doi:10.54302/mausam.v70i3.257

Zheng C., Abd-Elrahman, A., & Whitaker, V. (2021). Remote Sensing and Machine Learning in Crop Phenotyping and Management, with an Emphasis on Applications in Strawberry Farming. *Remote Sensing, 13*(3), 531, 1-28. doi:10.3390/rs13030531

Chapter 14
A Robust Method for Classification and Localization of Satellite Cyclonic Images Over the Bay of Bengal and the Arabian Sea Using Deep Learning

Manikyala Rao Tankala
https://orcid.org/0000-0002-5126-0588
IMD, Ministry of Earth Sciences, India

Samuel Stella
IMD, Ministry of Earth Sciences, India

Prayek Sandepogu
IMD, Ministry of Earth Sciences, India

Kondaveeti Nanda Gopal
IMD, Ministry of Earth Sciences, India

Ramesh Babu Mamillapalli
IMD, Ministry of Earth Sciences, India

Devarakonda Rambabu
IMD, Ministry of Earth Sciences, India

ABSTRACT

According to recent findings, deep learning algorithm outperforms in many tasks like image classification, image segmentation, image recognition, etc. in the field of computer vision. With the help of deep learning, classification tasks on remote sensing image data can attain better performance compared to traditional approaches. This chapter primarily demonstrates how residual neural networks are used to classify satellite images of cyclones in the Bay of Bengal (BoB) and the Arabian Sea (Arab Sea). The authors further discovered the cyclones' locations and investigated using satellite images in the infrared and visible bands of electromagnetic spectrum. From the evaluation metrics, the neural network looks to be capable of correctly identifying the cyclonic storm utilising Gradient Class Activation Mapping (Grad-CAM). Satellite images of both cyclone storm and non-cyclone storm are analysed for cyclonic storm recognition and classification.

DOI: 10.4018/978-1-6684-3981-4.ch014

INTRODUCTION

In the realm of weather forecasting, proper classification of cyclonic storm and non-cyclonic storm images has been imperative because of their status as one of the world's most damaging natural catastrophes. Seasonal tropical cyclones present difficulty in getting consistent observations as these storms spend most of their lives out at sea, far from any place on land. As a result, techniques like the Dvorak method, which measures cyclone severity exclusively by using visible and infrared satellite imagery, are very popular. There is a pressing need for improvements in satellite-based cyclone classifications to serve the needs of meteorologists and coastal community's alike (Melgani and Bruzzone, 2004; pp. 1778-1790). This improvement is of central importance in the field of atmospheric sciences, and the future development of these methods should be a top priority for research. The extraction of important aspects of cloud distributions relating to tropical cyclone strength is the challenge of the Dvorak approach. Experts rely on climatology, numerical weather prediction (NWP), satellite images, radar products, and their own experience for forecasting cyclones.

Deep Learning algorithm design has recently seen improvements due to the vast development in technologies and advancement in hardware resources for implementation (Bengio, Courville, and Vincent, 2013, p1799). Image recognition has been able to solve the problem of classifying ambiguous data, and hence it is important to remember that the application of deep learning algorithms may solve typical tasks like storm image classification and prediction (Chen et al., 2019). Also, deep learning may be used for the extraction of complex features from images, while the traditional approaches fail to do so in feature engineering (Boyo, Chen, and Lin 2018). Thus, deep learning may be useful at the user's choice, leading to specific task achievement. The preliminary objectives of this chapter are to reveal the facts about the classification of cyclone images and non-cyclone images along with the localisation of a cyclonic storm area with an eye or without an eye pattern using gradient mappings (Zhang et al., 2021, pp. 2070–2086). Also, this chapter discusses the different models' performances, which are trained on a dataset with images of recent cyclones across the Bay of Bengal and Arabian Sea. Furthermore, this chapter discusses evaluation metrics used for analysis in cyclone and non-cyclone image classification, detection of cyclones or non-cyclones in images, localisation using Grad-CAM in terms of heat map images, graphical plots, loss curves, comparison tabular forms and figures.

BACKGROUND

In the northern (southern) hemisphere, a cyclone is defined as a low-pressure system with winds rotating in an anticlockwise (clockwise) direction in the northern (southern) hemisphere with a minimum sustained wind speed of 34 knots (62kmph). Tropical cyclones are low-pressure systems that do not form a front and are centred over the tropical or sub-tropical oceans; they are also known as tropical depressions. There is clear wind circulation, or rotation, in these low-pressure systems, and there is also convection (i.e., thunderstorm activity). India is significantly reliant on agricultural production, with the majority of its food supply being dependent on rainfall. On an average, India experiences three to five tropical cyclones every year, and the India Meteorological Department (IMD) offers timely alerts and warnings to all government departments, the aviation sector, railways, NDRF, non-government departments, agricultural departments, etc., to ensure that they are fully informed of the impact of cyclones and measures to be taken accordingly.

The following sections discuss key terms and features of tropical cyclones.

Eye Wall/ Wall Cloud

1. An eye wall is defined by a ring of convective clouds around the eye of the cyclone.
2. Expect intense rain bands to spiral inwards, which are considered the most dangerous part of the Tropical Cyclone (TC).
3. A wall cloud's width is approximately 20–100 km of a tropical cyclone.
4. The region is experiencing maximum pressure gradients, maximum temperature gradients, and heaviest precipitation, with gale winds and storm surges.

Eye

The central part of the tropical cyclone is known as the Eye, with a diameter of 10 km to 100 km.

1. An eye of a cyclone is formed by air sinking from upper levels to lower levels and is characterised by calm winds, with a clear sky and the lowest pressure at the Centre.
2. When the eye passes or crosses over an area, there is a sudden cessation of precipitation.
3. The eye is circular or elliptical in shape, with a regular or diffused pupil, or a single or double pupil.
4. Also, the eye is warmer than the surrounding area of the cyclone.

The below Figure 1 indicates the anatomy of a tropical cyclone which comprises the eye, the eye-wall, Outflow cloud shield, clockwise winds, descending air, spiral winds.

The above Figure 2 depicts multiple tracks of tropical cyclones (TC) that were formed in the Bay of Bengal and Arabian Sea in the year 2020 period. They were also classified as Depression, Deep Depression, Cyclonic Storm, Severe Cyclonic Storm, Very Severe Cyclonic Storm, Extremely Severe Cyclonic Storm, and Super Cyclonic Storm based on wind velocities.

LITERATURE REVIEW

Pradhan et al. created a convolutional neural network (CNN) architecture to estimate hurricane strength categories (Pradhan et al., 2018; pp. 692-702). In a specific project, they did this in one task. They used the typical approach of machine learning and employed k-fold validation, which randomly partitioned the dataset, resulting in a higher than usual accuracy of 82%. To avoid data leakage, Cyclone photos and other time-series data should be divided in such a way that the training data has earlier timestamps than the test data. Random split validation works best with time-series data since it allows the model to incorporate future patterns not present in the training phase (Simonyan and Zisserman, 2014). When we used a sequential split instead of a random split in our experiment, a neural network learned that the accuracy of a typhoon intensity classification model built to be like Pradhan et al. drops to 55.7 percent. Shakya et al., proposed multiple CNN architectures which attained good accuracy (97%), but no localisation was achieved on cyclonic images (Shakya, Kumar, and Goswami, 2020, pp. 827-839). Higa et al. proposed a model using VGG-16 but attained lower accuracy. Pang et al.'s proposed model attained higher accuracy using DCGAN & YOLO v3, but the robustness of the model was not discussed.

Figure 1. Anatomy of tropical cyclone (image courtesy: IMD source)

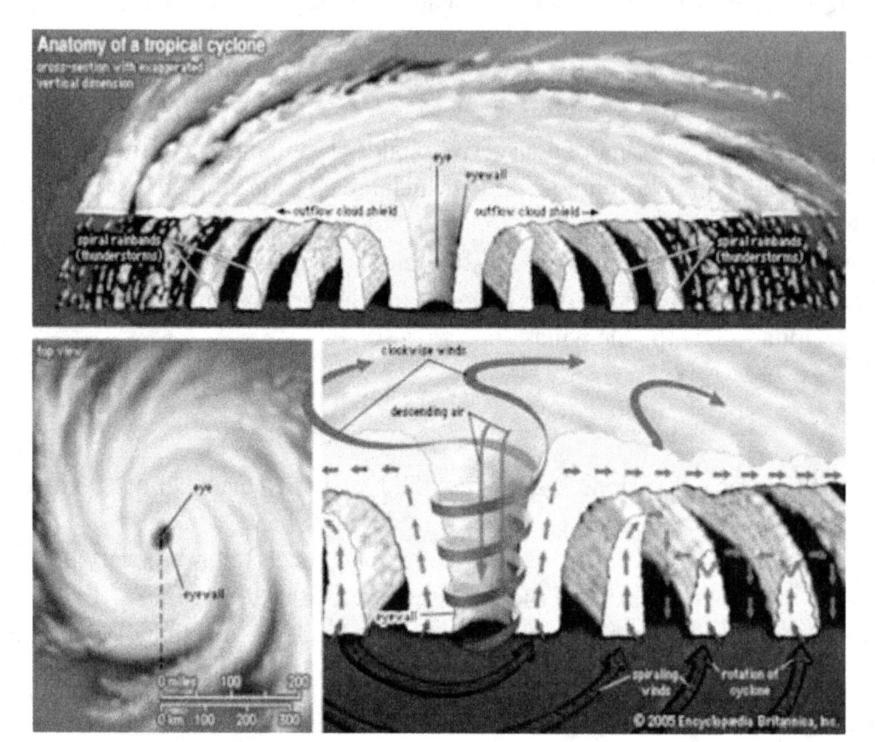

Figure 2. Tracks of cyclonic disturbances over north Indian Ocean and land region during 2020 year (image courtesy: IMD source)

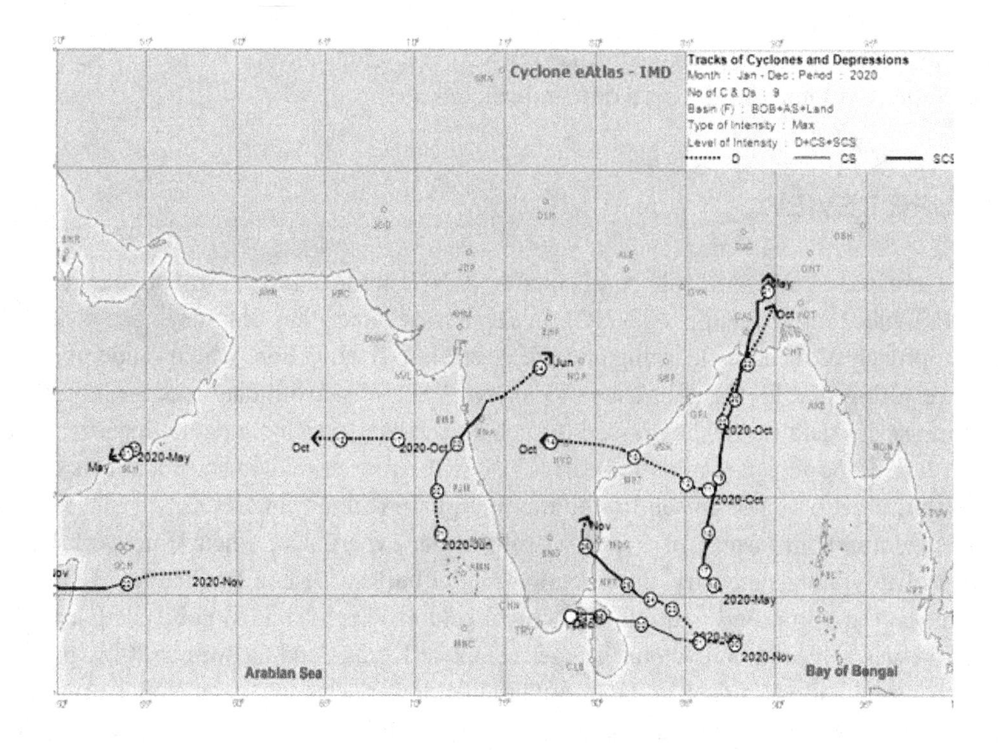

Table 1. Comparing the proposed algorithm to state-of-the-art methods

S.No	Authors	Deep Learning Methods employed	Performance metrics	Dataset	localisation
1	Shakya, Kumar and Goswami, 2020	Sequencial CNN, Xception,NasNetMobile, Mobilenet	Accuracy 97%	9,900 images	No
2	Hiag et al., 2021	VGG-16 model	Accuracy 77%	5,816 images	yes
3	Pang et al., 2021	DCGAN & YOLO v3 models	Accuracy 98.59%	2400 images	yes
4	Rajesh et al., 2019	DLR-FH model	Accuracy = 98.59%	10 to 100 images	No
5	**Proposed algorithm**	**ResNet model**	**Accuracy = 100%**	**3617 images**	**yes**

Also, Rajesh et. al. proposed a model using Logistic Regression with Fuzzy logic, but no localisation was achieved (Rajesh et al., 2019, p.688–696). Hence, in this chapter, a new model is being proposed for the robustness and localisation of cyclones along with classification accuracy metric. *Table 1* below indicates a comparison of different methods for detection and classification of cyclonic storm images.

MAIN FOCUS OF THE CHAPTER

The focus of this part is on the classification of cyclonic storm images, the detection of cyclone cloud distribution in images (Zhuge et al., 2015, pp. 5661–5676), and the localization of cyclonic storm distribution with the eye as the core. The following sections briefly discuss the objectives of the proposed method for image classification using a deep learning algorithm.

Classification and Prediction

The proposed method primarily focuses on the classification of satellite cyclonic storm images for the presence of cyclones or non-cyclonic cloud distributions. The proposed method performs image classification from steps 1 to 8 and is shown in Figure 3. The residual neural network model is trained using the Adam optimizer. The validation accuracy for classification is 100% and is achieved on all three models of ResNet with different times of training. Models from the Residual networks (ResNet) family, such as ResNet-50, ResNet-101, and ResNet-151, were applied to the satellite image dataset, and superior performance was obtained when compared to other state-of-the-art methodologies. The identical train-test split was applied to all experimental models. Precision, recall matrix, and categorical cross-entropy loss were calculated for 05 epochs, and their performance will be compared to find the best model for evaluation. While a model's correctness is questioned, when the gain is high and the loss is low, the model is deemed to be accurate on the dataset, showing that it has been well-trained.

Block diagram of proposed method

Figure 3. Proposed algorithm block diagram

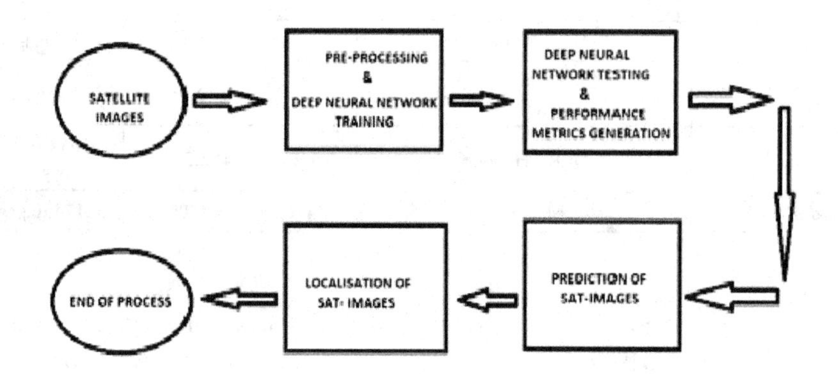

Steps involved in proposed method

The proposed algorithm in the block diagram from Figure 3, shown as steps from 1–10, explains the entire implementation clearly:

1. Reading a satellite image dataset containing infrared and visible spectrum images of cyclones and non-cyclones (Lin et al., 2005),
2. In the dataset, pre-processing techniques (image resizing, data augmentation, and segregation into two folders of cyclonic and non-cyclonic images) were used.
3. Train and Test phase on the given dataset.
4. Initialize the optimizer parameters and download of "image net" weights for the residual neural network algorithm.
5. Generating the model summary for each neural network.
6. Using "categorical cross-entropy" to calculate loss in the training phase, minimising the error function, plotting the training phase against total training epochs, and plotting loss curves vs. epochs.
7. Evaluation of metrics in terms of training accuracy, testing accuracy, and generation of a precision and recall matrix for each dataset generation of logs for the training and testing phases for each epoch.
8. Detection of cyclonic storm clouds using a model obtained after training on a satellite image dataset.
9. Localisation using class activation mapping for detection of cyclone areas in images.
10. Comparison of multiple models, i.e., ResNet-50, ResNet-101, and ResNet-151, against performance metrics and time of evaluation accordingly. *Table 2* indicates the pseudo code for the proposed algorithm.

Step 1: Reading the Satellite Image Dataset "Pi, qi" where i = 1 to n
Step 2: Resize all images to 224*224 dimensions and download the net weights of the images.
Step 2: Train the ResNet model with different tuning parameters for each epoch.

Step 3: Change the batch size to bs and epochs to 05.
Step 4: Choose the Adam optimizer (learning rate) and the image resize dimension.

- Set nb = n/bs as the number for mini-batch size.

Step 5: For each iteration, multiply by 1 to n.
batch 1 to batch n_b

- Choose a batch from the training dataset.
- Train the model for both cyclonic and non-cyclonic storm images.

By minimising the cross-entropy loss,

- Back-propagate the loss.
- Update the model parameters.

Step 6: Sort images into Cyclone and Non-Cyclone categories.
Step 7: Detection of cyclonic storms in satellite image dataset test image samples
Step 8: Using Grad-CAM, locate the Cyclone eye with or without an eye-pattern.

Table 2 above indicates the pseudo-code of algorithm implementation in steps for achieving the desired task of classification and localisation of cyclonic storm eyes using GRAD-CAM technology.

CLASSIFICATION USING RESNET-50 MODEL

Residual networks are deep convolutional networks that skip the blocks of convolutional layers (Krizhevsky et al., 2017, p. 84–90). They follow the basic design rules of rectangular blocks. The down-sampling is performed by layers that have batch normalization. The output and input are both of the same dimensions. The ResNet-50 in *Figure 4* is a distributed network with 50 weighted layers. The network's end-point is a 1,000 fully-connected (fc) layer. In this study, the ResNet-50 base model performs cyclone detection tasks on a satellite imagery dataset. Despite the significant differences between cyclonic storm and non-cyclonic storm images, this model can still be used to perform effective cyclonic storm image recognition tasks.

While designing, the first 49 layers of ResNet-50 are assigned to the classification task, and then trained fully-connected soft-max with the learned features extraction layers. The first 49 layers of ResNet50 were transferred using transfer learning techniques. These layers are usually referred to as "learned extraction layers of features." On the pre-processed dataset, a basic KERAS model was trained. The training was carried out using the NVIDIA GPU platform. The model was trained and validated on INSAT 3-D Satellite IR and visible spectrum images that were randomly jumbled and spit out. The table shows the model that was utilised. It was created using categorical cross-entropy. And the Adam optimizer was used for optimization. For identifying cyclonic storm and non-cyclonic storm images, the RESNET-50 Model achieves a peak validation accuracy of 100 per cent.

Figure 4. Architecture of ResNet50 Model with convolution and identity blocks

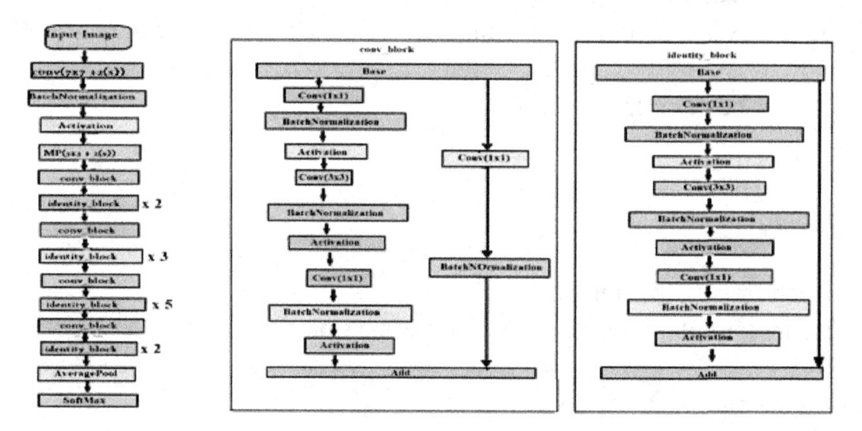

Figure 5. Residual Network identity block

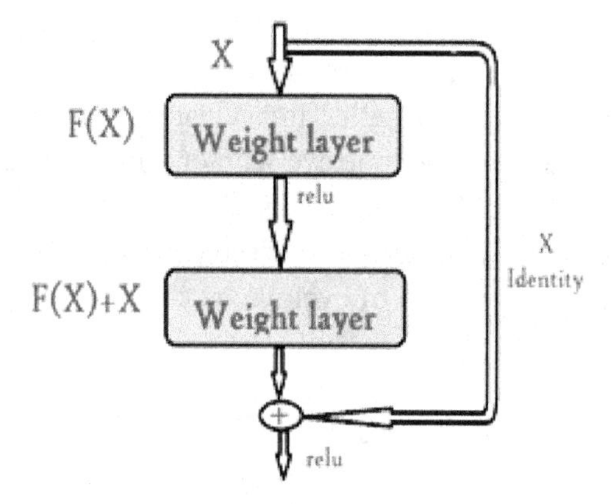

H(X) is a desired mapping function, and stacked non-linear mapping as F(X) = H(X)-x results in H(X) as F(X)+x, which is implemented by feed-forward neural networks with short-cut connections. Short-cut connections perform identity mappings as shown in *Figure (5)*. Short-cut connections are those which skip layers and are stacked to output layers during the process of output evaluation (LeCun, Bengio, and Hinton, 2015, 436-444).

Residual learning is adopted by defining a building block given by equation (1).

$$y = F(x_i, \{W_i\}) + \mathbf{x} \tag{1}$$

x and y are input and output with $F(x_i, Wi)$ as the residual learning unit. Figure (5) shows two layers with $F = W2. (W_1 x)$ *with* indicating that the RELU function and bias are omitted during computation. By using a shortcut connection and element-wise addition, you can get a function like (F(xi) +x). Performing linear projection W_s by short-cut connection to obtain a match in F and x is given by equation (2).

$$y = F(x_i, \{W_i\}) + W_s x \tag{2}$$

Figure 6. Deeper residual network function for IMAGENET

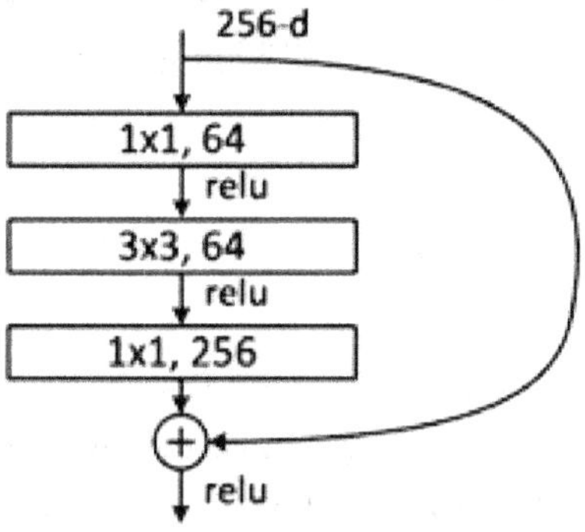

Residual Networks (ResNets) are convolutional networks with deep layers and with the core principle of exploiting shortcut connections to skip blocks of convolutional layers, as shown in Figure 6 (Bau et al., 2017, pp. 10; He et al., 2016, pp. 630–645). The "bottleneck" blocks have got two characteristic rules: (i) the layers have the same count of filters for the same output feature map size; and (ii) the number of filters is doubled if the feature map size is halved. And the layers of convolution have a stride of 2 to obtain direct down sampling, and batch normalisation is performed soon after each convolution and before RELU activation (Higa et al., 2021). The identity shortcut is used when both the input and output dimensions are the same. The projection shortcut is widely used to match dimensions through 1 x 1 convolutions as the dimensions grow larger. When performing shortcuts over feature maps of different sizes, a stride of 2 is used in both circumstances. Soft-max activation results in a neural network with 1,000 fully-connected (fc) layers.

Prediction Phase

A cyclone is a system of winds rotating around a low-pressure centre called an "eye" with wind speeds exceeding a particular threshold. Then A huge amount of data is required to create a robust model (Liu et al., 2016; Pang et al., 2021). Neural networks depend on augmentation to combat this. By artificially adding noise to the model, robustness is verified. Different augmentation techniques were used, such as scaling, translating, cropping, rotating, and adding Gaussian noise (Zhang, Liu, and Hang, 2020, pp. 586–597) (Szegedy et al., 2015, pp. 1–9). After the augmentation process, we were able to develop images for training and images for testing. The images from the satellite meteorological division in New Delhi were considered for the training and testing model. These satellite photos were taken during a half-hour period and in various spectrums (visible, infrared). The trained model is tested for prediction in images of cyclonic storms or not. In the end, an accuracy of 100 per cent was obtained.

Classification using RESNET-101/152 Model

ResNet of 101 and 152 layers are created by stacking several 3-layer blocks together. Surprisingly, the 152-layer ResNet (11.3 billion FLOPs) has less complexity than the VGG-16/19 nets (15.3/19.6 billion FLOPs) despite having a substantially higher depth. By a significant margin, the ResNet with 50/101/152 layers are more accurate than those with 34 layers. Further neural networks don't see any degradation, and as a result, the network is able to enjoy high accuracy and it increases from higher depth. The benefits of depth are obvious for all evaluation indicators (Li, Huang, and Gong 2019, Pages 1082–1086) (Wimmers, Velden, and Cussoth, 2019, p.2261–2282).

LOCALIZATION USING GRAD-CAM

Gradient Class Activation Mapping (Grad-CAM) uses the gradient information from the last convolutional layer to highlight significant image regions using a heat map. The Fully Connected layer translates feature maps from an input image into probability scores for each class. A gradient is a measurement of the likelihood score change from a given image portion alteration (Selvaraju et al., 2020, 336–359). The gradient is large for image components that have a major impact on class classification. For each cyclonic storm image localization, we used the Grad-CAM technique on the last convolution layer of each model and got a heat map of the image. Grad-CAM is a class-discriminative localization strategy for any CNN-based network that generates visual explanations without requiring architectural changes or retraining (Bazzini et al., 2016). Grad-CAM is evaluated for localisation and model faithfulness, where it outperforms baselines.

EVALUATION DATASET

The evaluation dataset consists of satellite infrared and visible images of cyclonic storm and non-cyclonic storm that occurred during the 2019–2021 period. All cyclonic storm and non-cyclone images were taken from the satellite image database of the India Meteorological Department and were pre-processed during the preparation of the dataset. Pre-processing techniques involve data augmentation with noise addition, rotation, and shift for each image as a part of image data enhancement. The number of non-cyclone images was 1812 and the number of cyclone images was 1805, which was considered. Non-cyclonic storm and cyclonic storm images of Yaas, Nisarga, Bureive, Amphan, Nivar, Tauktae, BOB were taken from the url address of the IMD website: https://www.satellite.imd.gov.in/. Cyclone Yaas, Cyclone Nisarga, Cyclone Bureive, Cyclone Amphan, and Cyclone Tauktae were thoroughly analysed in tests using INSAT 3-D satellite images to evaluate the proposed algorithm and whether it is competitive or superior to existing approaches. Furthermore, the outcomes of residual neural networks were compared to standard approaches for classification accuracy and cyclonic storm image localization.

Experimentation and Results

Experimental evaluation is carried out using the Python language in the Jupiter Notebook IDE with the usage of KERAS and TENSORFLOW neural network libraries. Classification results show better performance in terms of accuracy and false positives. Localisation of cyclone cloud images is obtained by Gradient Class Activation Mapping (Grad-CAM) and indicates well the detection of Cyclone-Eye or Cyclone without an Eye region. Using Keras with Tensorflow as a backend, the model of DL-based storm classification was applied to the dataset of storm signatures. Furthermore, training and testing of the images using the chosen neural network is done on the NVIDIA GTX 1050 GPU for effectiveness.

A tensor board is utilised for visualisation of the curve that displays how well the system can predict a value before and after it is delivered. This visualisation is applied for both training and testing, at the time when neural networks are utilising ResNet-50, ResNet-101, and ResNet-151. In addition, visualisations of the training and validation accuracy on the Tensor board are provided to display the graph that depicts training and validation accuracy as well as epoch loss. Promising results were acquired in the experiments, and it's beneficial to use these discoveries in the future. *Table 3 and Figure 7* indicate detection and localisation images obtained by using the models trained and tested for cyclonic storm and non-cyclonic storm images

Table 2. Detection of cyclonic storm /non-cyclonic storm images using deep learning algorithm

SNO	DEEP LEARNING RESIDUAL NEURAL NETWORK	Prediction by model	Prediction by model
1	ResNet50 -Model	 cyclone IR- image/TAUKTAE Cyclone/15.05.2021/0900 UTC	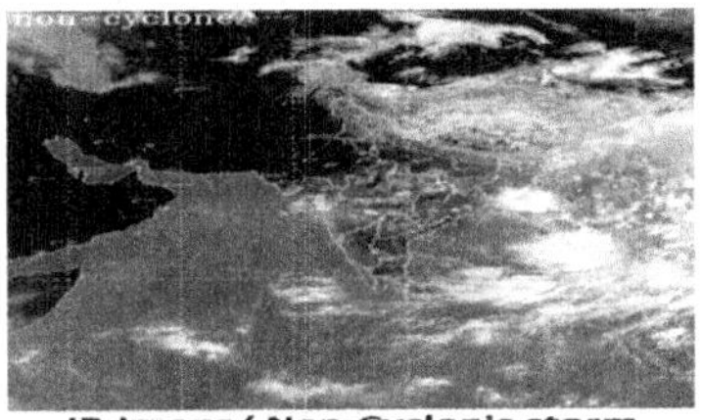 non-cyclone IR-Image/ Non-Cyclonic storm image/INSAT 3-D
2	ResNet101-Model	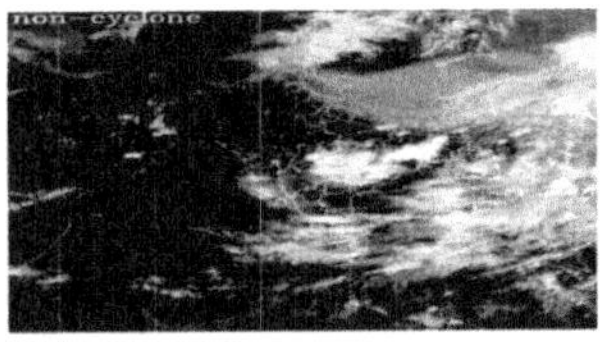 non-cyclone IR-Image/ Non-Cyclonic storm image/INSAT 3-D	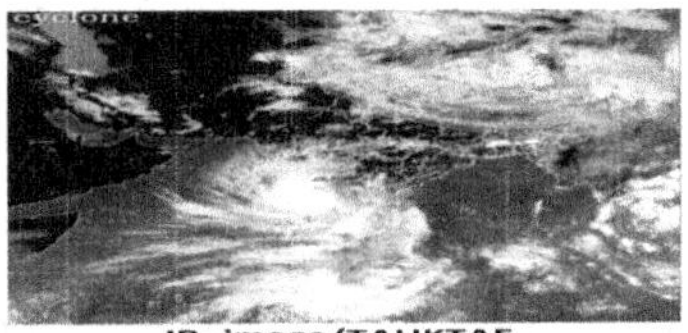 cyclone IR- image/TAUKTAE Cyclone/15.05.2021/0900 UTC
3	ResNet151- Model	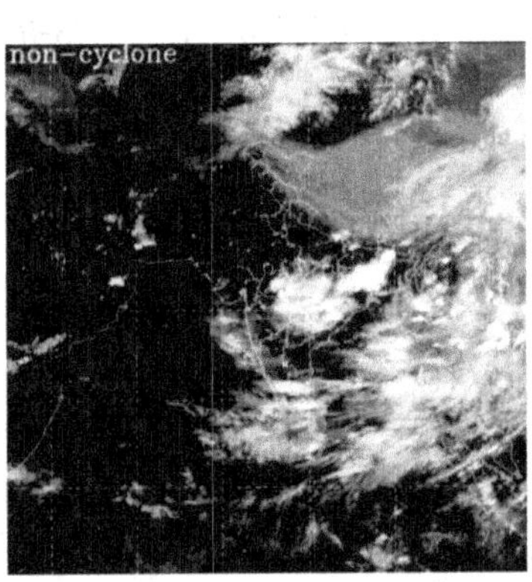 non-cyclone IR-Image/ Non-Cyclonic storm image/INSAT 3D satellite source	 cyclone IR- image/TAUKTAE Cyclone/15.05.2021/0900 UTC

Table 3. Model accuracy versus epochs trained for proposed method

Sno	Deep Learning Neural Network(DLNN)	Accuracy attained	Number of epochs
1	ResNet-50 Model	100 %	05
2	ResNet-101 Model	100%	05
3	Resnet-151 Model	100%	05
4	Grad-CAM	LOCALISATION OF CYCLONE-EYE **YAAS & TAUKTAE**	

Table 4 indicates the accuracies attained for multiple epochs trained on different variants of RESNET, which are used for the classification and localisation of cyclonic storm areas.

Table 4. Cyclone storm localisation using Grad-CAM method

Sno	Method used for localisation	Image of Cyclone YAAS	Image of Cyclone TAUKTAE
1.	Gradient Class Activation Mapping (Grad-CAM)	*IR-Image /YAAS cyclone/INSAT-3D/ 25.05.2021/0000 UTC*	*IR-Image /Tauktae cyclone/ INSAT-3D/17.05.2021/0030 UTC*

Graphical Plots of Deep Learning

The above *Figure 7* shows the accuracy and loss curves for the RESNET architecture for the total number of epochs. It shows good accuracy and minimum loss while training and testing the neural network. The neural network has been implemented on an NVIDIA graphics card for faster execution. *Table 5* indicates localisation using Grad-CAM. The Adam optimizer is used for training the three variants of the RESNET architecture in training. The ADAM optimizer reduces the error while training on the dataset, and specifically, it has two advantages. Faster optimization is achieved and it also combines the RMS propagation and adaptive gradient algorithms.

Figure 7. Training accuracy and loss curves for ResNet50 neural network vs epochs

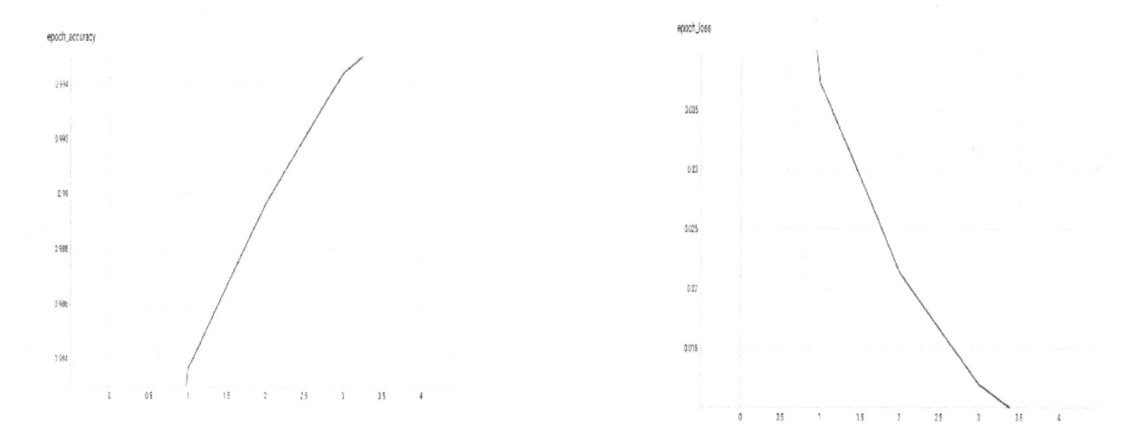

Advantages of Proposed Method

1. Satellite Tropical Cyclone image classification with an accuracy of 100%.
2. Detection of cyclonic storm clouds using an inference model obtained
3. Residual neural networks are used for the detection of cyclonic storms and are robust in the detection of the presence of noise, rotation, and shift in satellite images.
4. Localisation of a cyclone cloud using the Grad-CAM method.
5. Results exhibit better performance by the proposed method when compared with other methods that were employed earlier.

FUTURE RESEARCH DIRECTIONS

The Deep Learning (DL) algorithm used in the proposed Cyclonic Storm image classification and localization method can be implemented on hardware devices like the NVidia Jetson Nano board for faster implementation, and the proposed method can even be executed on cloud platforms using TPUs for faster processing. Thus, in the workplace, this technique has been expanded.

CONCLUSION

Thus, in this chapter, satellite infrared imagery was used to classify and locate cyclone cloud distribution using the suggested method, and the results showed that this method outperformed others in terms of assessment metrics and localization. During the implementation process, the model was found to be resilient to noise, rotation, and shift in images. A classification accuracy of 100% is obtained for all three models, i.e., ResNet-50, ResNet-101, and ResNet-151. Implementation of training and validation is well achieved using NVIDIA GPU for faster execution Also, with less time (a few seconds), the detection process for the presence of a cyclonic storm in satellite images was achieved. Furthermore, the model is robust in the detection phase for the presence of noise, shift, and rotation in test sample images of the dataset. Therefore, the proposed method has attained the objectives that were important during the phase of evaluation.

ACKNOWLEDGMENT

We express our sincere gratitude to Dr.Mrutyunjay Mohapatra, Director General of Meteorology (DGM), India Meteorological Department (IMD) for encouraging and motivating us in research, development, and implementation of this chapter. We also express our sincere thanks to Dr.S.Balachandran, DDGM, RMC-Chennai, India Meteorological Department and Satellite Meteorology Division, New Delhi for helping us with getting image data and obtaining fruitful results.

REFERENCES

Bau, D., Zhou, B., Khosla, A., Oliva, A., & Torralba, A. (2017). Network dissection: Quantifying interpretability of deep visual representations. Computer Vision and Pattern Recognition.

Bazzani, L., Bergamo, A., Anguelov, D., & Torresani, L. (2016). Self-taught object localization with deep networks. WACV.

Bengio, Y., Courville, A., & Vincent, P. (2013). Representation learning: A review and new perspectives. *IEEE Transactions on Pattern Analysis and Machine Intelligence*, *35*(8), 1798–1828.

Chen, B., Chen, B.-F., & Lin, H.-T. (2018). Rotation-blended CNNs on a New Open Dataset for Tropical Cyclone Image-to-intensity Regression. In Proceedings of the 24th ACM SIGKDD International Conference on Knowledge Discovery & Data Mining (KDD '18) (pp. 90–99). Association for Computing Machinery. doi.org/10.1145/3219819.3219926.

Chen, B.-F., Chen, B., Lin, H.-T., & Elsberry, R. L. (2019). Estimating Tropical Cyclone Intensity by Satellite Imagery Utilizing Convolutional Neural Networks. *Weather and Forecasting*. doi:10.1175/waf-d-18-0136

He, K., Zhang, X., Ren, S., & Sun, J. (2016). Identity mappings in deep residual networks. In *European conference on computer vision* (pp. 630-645). Springer.

Higa, M., Tanahara, S., Adachi, Y., Ishiki, N., Nakama, S., Yamada, H., ... Miyata, R. (2021). Domain knowledge integration into deep learning for typhoon intensity classification. *Scientific Reports*, *11*(1), 1–10.

Krizhevsky, A., Sutskever, I., & Hinton, G. E. (2017, June). ImageNet classification with deep convolutional neural networks. *Communications of the ACM*, *60*(6), 84–90. doi:10.1145/3065386

LeCun, Y., Bengio, Y., & Hinton, G. (2015). Deep learning. *Nature, 521*(7553), 436-444.

Li, J., Huang, X., & Gong, J. (2019). Deep neural network for remote-sensing image interpretation: Status and perspectives. *National Science Review*, *6*(6), 1082–1086.

Lin, W.-C., Liao, D.-Y., Liu, C.-Y., & Lee, Y.-Y. (2005). Daily imaging scheduling of an Earth observation satellite. *IEEE Transactions on Systems, Man, and Cybernetics. Part A, Systems and Humans*, *35*(2), 213–223. doi:10.1109/TSMCA.2005.843380

Liu, Q., Wu, S., Wang, L., & Tan, T. (2016). *Predicting the Next Location: A Recurrent Model with Spatial and Temporal Contexts*. AAAI.

Melgani, F., & Bruzzone, L. (2004). Classification of hyperspectral remote sensing images with support vector machines. *IEEE Transactions on Geoscience and Remote Sensing*, *42*(8), 1778–1790.

Pang, S., Xie, P., Xu, D., Meng, F., Tao, X., & Li, B. ... Song, T. (2021). NDFTC: A New Detection Framework of Tropical Cyclones from Meteorological Satellite Images with Deep Transfer Learning. *Remote Sensing*, *13*(9). Advance online publication. doi:10.3390/rs13091860

Pradhan, R., Aygun, R. S., Maskey, M., Ramachandran, R., & Cecil, D. J. (2018). *Tropical Cyclone Intensity Estimation Using a Deep Convolutional Neural Network. IEEE Transactions on Image Processing*, *27*(2), 692–702.

Rajesh, K., Ramaswamy, V., Kannan, K., & Arunkumar, N. (2019). Satellite cloud image classification for cyclone prediction using Dichotomous Logistic Regression Based Fuzzy Hypergraph model. *Future Generation Computer Systems*, *98*, 688–696. https://doi.org/10.1016/j.future.2018.12.042

Selvaraju, R. R., Cogswell, M., & Das, A. (2020). Grad-CAM: Visual Explanations from Deep Networks via Gradient-Based Localization. *International Journal of Computer Vision, 128,* 336–359. doi:10.1007/s11263-019-01228-7

Shakya, S., Kumar, S., & Goswami, M. (2020). Deep learning algorithm for satellite imaging based cyclone detection. *IEEE Journal of Selected Topics in Applied Earth Observations and Remote Sensing*, *13*, 827–839.

Simonyan, K., & Zisserman, A. (2014). *Very deep convolutional networks for large-scale image recognition.* arXiv preprint arXiv:1409.1556.

Szegedy, C., Liu, W., Jia, Y., Sermanet, P., Reed, S., Anguelov, D., ... Rabinovich, A. (2015). Going deeper with convolutions. In *Proceedings of the IEEE conference on computer vision and pattern recognition* (pp. 1-9). IEEE.

Wimmers, A. J., Velden, C. S., & Cossuth, J. (2019). Using Deep Learning to Estimate Tropical Cyclone Intensity from Satellite Passive Microwave Imagery. *Monthly Weather Review*, *147*, 2261–2282. https://doi.org/10.1175/mwr-d-18-0391.1

Zhang, C.-J., Wang, X.-J., Ma, L.-M., & Lu, X.-Q. (2021). Tropical Cyclone Intensity Classification and Estimation Using Infrared Satellite Images With Deep Learning. *IEEE Journal of Selected Topics in Applied Earth Observations and Remote Sensing*, *14*, 2070–2086. doi:10.1109/JSTARS.2021.3050767

Zhang, R., Liu, Q., & Hang, R. (2020). Tropical Cyclone Intensity Estimation Using Two-Branch Convolutional Neural Network From Infrared and Water Vapor Images. *IEEE Transactions on Geoscience and Remote Sensing, 58*(1), 586-597. doi:10.1109/TGRS.2019.2938204

Zhuge, X., Guan, J., Yu, F., & Wang, Y. (2015, October). A New Satellite-Based Indicator for Estimation of the Western North Pacific Tropical Cyclone Current Intensity. *IEEE Transactions on Geoscience and Remote Sensing*, *53*(10), 5661–5676. https:// doi.org/ 10.1109/TGRS.2015.2427035

ADDITIONAL READING

Abadi, M., Barham, P., Chen, J., Chen, Z., Davis, A., Dean, J., Devin, M., Ghemawat, S., Irving, G., Isard, M., Kudlur, M., Levenberg, J., Monga, R., Moore, S., Murray, D. G., Steiner, B., Tucker, P., Vasudevan, V., Warden, P., ... Zheng, X. (2016). TensorFlow: a system for large-scale machine learning. In *Proceedings of the 12th USENIX conference on Operating Systems Design and Implementation (OSDI'16)* (pp. 265–283). USENIX Association. https://www.usenix.org/conference/osdi16/technical-sessions/presentation/abadi

Cinbis, R. G., Verbeek, J., & Schmid, C. (2016). Weakly supervised object localization with multi-fold multiple instance learning. *IEEE Transactions on Pattern Analysis and Machine Intelligence*. PMID:26930676

Das, A., Datta, S., Gkioxari, G., Lee, S., Parikh, D., & Batra, D. (2018). Embodied question answering. *Proceedings of the IEEE conference on computer vision and pattern recognition (CVPR)*.

Lin, W.-C., Liao, D.-Y., Liu, C.-Y., & Lee, Y.-Y. (2005). Daily imaging scheduling of an Earth observation satellite. *IEEE Transactions on Systems, Man, and Cybernetics. Part A, Systems and Humans, 35*(2), 213–223. doi:10.1109/TSMCA.2005.843380

Szegedy, C., Liu, W., Jia, Y., Sermanet, P., Reed, S., Anguelov, D., ... Rabinovich, A. (2015). Going deeper with convolutions. In *Proceedings of the IEEE conference on computer vision and pattern recognition* (pp. 1-9). IEEE.

Wimmers, A. J., Velden, C. S., & Cossuth, J. (2019). Using Deep Learning to Estimate Tropical Cyclone Intensity from Satellite Passive Microwave Imagery. *Monthly Weather Review, 147*(6), 2261–2282. doi:10.1175/MWR-D-18-0391.1

Zhuge, X., Guan, J., Yu, F., & Wang, Y. (2015, October). A New Satellite-Based Indicator for Estimation of the Western North Pacific Tropical Cyclone Current Intensity. *IEEE Transactions on Geoscience and Remote Sensing*, *53*(10), 5661–5676. doi:10.1109/TGRS.2015.2427035

KEY TERMS AND DEFINITIONS

Cyclone: A low pressure system with winds rotating in an anticlockwise (clockwise) direction in the northern (southern) hemisphere with a minimum sustained wind speed of 34 knots (62kmph).

Epochs: The number of epochs used to train a deep learning model.

Gradient Mapping: Activation Mapping is used for the localisation of images of interest.

Non-Cyclonic Storm: When there is no cyclogenesis or depression formation, a storm is classified as non-cyclonic.

Precision: Is defined as the ratio of true positives to the sum of true and false positives.

Recall: The ratio of true positives is equal to the sum of true and false positives.

Testing Accuracy: Testing accuracy is obtained on a test dataset having images.

Chapter 15
Review of Weather-Affected Urban Air Pollution Forecast Models

Ankit Didwania
https://orcid.org/0000-0003-3141-054X
Gujarat Technological University, India

Vibha Patel
Gujarat Technological University, India

ABSTRACT

Weather affects air quality globally since different aspects of the weather like humidity, temperature, wind speed, and direction essentially affect the movement, creation, and concentration of various major air pollutants like surface ozone, PM 2.5, methane, carbon dioxide, etc. Air pollution is caused when an excessive amount of harmful substances like gases, particles, etc. are poured into our atmosphere which can severely affect the health of any living organisms. In this chapter, the most relevant weather affected urban air quality prediction papers are studied along with recent IoT systems developed for air pollution, and the authors observed that modern artificial intelligence algorithms are better than traditional statistical models. However, artificial intelligence-based algorithms cannot be directly compared effectively due to the hybrid nature of data sources used. Also, a need is identified to develop a powerful end-to-end model based on artificial intelligence algorithms and IoT systems.

INTRODUCTION

World Health Organization (WHO) confirms that "the primary cause of air pollution (burning fossil fuels) is also a major contributor to climate change, which impacts people's health in different ways" (WHO, 2019). WHO shows that air pollution is the topmost environmental threat to global health (WHO, 2019). WHO report says "9 out of 10 people breathe polluted air every day!" (WHO, 2019) Although the quality of air majorly affects urban and industrial areas, still due to its unique characteristics of global-scale

DOI: 10.4018/978-1-6684-3981-4.ch015

pollution dispersal capability, it also affects rural and forest areas far away from the source of pollution. So, air pollution is the single biggest environmental health crisis everyone faces daily.

In 2021, India, which has high air pollution levels in many major cities, has given a challenging promise to cut its emissions to net-zero by 2070 during the COP26 (Glasgow) summit. In 2015, the United Nations (UN) had adopted improvement in air quality as the 11th of its 17 Sustainable Development Goals (SDG-11) for the 2030 Agenda for Sustainable Development. The global targets for Sendai Framework for Disaster Risk Reduction (SFDRR) also have synergy with SDG 11. So, by studying the various existing models/algorithms for predicting urban air quality, we can provide future directions to improve the air quality forecast algorithms to achieve the UN SDG-11 goals and COP26 promise by preparing for pollution mitigations strategies.

Air pollution can cause climate change globally and also majorly contribute to the formation of meteorological hazards like fog or smog in urban areas. As, urban areas release lots of smaller particles of pollutants including nitrogen oxides (NOx) like nitrogen dioxide (NO2), methane (CH4), Volatile organic compounds (VOC) like benzene (C6H6), carbon oxides like carbon monoxide (CO) and carbon dioxide (CO2), sulfur oxides (SOx) like sulfur dioxide (SO2), ozone (O3) and particulate matters (PM 0.1, PM 2.5, PM 10). The major sources of such pollutants in urban areas include incomplete vehicular fuel combustion from old vehicles, vehicles traffic density and traffic jams releasing majorly nitrogen oxides, particulate matter and VOCs; inefficient burning of fossil fuels for cooking, heating or lighting purposes releasing majorly sulfur oxides and particulate matters and other harmful pollutants, inefficient disposal of human and animal waste releasing majorly methane, open garbage burning releasing majorly particulate matters and carbon oxides, increased usage of beauty products releasing majorly VOCs, high energy demand from coal-powered power plants releasing majorly nitrogen oxides and particulate matters and industrial toxic emissions due to use of harmful chemicals. The chemical reaction of NOx and VOCs under sunlight creates ground level harmful ozone particles so its major sources are also vehicular and power plant pollution and the burning of fossil fuels.

Such smaller pollutant particles produced due to human activities forms aerosols which affect the physical properties of water droplets in our atmosphere at the micro-level and even boost the formation of fog in urban areas. Such smoky fog is also known as smog. It is a life-threatening hazard as it decreases the visibility of roads causing major accidents and also weakens the health of every living being who breathes such air. It may be fatal to asthma, respiratory and cardiovascular patients and other vulnerable groups like children, senior citizens, pregnant women, etc.

The use of IoT based sensor network system like pollution control boards sensor systems in urban areas helps to augment traditional observations like ground-based weather monitoring stations. It helps to fill the gaps in observation caused due to the limited capacity and uneven distribution of weather monitoring stations. Although IoT based sensor network is not as reliable and easy to maintain as traditional weather monitoring stations. The use of earth observation/satellite (EO) based Artificial Intelligence/ Machine Learning (AI/ML) techniques has been increased in early warning and detection of major air pollutants globally due to their low observation gaps, global view and quick computation. Although their accuracy still needs improvement.

BACKGROUND

Air quality / Air pollution is complex compared to weather data as it can be different for nearby locations and can dramatically change quickly. Also, health problems like asthma, respiratory and cardiovascular diseases, cancer, etc. are known to be worsening due to air pollution. So, it can cause fatal problems to vulnerable groups like children, pregnant women, senior citizens, patients, etc.

So, real-time air quality monitoring and near-future prediction will help the public and government authorities to plan mitigation strategies like the control on generation and transportation of air pollutants to reduce the financial burden caused by air pollution-related health diseases. For larger public consumption this complex air quality data is converted to a numerical value known as Air Quality Index (AQI). Although this index is not the same for all countries, generally higher numbers indicate high air pollution levels. Among the various air pollutants; PM 2.5, PM 10, SO2, O3 and NO2 are significantly observed globally. Fine Particulate Matter like PM 2.5 causes many adverse health effects globally. In-situ sites like ground-based pollution monitoring networks and IoT systems give exact estimates of that particular small area only, so the prediction cannot be generalized.

Aerosols in the atmosphere are good indicators of PM based air pollution. It can be natural like fog, dust, etc. or anthropogenic like smoke, particulate matter in the air etc. Aerosol data like Aerosol Optical Depth (AOD) values retrieved from satellite data are majorly used for establishing a correlation with PM 2.5 and other major pollutants. Meteorological stations data like wind speed, wind direction, temperature, etc are also used as parameters for enhancing this correlation. Highly accurate data from various in-situ (ground) systems such as pollution monitoring networks and IoT systems are majorly used as ground truth data for establishing this correlation. In this manner, a highly positive correlation is generally established between PM 2.5 and other major pollutants, which are used for monitoring and displaying real-time air quality data. So, AOD values allow estimating PM 2.5 globally.

SENDAI FRAMEWORK SYNERGY WITH SDG-11

Due to the global increase in urban areas and population, there was a need for global policies for sustainable urban living. Due to the global nature of air pollution, globally agreed frameworks and goals were developed having measurable indicators and attainable targets.

The Sustainable Development Goals (SDG) under the United Nations Department of Economic and Social Affairs, also known as the Global Goals, were adopted by UN member states in the year 2015 ("UN SDGs," 2015). SDGs are a combination of 17 interlinked objectives designed to achieve a sustainable future globally. The 11th goal of SDG (SDG-11) also focuses on making urban areas more sustainable by improving their air quality. It provides well-defined SDG indicators to continuously measure the progress of the goals achieved by various countries, as the goal is agreed to be fulfilled by the year 2030.

The Sendai Framework for Disaster Risk Reduction (SFDRR) under the United Nations Office for Disaster Risk Reduction (UNDRR), also known as the Sendai Framework, was adopted by UN member states in the Sendai city of Japan in the year 2015 and the framework is adopted till the year 2030 similarly as UN Sustainable Development Goals (SDG) ("Sendai Framework," 2015). This framework was developed with the help of the United Nations International Strategy for Disaster Reduction to set common standards among member states and give them achievable targets along with legally based instruments for disaster risk reduction and climate change adaptation. This framework also helps in monitoring the progress of SDG-11 goals.

The below-mentioned Sendai framework global targets (A to E) related to a sustainable reduction in air pollution have much synergy with the 11th UN Sustainable Development Goals (SDG-11) whose goal is to make human settlements more safe and sustainable ("Sendai Framework," 2015):

1. Reduce global disaster mortality
    a. Sendai indicator-A1 is similar to SDG indicator-11.5.1
    b. Sendai indicator-A1: It indicates the number of deaths and missing persons due to disasters per lakh population
    c. SDG indicator-11.5.1: It indicates the number of deaths, missing persons and directly affected persons due to disasters per lakh population
2. Reduce the number of affected people globally
    a. Sendai indicator-B1 is similar to SDG indicator-11.5.1
    b. Sendai indicator-B1: It indicates the number of directly affected people due to disasters per lakh population
    c. SDG indicator-11.5.1: It indicates the number of deaths, missing persons and directly affected persons due to disasters per lakh population
3. Reduce direct economic loss in relation to GDP
    a. Sendai indicator-C1 is similar to SDG indicator-11.5.2
    b. Sendai indicator-C1: It indicates direct financial loss due to disasters concerning the global GDP
    c. SDG indicator-11.5.2: It indicates direct financial loss concerning global GDP and harm to significant infrastructure and basic services due to disasters
4. Reduce disaster damage to critical infrastructure and disruption of basic services
    a. Sendai indicators-D1, D5 are similar to SDG indicator-11.5.2
    b. Sendai indicator-D1: It indicates damage to important infrastructure due to disasters
    c. Sendai indicator-D5: It indicates the number of disruptions to basic services due to disasters
    d. SDG indicator-11.5.2: It indicates direct financial loss concerning global GDP and harm to significant infrastructure and basic services due to disasters
5. Increase the number of countries with national and local disaster risk reduction strategies
    a. Sendai indicators-E1, E2 is similar to SDG indicators-11. b.1 and 11.b.2 respectively
    b. Sendai indicator-E1: It indicates the number of countries that adopt and implement national disaster risk reduction strategies in line with the Sendai Framework for Disaster Risk Reduction 2015-2030
    c. Sendai indicator-E2: It indicates the percentage of local governments that adopt and implement local disaster risk reduction strategies in line with national strategies
    d. SDG indicator-11. b.1: It indicates the number of countries that adopt and implement national disaster risk reduction strategies in line with the Sendai Framework for Disaster Risk Reduction 2015–2030
    e. SDG indicator-11. b.2: It indicates the proportion of local governments that adopt and implement local disaster risk reduction strategies in line with national disaster risk reduction strategies

WEATHER AFFECTED URBAN AIR POLLUTION MONITORING AND PREDICTION

IoT Systems for Air Pollution Monitoring

There have been many recent developments in IoT systems also, for detecting air pollutants at micro-local levels. Table 1 describes the study of recent IoT systems developed for air pollution (Arora, Pandya, Shah, & Doshi, 2019).

Table 1. Study of recent IoT systems developed for air pollution (Arora et al., 2019)

Sr. No.	Measurement Capabilities	Connectivity	Sensors used	Research Gaps
1	General pollutants like Ammonia, Benzene, Carbon Monoxide, Smoke, Temperature, Humidity, UV rays	Bluetooth Module (HC-06)	MQ-135, MQ-2, UV-01, DHT11	Need to study UV effects on the atmosphere
2	Vehicular pollution like Ammonia, Benzene, Smoke Monitor	Wired Ethernet of Raspberry- Pi	MQ-135	Elimination of interference of other gases from the atmosphere
3	Air and noise pollutants like Ammonia, Benzene, Smoke, Noise	ESP8266 Wi-Fi Module	MQ-135	Efficient classification of pollutant causes and categories.
4	Air and noise pollutants like Ammonia, Benzene, Carbon Monoxide, Smoke, Temperature, Humidity, Noise, LPG, Propane, Hydrogen	ESP8266 Wi-Fi Module	MQ-135, MQ-7, MQ-2, DHT11	Deployment feasibility and Mechanism to classify causes and categories
5	Air pollutants like LPG, Propane, Hydrogen, Ozone, Sulphur Dioxide, Carbon Monoxide	ESP8266-12E Wi-Fi modules	MQ-2	Causes need to be predicted by ML methods

The availability of a large amount of Earth Observation/Satellite (EO) data and advancement in Artificial Intelligence (AI) techniques has speedup the analysis of EO data and opened new frontiers for the application of EO data. But, freely available EO data has a low spatial resolution which affects their accuracy. So, AI-based models like variants of reinforcement learning, GAN's, etc are created for improving the spatial resolution (super-resolution) of major pollutants like PM 2.5, NO2, etc. and accurately predicting the air quality up to the local level. Monitoring and prediction up to the local level will help in improving the planning of smart cities for smarter and healthier living.

Many predictions and monitoring systems are already present for air quality, but they have a trade-off between accuracy, resolution and cost. Ground-based in-situ monitoring is cheap to install but overall costly to operate and maintain due to its limited operational lifetime. Although they have partial and uneven spatial distribution due to limited installation possible, their accuracy and temporal resolution are very high. Satellite-based remote sensing monitoring is costly to install but overall cheap to operate and maintain due to its high operational lifetime. They have high and even spatial distribution but low accuracy and low temporal resolution. Many AI-based prediction algorithms for air quality prediction also exist but they have low accuracy for local levels due to the usage of low-resolution data.

Models for Air Pollution Prediction

There are two major models to predict air quality information:

Statistical Models

Table 2 shows the various statistical models.

Table 2. Statistical models

Sr. No.	Statistical Models	Working
1	Simple Linear Regression Model or Two-Valued Model (TVM) (Chu et al., 2003)	This paper uses a very simplified regression model. It uses AOD as the only predictor. Its implementation is also very easy.
2	Multi-Valued Model (MVM) or Generalized /Multiple Linear Regression Model (Gupta & Christopher, 2009)	This paper uses a slightly complicated regression model. Meteorological data are used as additional parameters along with AOD to increase accuracy.
3	Geographically Weighted Regression model (GWR) (Wu, Yao, Li, & Si, 2016)	This paper pays attention to the local variations affecting the AOD-PM 2.5 relationship by using a complicated regression model. Parameters from each monitoring site are used to model such spatially varying relationships.

The advantages and disadvantages of the statistical models mentioned above are as follows:

- Advantages:
  - Once a statistical model is developed, it can be used repetitively quickly and easily
  - Can be used as a benchmark model for future models
- Disadvantages:
  - Not able to model the non-linearity effectively
  - Does not reveal the complex relationships efficiently between PM 2.5 and the predictors

Machine Learning (ML) models

Table 3 shows the various AI/ML models

The detailed review of ML models mentioned above is as follows:

1. For air pollution prediction in major polluted cities of China, (Ameer et al., 2019) demonstrated that the time interval required to execute the decision tree and random forest algorithms are less compared to multi-layer perceptron and gradient boosting algorithms. Based on the processing time and error rate parameters, (Ameer et al., 2019) recommends using the random forest for air

quality prediction as after hyperparameter tuning, random forest helps to reduce over-fitting also. But, its author would need to use additional parameters for improving prediction. Table 4 shows the comparison of various AI/ML regression techniques for air pollution prediction.

Table 3. Various Machine Learning (ML) models

Sr. No.	Major Pollutant	Data Source	Algorithm Type	Methods	Period	Evaluation Metrics	Data Rate
1 (Ameer et al., 2019)	PM2.5	AQ, MET	Ensemble, Regression	DTR, RFR, MLP, GBR	1 week	MAE, RMSE	Not Specified
2 (Martínez-España et al., 2018)	O3	AQ, MET	Ensemble, Regression	Bagging, RC(RT), RF,DT(M5P), KNN	24 h	MAE, RMSE, R2	Hourly
3 (Eldakhly, Aboul-Ela, & Abdalla, 2017)	PM10	AQ, MET, Temporal	Regression	chWSVR	1 h	RMSE, R, z', t-value	Hourly
4 (Abu Awad, Koutrakis, Coull, & Schwartz, 2017)	BC	AQ, MET, Spatial, Temporal	Regression	nu-SVM	24 h	R2	Daily
5 (Oprea, Dragomir, Popescu, & Mihalache, 2016)	PM10	AQ, MET	Regression	M5P, REPTree,	24 h, 48 h, 72 h	R, MAE, RMSE	Daily
6 (Contreras & Ferri, 2016)	SO2, O3, NO, NO2	AQ, MET, TIF, Temporal	Regression, Ensemble	LR, QR, IBKreg, M5P, RF	3 h	RMSE	Hourly
7 (Sayegh, Munir, & Habeebullah, 2014)	PM10	AQ, MET	Regression	MLR, QR, GAM, BRT1, BRT2	1 h	MBE, MAE, RMSE, FACT2,R, IA	Hourly
8 (Ip, Vong, Yang, & Wong, 2010)	SPM, SO2, NO2,O3	AQ, MET	Regression	LSSVM	24 h	Relative Error	Daily
9 (W. Wang, Men, & Lu, 2008)	PM10, NOx, SO2	AQ, MET	Regression	online SVM	24 h, 1 week	MAE, RMSE, WIA	Hourly
10 (Weizhen Lu et al., 2002)	PM10	AQ, MET	Regression	SVM	24 h, 1 week	MAE	Hourly
11 (Y. Zhang et al., 2019)	PM2.5	AQ, MET, WFD, Spatial	Ensemble	LightGBM	24 h	SMAPE, MSE, MAE	Not Specified
12 (Zheng, Li, Lu, & Ruan, 2018)	PM2.5	AQ, MET, Temporal	Ensemble	MKSVC	1 h, 3 h, 6 h, 9 h,12 h	ACC, MSE, WP, WR, WF	Hourly
13 (Eslami, Salman, Choi, Sayeed, & Lops, 2020)	O3	AQ, MET	Ensemble	DNN(extra-trees)	24 h	IA	Hourly
14 (J. Wang & Song, 2018)	PM2.5	AQ, WFD, Spatial	Ensemble	STE	6 h, 12 h, 24 h, 48h	RMSE, MAE, ACC	Hourly
15 (C. Zhang et al., 2017)	PM2.5, PM10, SO2, CO, NOx, O3	AQ, MET, WFD	Ensemble	MELSA	72 h	RAE, RSE, R	Not Specified
16 (Debry & Mallet, 2014)	O3, NO2, PM10	AQ, MET	Ensemble	DRR	24 h, 48 h, 72 h	RMSE	Hourly
17 (Kleinert, Leufen, & Schultz, 2020)	near-surface O3 concentrations	AQ, MET	Regression	CNN	8 h	MSE, skill score	Hourly

Table 4. Various ML regression techniques for air pollution prediction (Ameer et al., 2019)

Sr. No.	ML Regression Techniques (Ameer et al., 2019)	Processing Time	Error Rate
1	Decision Tree	Least among all four techniques	High
2	Random Forest	Much lower than Multi-Layer Perceptron and Gradient Boosting	Least among all
3	Multi-Layer Perceptron	High	High
4	Gradient Boosting	Highest among all four techniques	Highest among all

2. (Martínez-España et al., 2018) proves that random forest algorithm is better for ozone level prediction in Murcia which is a smart city in Spain, as it has comparatively lower RMSE and MAE values. Ozone is just one of the various major air pollutants, so its author would need to experiment with other major pollutants also for getting overall air quality prediction of smart cities.
3. (Eldakhly et al., 2017) proposes a model using the chance theory that handles PM 10 interval variations better than random forest and bootstrap aggregating approaches. But, its author would need to demonstrate their model efficiency on other air pollutants also.
4. (Abu Awad et al., 2017) propose a model based on SVM capable of black carbon prediction in three New England states. This model uses additional attributes of topography and transportation but is deficient in monitors.
5. (Oprea et al., 2016) used inductive learning for air pollutants forecasting and its author observed that the M5P method gives better PM 10 prediction than the REPTree method.
6. Wind flow in an urban area is different from rural areas due to the various building elevations, so (Contreras & Ferri, 2016) considered wind as an important parameter for the prediction of nitrogen oxides, ozone and sulphur dioxides in the urban area of Valencia. Its author also interpolates real-time forecasts and proves that random forest is comparatively good in urban air quality prediction, but they overlook the local characteristics of the prediction area.
7. (Sayegh et al., 2014) argue that the QR method is comparatively better in predicting hourly PM 10 concentrations for the holy city of Makkah in Saudi Arabia as it realizes covariates significance at different quantiles. But it overlooks traffic attributes and also needs to include many historical data from various monitoring locations of different same and different cities.
8. (Ip et al., 2010) conclude that its proposed method using least-squares SVM is better than the MLP method for air quality prediction in China. As the study was done on data from 2003-06, more years of data should be considered.
9. (W. Wang et al., 2008) argue online SVM gives a superior air quality prediction model for the city of Hong Kong in china against conventional SVM due to its dynamic nature. But it needs more computation capabilities due to its high dimensionality.
10. (Weizhen Lu et al., 2002) shows that the support vector machine gives enhanced generalization performance than Radial Basis Function. It predicts hourly air quality with higher accuracy for a particular year in Hong Kong, China. More data should be considered to get a continuous accurate prediction.
11. (Y. Zhang et al., 2019) proposes the LightGBM model which can perform parallel learning on data with high dimensions. Data from only Beijing city, China was used from 2017-18, so data

augmentation on predictive and historical data was done using the sliding window mechanism to achieve higher accuracy. To increase the training speed, PCA was used to choose only important features.

12. (Zheng et al., 2018) believe that their proposed multiple kernel-based SVC model can forecast extreme air pollution events in cities of China as the model ensembles various methods, prunes unwanted attributes and performs metric learning for the prediction. Although the model performs better than LSTM, ARIMA, RF, MLP and SVM still further study is needed.
13. (Eslami et al., 2020) believes using an ensemble of extra trees method and the deep neural network would give a quicker real-time estimate of ozone peak levels, which is one of the major air pollutants in South Korea. Still, higher ozone peak levels were not estimated correctly.
14. (J. Wang & Song, 2018) conclude its proposed deep STE model is comparatively effective in Beijing, China for air quality prediction as its dataset contains major pollutant values and meteorological events for a few years of multiple monitoring stations. It also considers the distance of the monitoring stations and the meteorological trends in its model.
15. (C. Zhang et al., 2017) believes its proposed MELSA algorithm gives better air quality index values for 1-3 days in advance for Beijing, China via web service. It uses weather events and dominant air pollutants as the major data source in its ensemble learning.
16. (Debry & Mallet, 2014) writes that daily air quality prediction given by the Prev'Air platform is enhanced by using a discounted ridge regression model. Only a few years' data are used so more data should be included to prove the effectiveness of the model.
17. (Kleinert et al., 2020) confirms that its manifold CNN layers based IntelliO3-ts model predicts daily ozone pollution in Germany better than other models from 8 hours up to 2 days in advance. But author confirms that the predictions for lead times longer than 2 days are not valuable. Also, it overlooks the geographic information and relation between monitoring stations.

The advantages and disadvantages of the ML models mentioned above are as follows:

- Advantages:
  - Powerful ability to model the non-linearity
  - Better performance
  - Can be used as a benchmark for future models
- Disadvantages:
  - Not easy to implement and use
  - Need careful consideration of handcrafted features

The Major Learning's from The Above Discussed AI/ML Models

- Major Evaluation Metric Used: MAE and RMSE
- Major Time Granularity: 24 Hours (next day prediction)
- Major Prediction Target: PM 2.5, PM 10
- Major Limitation: Lack of Data
- Other important data sources: weather events and transportation trends

FUTURE RESEARCH DIRECTIONS

In future, there is a need to further study and compare all the relevant algorithms using similar data sources which have the same spatial and temporal resolutions and develop powerful end-to-end models based on machine learning architecture and IoT systems.

CONCLUSION

Based on the review of the above algorithms, in this chapter per the authors can observe that machine learning-based models are better than statistical models in air quality prediction. Still, these various machine learning algorithms discussed above have analyzed data having different data rates, types, sources, etc. So it is difficult to compare their accuracy efficiently. Weather affected air pollution impacts everyone's life. In this chapter per the authors, a detailed review is conducted of various AI-based models and the latest IoT technologies for monitoring and forecasting air quality. Traditional air quality prediction methods heavily depend on high computing power and numerical data availability, which reduces their accuracy and usage. Satellite-based estimates can provide complete global coverage compared to ground-based air quality monitors. The IoT system can be used to gather micro surface-level air pollution data for the region of interest taken at regular intervals.

REFERENCES

Abu Awad, Y., Koutrakis, P., Coull, B. A., & Schwartz, J. (2017). A spatio-temporal prediction model based on support vector machine regression: Ambient Black Carbon in three New England States. *Environmental Research, 159*, 427–434. doi:10.1016/j.envres.2017.08.039 PMID:28858756

Ameer, S., Shah, M. A., Khan, A., Song, H., Maple, C., Islam, S. U., & Asghar, M. N. (2019). Comparative Analysis of Machine Learning Techniques for Predicting Air Quality in Smart Cities. *IEEE Access: Practical Innovations, Open Solutions, 7*, 128325–128338. doi:10.1109/ACCESS.2019.2925082

Arora, J., Pandya, U., Shah, S., & Doshi, N. (2019). Survey- Pollution monitoring using IoT. *Procedia Computer Science, 155*, 710–715. doi:10.1016/j.procs.2019.08.102

Chu, D. A., Kaufman, Y. J., Zibordi, G., Chern, J. D., Mao, J., Li, C., & Holben, B. N. (2003). Global monitoring of air pollution over land from the Earth Observing System-Terra Moderate Resolution Imaging Spectroradiometer (MODIS). *Journal of Geophysical Research, 108*(21), 1–18. doi:10.1029/2002JD003179

Contreras, L., & Ferri, C. (2016). Wind-sensitive Interpolation of Urban Air Pollution Forecasts. *Procedia Computer Science, 80*, 313–323. doi:10.1016/j.procs.2016.05.343

Debry, E., & Mallet, V. (2014). Ensemble forecasting with machine learning algorithms for ozone, nitrogen dioxide and PM10 on the Prev'Air platform. *Atmospheric Environment, 91*, 71–84. doi:10.1016/j.atmosenv.2014.03.049

Eldakhly, N. M., Aboul-Ela, M., & Abdalla, A. (2017). Air Pollution Forecasting Model Based on Chance Theory and Intelligent Techniques. *International Journal of Artificial Intelligence Tools*, *26*(6), 1–30. doi:10.1142/S0218213017500245

Eslami, E., Salman, A. K., Choi, Y., Sayeed, A., & Lops, Y. (2020). A data ensemble approach for real-time air quality forecasting using extremely randomized trees and deep neural networks. *Neural Computing & Applications*, *32*(11), 7563–7579. doi:10.100700521-019-04287-6

Framework, S. (2015). Retrieved April 6, 2022, from https://www.undrr.org/implementing-sendai-framework/what-sendai-framework

Gupta, P., & Christopher, S. A. (2009). Particulate matter air quality assessment using integrated surface, satellite, and meteorological products: Multiple regression approach. *Journal of Geophysical Research*, *114*(D14), D14205. Advance online publication. doi:10.1029/2008JD011496

Ip, W. F., Vong, C. M., Yang, J. Y., & Wong, P. K. (2010). Forecasting daily ambient air pollution based on least squares support vector machines. *2010 IEEE International Conference on Information and Automation, ICIA 2010*, 571–575. 10.1109/ICINFA.2010.5512401

Kleinert, F., Leufen, L. H., & Schultz, M. G. (2020). IntelliO3-ts v1.0: A neural network approach to predict near-surface ozone concentrations in Germany. *Geoscientific Model Development Discussions*, *2020*, 1–69. doi:10.5194/gmd-2020-169

Lu, W., Wang, W., Leung, A. Y. T., Lo, S-Y., Yuen, R. K. K., Xu, Z., & Fan, H. (2002). Air pollutant parameter forecasting using support vector machines. *IEEE Proceedings of the 2002 International Joint Conference on Neural Networks, 1*, 630–635. 10.1109/IJCNN.2002.1005545

Martínez-España, R., Bueno-Crespo, A., Timón, I., Soto, J., Muñoz, A., & Cecilia, J. M. (2018). Air-pollution prediction in smart cities through machine learning methods: A case of study in Murcia, Spain. *Journal of Universal Computer Science*, *24*(3), 261–276.

Oprea, M., Dragomir, E. G., Popescu, M., & Mihalache, S. F. (2016). Particulate matter air pollutants forecasting using inductive learning approach. *Revista de Chimie*, *67*(10), 2075–2081.

Sayegh, A. S., Munir, S., & Habeebullah, T. M. (2014). Comparing the Performance of Statistical Models for Predicting PM10 Concentrations. *Aerosol and Air Quality Research*, *14*(3), 653–665. doi:10.4209/aaqr.2013.07.0259

UN SDGs. (2015). Retrieved April 6, 2022, from https://sdgs.un.org/goals

Wang, J., & Song, G. (2018). A Deep Spatial-Temporal Ensemble Model for Air Quality Prediction. *Neurocomputing*, *314*, 198–206. doi:10.1016/j.neucom.2018.06.049

Wang, W., Men, C., & Lu, W. (2008). Online prediction model based on support vector machine. *Neurocomputing*, *71*(4–6), 550–558. doi:10.1016/j.neucom.2007.07.020

WHO. (2019). https://www.who.int/emergencies/ten-threats-to-global-health-in-2019

Wu, J., Yao, F., Li, W., & Si, M. (2016). VIIRS-based remote sensing estimation of ground-level PM 2.5 concentrations in Beijing–Tianjin–Hebei: A spatiotemporal statistical model. *Remote Sensing of Environment, 184*, 316–328. doi:10.1016/j.rse.2016.07.015

Zhang, C., Yan, J., Li, Y., Sun, F., Yan, J., Zhang, D., ... Bie, R. (2017). Early Air Pollution Forecasting as a Service: An Ensemble Learning Approach. *IEEE International Conference on Web Services*, 636–643. 10.1109/ICWS.2017.76

Zhang, Y., Wang, Y., Gao, M., Ma, Q., Zhao, J., Zhang, R., Wang, Q., & Huang, L. (2019). A Predictive Data Feature Exploration-Based Air Quality Prediction Approach. *IEEE Access: Practical Innovations, Open Solutions, 7*, 30732–30743. doi:10.1109/ACCESS.2019.2897754

Zheng, H., Li, H., Lu, X., & Ruan, T. (2018). A Multiple Kernel Learning Approach for Air Quality Prediction. *Advances in Meteorology, 2018*, 1–15. doi:10.1155/2018/3506394

ADDITIONAL READING

Appel, K. W., Bash, J. O., Fahey, K. M., Foley, K. M., Gilliam, R. C., Hogrefe, C., Hutzell, W. T., Kang, D., Mathur, R., Murphy, B. N., Napelenok, S. L., Nolte, C. G., Pleim, J. E., Pouliot, G. A., Pye, H. O. T., Ran, L., Roselle, S. J., Sarwar, G., Schwede, D. B., ... Wong, D. C. (2021). The Community Multiscale Air Quality (CMAQ) model versions 5.3 and 5.3.1: System updates and evaluation. *Geoscientific Model Development, 14*(5), 2867–2897. doi:10.5194/gmd-14-2867-2021 PMID:34676058

Ciais, P., Bastos, A., Chevallier, F., Lauerwald, R., Poulter, B., Canadell, J. G., Hugelius, G., Jackson, R. B., Jain, A., Jones, M., Kondo, M., Luijkx, I. T., Patra, P. K., Peters, W., Pongratz, J., Petrescu, A. M. R., Piao, S., Qiu, C., Von Randow, C., ... Zheng, B. (2022). Definitions and methods to estimate regional land carbon fluxes for the second phase of the REgional Carbon Cycle Assessment and Processes Project (RECCAP-2). *Geoscientific Model Development, 15*(3), 1289–1316. doi:10.5194/gmd-15-1289-2022

Feng, X., Lin, H., Fu, T.-M., Sulprizio, M. P., Zhuang, J., Jacob, D. J., Tian, H., Ma, Y., Zhang, L., Wang, X., Chen, Q., & Han, Z. (2021). WRF-GC (v2.0): Online two-way coupling of WRF (v3.9.1.1) and GEOS-Chem (v12.7.2) for modeling regional atmospheric chemistry–meteorology interactions. *Geoscientific Model Development, 14*(6), 3741–3768. doi:10.5194/gmd-14-3741-2021

Huang, L., Liu, S., Yang, Z., Xing, J., Zhang, J., Bian, J., Li, S., Sahu, S. K., Wang, S., & Liu, T.-Y. (2021). Exploring deep learning for air pollutant emission estimation. *Geoscientific Model Development, 14*(7), 4641–4654. doi:10.5194/gmd-14-4641-2021

Huang, L., & Topping, D. (2021). JlBox v1.1: A Julia-based multi-phase atmospheric chemistry box model. *Geoscientific Model Development, 14*(4), 2187–2203. doi:10.5194/gmd-14-2187-2021

Humphreys, M. P., Lewis, E. R., Sharp, J. D., & Pierrot, D. (2022). PyCO2SYS v1.8: Marine carbonate system calculations in Python. *Geoscientific Model Development, 15*(1), 15–43. doi:10.5194/gmd-15-15-2022

Leach, N. J., Jenkins, S., Nicholls, Z., Smith, C. J., Lynch, J., Cain, M., Walsh, T., Wu, B., Tsutsui, J., & Allen, M. R. (2021). FaIRv2.0.0: A generalized impulse response model for climate uncertainty and future scenario exploration. *Geoscientific Model Development*, *14*(5), 3007–3036. doi:10.5194/gmd-14-3007-2021

Wu, D., Lin, J. C., Duarte, H. F., Yadav, V., Parazoo, N. C., Oda, T., & Kort, E. A. (2021). A model for urban biogenic CO2 fluxes: Solar-Induced Fluorescence for Modeling Urban biogenic Fluxes (SMUrF v1). *Geoscientific Model Development*, *14*(6), 3633–3661. doi:10.5194/gmd-14-3633-2021

KEY TERMS AND DEFINITIONS

Air Pollution: The presence of harmful substances like PM 2.5, CO2, etc. in the atmosphere.

Air Quality Index: An index that indicates the quality of air. It can vary for different countries.

Earth Observation: To observe or collect information regarding various characteristics of the earth from our sky using air-borne instruments like drones, satellites, etc.

In-Situ: To collect data from the ground level of a remotely sensed location. It is used to verify remote sensing data.

IoT: The Internet of Things (IoT) is a network of ground-based sensors in a particular location.

Meteorology: The study of the earth's weather and climate events and trends.

Pollution Forecast: To predict pollution events or trends in future. It can be from minutes to years.

Remote Sensing: The act of sensing or collecting data like images, sound, etc. from a distance.

Spatial Resolution: The number of pixels available in the satellite image of a certain area. It can be from meters to kilometers.

Temporal Resolution: The revisit duration of a satellite to observe the same location again. It can be from hours to days.

Compilation of References

Abdalla, A. M., Ghaith, I. H., & Tamimi, A. A. (2021). Deep learning weather forecasting survey. *International Conference on Information Technology (ICIT).*

Abdikan, S., Balik Sanli, F., Bektas Balcik, F., & Goksel, C. (2008). Fusion of SAR images (pulsar and radarsat-1) with multispectral spot image: A comparative analysis of resulting images. *International Archives of the Photogrammetry, Remote Sensing and Spatial Information Sciences - ISPRS Archives, 37*, 1197–1202.

Abu Awad, Y., Koutrakis, P., Coull, B. A., & Schwartz, J. (2017). A spatio-temporal prediction model based on support vector machine regression: Ambient Black Carbon in three New England States. *Environmental Research, 159*, 427–434. doi:10.1016/j.envres.2017.08.039 PMID:28858756

Afan, Salman, Kanigoro, & Heryadi. (2015). Weather forecasting using deep learning techniques. Proc. Int'l. Conf. Advanced Computer Science and Info. Systems (ICACSIS), 281-285.

Ahmad, H. F., Alam, A., Bhat, M. S., & Ahmad, S. (2016). One Dimensional Steady Flow Analysis Using HECRAS – A case of River Jhelum, Jammu and Kashmir. *European Scientific Journal, ESJ, 12*(32), 340. doi:10.19044/esj.2016.v12n32p340

Ahmed, A. N., Othman, F. B., Afan, H. A., Ibrahim, R. K., Fai, C. M., Hossain, M. S., Ehteram, M., & Elshafie, A. (2019). Machine learning methods for better water quality prediction. *Journal of Hydrology (Amsterdam), 578*, 124084. doi:10.1016/j.jhydrol.2019.124084

Ahmed, U., Mumtaz, R., Anwar, H., Shah, A. A., Irfan, R., & García-Nieto, J. (2019). Efficient water quality prediction using supervised Machine Learning. *Water (Basel), 11*(11), 1–14. doi:10.3390/w11112210 PMID:32021704

Aktar, M. W., Paramasivam, M., Ganguly, M., Purkait, S., & Sengupta, D. (2010). Assessment and occurrence of various heavy metals in surface water of Ganga river around Kolkata: A study for toxicity and ecological impact. *Environmental Monitoring and Assessment, 160*(1), 207–213. doi:10.100710661-008-0688-5 PMID:19101812

Alawlaki, A. A. (2006). *Statistical methods of classification in GIS*. HTTP: //hdl.handle.net/10603/126331

Alley, W. M. (1984, July). The Palmer Drought Severity Index: Limitations and Assumptions. *Journal of Applied Meteorology, 23*, 110–1109. doi:10.1175/1520-0450(1984)023<1100:TPDSIL>2.0.CO;2

Al-Matarneh, Sheta, Bani-Ahmad, Alshaer, & Al-Oqily. (2014). Development of temperature-based weather forecasting models using neural networks and fuzzy logic. *Int'l. J. of Multimedia and Ubiquitous Engineering, 9*(12), 343-366.

Ameer, S., Shah, M. A., Khan, A., Song, H., Maple, C., Islam, S. U., & Asghar, M. N. (2019). Comparative Analysis of Machine Learning Techniques for Predicting Air Quality in Smart Cities. *IEEE Access: Practical Innovations, Open Solutions, 7*, 128325–128338. doi:10.1109/ACCESS.2019.2925082

Anjana, E. N. S. S., & Student, B. (n.d.). *Review IoT Sensors Classification and Applications in Weather Monitoring*. Academic Press.

An, Q., & Zhao, M. (2017). Time Series Analysis in the Prediction of Water Quality. *Proc. 7th Int. Conf. on Education, Management, Information and Mechanical Engineering (EMIM 2017)*. 10.2991/emim-17.2017.11

Arora, J., Pandya, U., Shah, S., & Doshi, N. (2019). Survey- Pollution monitoring using IoT. *Procedia Computer Science, 155*, 710–715. doi:10.1016/j.procs.2019.08.102

Arya, L. M., Richter, J. C., & Paris, J. F. (1983). Estimating Profile Water Storage from Surface zone Soil Moisture Measurements Under Bare Field Conditions. *Water Resources Research, 19*, 403–1.

Asghar, M. H., Negi, A., & Mohammadzadeh, N. (2015, May). Principle application and vision in Internet of Things (IoT). In *International Conference on Computing, Communication & Automation* (pp. 427-431). IEEE.

Atkinson, P. M., & Tatnall, A. (1997). Introduction neural networks in remote sensing. *International Journal of Remote Sensing, 18*(4), 699–709. doi:10.1080/014311697218700

Baboo, S. S., & Shereef, I. K. (2010). An efficient weather forecasting system using artificial neural network. *International Journal of Environmental Sciences and Development, 1*(4), 321–326. doi:10.7763/IJESD.2010.V1.63

Baker, J. M. (1990). Measurement of soil water content. *Remote Sensing Reviews, 5*, 263–279.

Ball, J. E., Anderson, D. T., & Chan, C. S. (2017). Comprehensive survey of deep learning in remote sensing: Theories, tools, and challenges for the community. *Journal of Applied Remote Sensing, 11*(04), 1. doi:10.1117/1.JRS.11.042609

Banara, S., Singh, T., & Chauhan, A. (2022, January). IoT Based Weather Monitoring System for Smart Cities: A Comprehensive Review. In *2022 International Conference for Advancement in Technology (ICONAT)* (pp. 1-6). IEEE.

Baste, P., & Dighe, D. (2017). Low Cost Weather Monitoring Station Using Raspberry PI. *International Research Journal of Engineering and Technology, 4*(5).

Battineni, G., Chintalapudi, N., & Amenta, F. (2020). Forecasting of COVID-19 epidemic size in four high hitting nations (USA, Brazil, India and Russia) by Fb-Prophet machine learning model. *Applied Computing and Informatics.*

Bau, D., Zhou, B., Khosla, A., Oliva, A., & Torralba, A. (2017). Network dissection: Quantifying interpretability of deep visual representations. Computer Vision and Pattern Recognition.

Bauer, P., Thorpe, A., & Brunet, G. (2015). The quiet revolution of numerical weather prediction. *Nature, 525*(7567), 47–55. doi:10.1038/nature14956 PMID:26333465

Bazzani, L., Bergamo, A., Anguelov, D., & Torresani, L. (2016). Self-taught object localization with deep networks. WACV.

Beer. (1852). Bestimmung der Absorption des rothen Lichts in farbigen Flüssigkeiten [Determination of the absorption of red light in colored liquids]. *Annalen der Physik und Chemie, 86*, 78–88.

Benediktsson, J. A., Swain, P. H., & Ersoy, O. K. (1993). Conjugate-gradient neural networks in multisource and very-high- dimensional remote sensing data classification. *International Journal of Remote Sensing, 14*(15), 2883–2903. doi:10.1080/01431169308904316

Bengio, Y., Courville, A., & Vincent, P. (2013). Representation learning: A review and new perspectives. *IEEE Transactions on Pattern Analysis and Machine Intelligence, 35*(8), 1798–1828.

Benyezza, H., Bouhedda, M., Djellout, K., & Saidi, A. (2018). Smart irrigation system based Thingspeak and Arduino. *IEEE International Conference on Applied Smart Systems (ICASS2018) Proceedings,* 1–4.

Bergmier, C., & Benitez, J. (2012). Neural networks in R using the stuttgart neural network simulator: RSNNS. *Journal of Statistical Software*, *46*(7), 1–26.

Bigah, Y., Rousseau, A. N., & Gumiere, S. J. (2019). Development of a steady-state model to predict daily water table depth and root zone soil matric potential of a cranberry field with a subirrigation system. *Agricultural Water Management*, *213*, 1016–1027.

Bochenek, B., & Ustrnul, Z. (2022). Machine Learning in Weather Prediction and Climate Analyses—Applications and Perspectives. *Atmosphere*, *13*(2), 180. doi:10.3390/atmos13020180

Brenowitz, N. D., & Bretherton, C. S. (2019). Spatially extended tests of a neural network parametrization trained by coarse-graining. *Journal of Advances in Modeling Earth Systems*, *11*(8), 2728–2744. https://onlinelibrary.wiley.com/doi/ abs/10.1029/2019MS001711

Budiarti, R. P. N., Tjahjono, A., Hariadi, M., & Purnomo, M. H. (2019). Development of IoT for Automated Water Quality Monitoring System. *Proc. IEEE Int. Conf. on Computer Science, Information Technology, and Electrical Engineering (ICOMITEE)*, 211–216. 10.1109/ICOMITEE.2019.8920900

Bulusu, N., Estrin, D., & Girod, L. (2001). *Scalable coordination for wireless sensor networks: self-configuring localization systems*. In *International Symposium on Communication Theory and Applications (ISCTA 2001)*, Ambleside, UK.

Campbell, G. S., & Gee, G. W. (1986). Water potential: miscellaneous methods. In A. Klute (Ed.), Methods of Soil Analysis, Part 1. Physical and Mineralogical Methods. Academic Press.

Campbell, G. S. (1988). Soil water potential measurement: An overview. *Irrigation Science*, *9*, 265–273.

Cao, S., & Wang, S. (2018). Design of River Water Quality Assessment and Prediction Algorithm. *2018 Eighth International Conference on Instrumentation \& Measurement, Computer, Communication and Control (IMCCC)*, 1625–1631. 10.1109/IMCCC.2018.00335

Carpenter, G. A., Gopal, S., Macomber, S., Martens, S., Woodcock, C. E., & Franklin, J. (1999). A neural network method for efficient vegetation mapping. *Remote Sensing of Environment*, *70*(3), 326–338. doi:10.1016/S0034-4257(99)00051-6

Cassel, D. K., & Klute, A. (1986). Water potential: Tensiometry. In A. Klute (Ed.), *Methods of soil analysis*. ASA and SSSA.

Cerpa, A., & Estrin, D. (2004). ASCENT: Adaptive self-configuring sensor networks topologies. *IEEE Transactions on Mobile Computing*, *3*(3), 272–285.

Chang, T.-C. (2019). The Performance of Grey Model and Auto-Regressive Integrated Moving Average for Human Resources Prediction in China. *2019 IEEE International Conference on Computation, Communication and Engineering (ICCCE)*, 245–248. 10.1109/ICCCE48422.2019.9010801

Chauhan, D., & Thakur, J. (2014). Data mining techniques for weather prediction: A review. *International Journal on Recent and Innovation Trends in Computing and Communication*, *2*(8), 2184–2189.

Chawla, P., Kumar, P., Singh, M., Hasteer, N., & Ghanshyam, C. (2015). Prediction of pollution potential of Indian rivers using empirical equation consisting of water quality parameters. *Proc. IEEE Technological Innovation in ICT for Agriculture and Rural Development (TIAR)*, 214–219. 10.1109/TIAR.2015.7358560

Chen, B., Chen, B.-F., & Lin, H.-T. (2018). Rotation-blended CNNs on a New Open Dataset for Tropical Cyclone Image-to-intensity Regression. In Proceedings of the 24th ACM SIGKDD International Conference on Knowledge Discovery & Data Mining (KDD '18) (pp. 90–99). Association for Computing Machinery. doi.org/10.1145/3219819.3219926.

Chen, B.-F., Chen, B., Lin, H.-T., & Elsberry, R. L. (2019). Estimating Tropical Cyclone Intensity by Satellite Imagery Utilizing Convolutional Neural Networks. *Weather and Forecasting*. doi:10.1175/waf-d-18-0136

Chen, B.-J., Chang, M.-W., & Lin, C.-J. (2004). Load forecasting using support vector Machines: A study on EUNITE competition 2001. *IEEE Transactions on Power Systems*, *19*(4), 1821–1830. doi:10.1109/TPWRS.2004.835679

Chen, C.-T., Chen, K.-S., & Lee, J.-S. (2003). The use of fully polarimetric information for the fuzzy neural classification of SAR images. *IEEE Transactions on Geoscience and Remote Sensing*, *41*(9), 2089–2100. doi:10.1109/TGRS.2003.813494

Chen, G., & Hou, R. (2007). A new machine double-layer learning method and its application in nonlinear time series forecasting. *2007 International Conference on Mechatronics and Automation*, 795–799. 10.1109/ICMA.2007.4303646

Chen, Y., Song, L., Liu, Y., Yang, L., & Li, D. (2020). A review of the artificial neural network models for water quality prediction. *Applied Sciences (Basel, Switzerland)*, *10*(17), 5776. doi:10.3390/app10175776

Chevallier, F., Chéruy, F., Scott, N. A., & Chédin, A. (1998). A neural network approach for a fast and accurate computation of a longwave radiative budget. *Journal of Applied Meteorology*, *37*(11), 1385–1397. doi:10.1175/1520-0450(1998)037<1385:ANNAFA>2.0.CO;2

Chien, H.-F., Lee, S.-H., Lee, W., & Tsai, Y. (2007). Forecasting Monthly Sales of Cell-phone Companies - the Use of VAR Model. *Second International Conference on Innovative Computing, Informatio and Control (ICICIC 2007)*, 459. 10.1109/ICICIC.2007.314

Chiuderi, A. (1997). Multisource and multitemporal data in land cover classification tasks: The advantage offered by neural networks. In Geoscience and Remote Sensing, 1997. IGARSS'97. Remote Sensing-A Scientific Vision for Sustainable Development., 1997 IEEE International, 4, 1663–1665.

Chow, V. T., Maidment, D. R., & Mays, L. W. (1998). *Applied Hydrology Chow 1988*. http://ponce.sdsu.edu/Applied_Hydrology_Chow_1988.pdf

Christopher, M. B. (2016). *Pattern Recognition and Machine Learning*. Springer-Verlag.

Chu, D. A., Kaufman, Y. J., Zibordi, G., Chern, J. D., Mao, J., Li, C., & Holben, B. N. (2003). Global monitoring of air pollution over land from the Earth Observing System-Terra Moderate Resolution Imaging Spectroradiometer (MODIS). *Journal of Geophysical Research*, *108*(21), 1–18. doi:10.1029/2002JD003179

Chu, P. S., & Zhao, X. (2007). A Bayesian Regression Approach for Predicting Seasonal Tropical Cyclone Activity Over the Central North Pacific. *Journal of Climate*, *20*(15), 4002–4013. doi:10.1175/JCLI4214.1

Cire¸san, D. C., Meier, U., Gambardella, L. M., & Schmidhuber, J. (2010). Deep, big, simple neural nets for handwritten digit recognition. *Neural Computation*, *22*(12), 3207–3220. doi:10.1162/NECO_a_00052 PMID:20858131

Civco, D. L. (1993). Artificial neural networks for land-cover classification and mapping. *International Journal of Geographical Information Science*, *7*(2), 173–186.

Colman, E. A., & Hendrix, T. M. (1949). The fibreglass electrical soil-moisture instrument. *Soil Science*, *67*, 425–438.

Contreras, L., & Ferri, C. (2016). Wind-sensitive Interpolation of Urban Air Pollution Forecasts. *Procedia Computer Science*, *80*, 313–323. doi:10.1016/j.procs.2016.05.343

Cummings, R. W., & Chandler, R. F. (1940). A field comparison of the electrothermal and gypsum block electrical resistance methods with the tensiometer method for estimating soil moisture in situ. *Soil Science Society of America Proceedings*, *5*, 80–85.

Daris, Fente, & Singh. (2018). Weather forecasting using artificial neural networks. *2018 2nd Int'l. Conf. on Inventive Communication and Computational Technologies (ICICCT)*, 1757-1761.

Das, L., Kumar, A., Singh, S., Ashar, A. R., & Jangu, R. (2021, January). IoT Based Weather Monitoring System Using Arduino-UNO. In *2021 2nd International Conference on Computation, Automation and Knowledge Management (ICCAKM)* (pp. 260-264). IEEE.

Dataset Sources. (2021). https://gisgeography.com/free-satellite-imagery-data-list/

Datasets. (2017). https://geoawesomeness.com/list-of-top-10-sources-of-free-remote-sensing-data/

De Oliveira, L., & Talamini, E. (2010). Water resources management in the Brazilian agricultural irrigation. *Journal of Ecology and the Natural Environment*, *2*, 123–133.

Dean, T. J., Bell, J. P., & Baty, A. J. B. (1987). Soil Moisture Measurement by an Improved Capacitance Technique. Part I. Sensor Design and Performance. *Journal of Hydrology (Amsterdam)*, *93*, 67–78.

Debry, E., & Mallet, V. (2014). Ensemble forecasting with machine learning algorithms for ozone, nitrogen dioxide and PM10 on the Prev'Air platform. *Atmospheric Environment*, *91*, 71–84. doi:10.1016/j.atmosenv.2014.03.049

Del Pra. (2020). *Time series forecasting with deep learning and attention mechanism*. Towards Data Science. https://towardsdatascience.com/time-series-forecasting-with-deep-learning-and-attention-mechanism-2d001fc871fc

Deng, W., Wang, G., Zhang, X., Guo, Y., & Li, G. (2014). Water quality prediction based on a novel hybrid model of ARIMA and RBF neural network. *2014 IEEE 3rd International Conference on Cloud Computing and Intelligence Systems*, 33–40.

Ding, J., Chen, B., Liu, H., & Huang, M. (2016). Convolutional Neural Network with Data Augmentation for SAR Target Recognition. *IEEE Geoscience and Remote Sensing Letters*, *13*(3), 364–368. doi:10.1109/LGRS.2015.2513754

Doorenbos, J., & Pruitt, W. O. (1984). *Crop water requirements. FAO Irrigation and Drainage, Paper No. 24*. FAO.

Dunkel, Z. (2009). Brief surveying and discussing of drought indices used in agricultural meteorology. *Quarterly Journal of the Hungarian Meteorological Service*, *113*, 23–37.

Earth data by NASA. (2021). https://earthdata.nasa.gov/learn/backgrounders/remote-sensing

Ebert-Uphoff, I., Thompson, D. R., Demir, I., Gel, Y. R., Karpatne, A., Guereque, M., ... Smyth, P. (2017, September). A vision for the development of benchmarks to bridge geoscience and data science. *17th International Workshop on Climate Informatics*.

Eccles, R., Zhang, H, & Hamilton, D. P. (2019). *A review of the effects of climate change on riverine flooding in subtropical and tropical regions*. doi:10.2166/wcc.2019.175

Edwards, D. C., & McKee, T. B. (1997). *Characteristics of 20th century drought in the United States at multiple time scales*. Colorado State Univ., Climatology Report No. 97-2.

Eldakhly, N. M., Aboul-Ela, M., & Abdalla, A. (2017). Air Pollution Forecasting Model Based on Chance Theory and Intelligent Techniques. *International Journal of Artificial Intelligence Tools*, *26*(6), 1–30. doi:10.1142/S0218213017500245

Eniolorunda, N. (2014). Climate change analysis and adaptation: The role of remote sensing (Rs) and geographical information system (Gis). *International Journal of Computational Engineering Research, 4*(1), 41–51.

Eslami, E., Salman, A. K., Choi, Y., Sayeed, A., & Lops, Y. (2020). A data ensemble approach for real-time air quality forecasting using extremely randomized trees and deep neural networks. *Neural Computing & Applications, 32*(11), 7563–7579. doi:10.100700521-019-04287-6

European Space Agency. (2020). https://www.esa.int/ESA_Multimedia/Images/2020/03/Low_Earth_orbit

Evett, S. R., Colaizzi, P. D., O'Shaughnessy, S. A., Hunsaker, D. J., & Evans, R. G. (2014). Irrigation Management. In *Encyclopedia of Remote Sensing.* Springer. doi:10.1007/978-0-387-36699-9_73

Evett, S. R. (2008). Neutron moisture meters. In S. R. Evett, L. K. Heng, P. Moutonnet, & M. L. Nguyen (Eds.), *Field estimation of soil water content: A practical guide to methods, instrumentation, and sensor technology. IAEA-TCS-30* (pp. 39–54). International Atomic Energy Agency.

Faruk, D. Ö. (2010). A hybrid neural network and ARIMA model for water quality time series prediction. *Engineering Applications of Artificial Intelligence, 23*(4), 586–594. doi:10.1016/j.engappai.2009.09.015

Fletcher, S. J. (2017). *Data assimilation for the geosciences: From theory to application.* Elsevier.

Foody, G. (1995a). Using prior knowledge in artificial neural network classification with a minimal training set. *Remote Sensing, 16*(2), 301–312. doi:10.1080/01431169508954396

Foody, G. M. (1995b). Land cover classification by an artificial neural network with ancillary information. *International Journal of Geographical Information Systems, 9*(5), 527–542. doi:10.1080/02693799508902054

Framework, S. (2015). Retrieved April 6, 2022, from https://www.undrr.org/implementing-sendai-framework/what-sendai-framework

Gagne, D. J. II, McGovern, A., & Xue, M. (2014). Machine learning enhancement of storm-scale ensemble probabilistic quantitative precipitation forecasts. *Weather and Forecasting, 29*(4), 1024–1043. doi:10.1175/WAF-D-13-00108.1

Gangopadhyay, S., & Mondal, M. K. (2016, January). A wireless framework for environmental monitoring and instant response alert. In *2016 international conference on microelectronics, computing and communications (MicroCom)* (pp. 1-6). IEEE.

Gardner, W. H. (1986). Water content. In A. Klute (Ed.), Methods of Soil Analysis, Part 1. Physical and Mineralogical Methods. Academic Press.

Gardner, W., Israelsen, O. W., Edlefsen, N. E., & Clyde, D. (1922). The capillary potential function and its relation to irrigation practice. *Physical Review, 20*, 196–204.

Geetha, S., & Gouthami, S. (2016). Internet of things enabled real time water quality monitoring system. *Smart Water, 2*(1), 1–19. doi:10.118640713-017-0005-y

Geng, J., Fan, J., Wang, H., Ma, X., Li, B., & Chen, F. (2015). High- Resolution SAR Image Classification via Deep Convolutional Autoencoders. *IEEE Geoscience and Remote Sensing Letters, 12*(11), 2351–2355. doi:10.1109/LGRS.2015.2478256

Genuer, R., Poggi, J. M., & Tuleau-Malot, C. (2010). Variable selection using random forests. *Pattern Recognition Letters, 31*(14), 2225–2236.

Ghaderi, Sanandaji, & Ghaderi. (2017). *Deep forecast: Deep learning-based spatio-temporal forecasting*. arXiv preprint arXiv:1707.08110

Ghanbarzadeh, A., Noghrehabadi, A., Assareh, E., & Behrang, M. (2009). Solar radiation forecasting based on meteorological data using artificial neural networks. *Proc. 7th IEEE International Conference on Industrial Informatics*, 227-231. 10.1109/INDIN.2009.5195808

Girshick, R., Donahue, J., Darrell, T., & Malik, J. (2016). Region- based convolutional networks for accurate object detection and segmentation. *IEEE Transactions on Pattern Analysis and Machine Intelligence*, *38*(1), 142–158. doi:10.1109/TPAMI.2015.2437384 PMID:26656583

Gocheva-Ilieva, S. G., Ivanov, A. V., Voynikova, D. S., & Boyadzhiev, D. T. (2014). Time series analysis and forecasting for air pollution in small urban area: An SARIMA and factor analysis approach. *Stochastic Environmental Research and Risk Assessment*, *28*(4), 1045–1060. doi:10.100700477-013-0800-4

Goodfellow, I., Bengio, Y., & Courville, A. (2016). *Deep Learning*. MIT Press. Available from http://www.deeplearningbook.org

Gopal, S., Woodcock, C. E., & Strahler, A. H. (1999). Fuzzy neural network classification of global land cover from a 1 avhrr data set. *Remote Sensing of Environment*, *67*(2), 230–243. doi:10.1016/S0034-4257(98)00088-1

Grönquist, P., Yao, C., Ben-Nun, T., Dryden, N., Dueben, P., Li, S., & Hoefler, T. (2021). Deep learning for post-processing ensemble weather forecasts. *Philosophical Transactions of the Royal Society A*, *379*(2194), 20200092.

Grover, Kapoor, & Horvitz. (2015). A deep hybrid model for weather forecasting. *Proceedings of the 21th ACM SIGKDD Int'l.Conf. Knowledge Discovery and Data Mining*, 379-386. 10.1145/2783258.2783275

Gupta, P., & Christopher, S. A. (2009). Particulate matter air quality assessment using integrated surface, satellite, and meteorological products: Multiple regression approach. *Journal of Geophysical Research*, *114*(D14), D14205. Advance online publication. doi:10.1029/2008JD011496

Guttman, N. B. (1998). Comparing the Palmer Drought Index and the Standardized Precipitation Index. *Journal of the American Water Resources Association*, *34*(1), 113–121.

Guttman, N. B. (1999). Accepting the Standardized Precipitation Index: A calculation algorithm. *Journal of the American Water Resources Association*, *35*(2), 311–322.

Haghiabi, A. H., Nasrolahi, A. H., & Parsaie, A. (2018). Water quality prediction using machine learning methods. *Water Quality Research Journal*, *53*(1), 3–13. doi:10.2166/wqrj.2018.025

Halim, S., & Bisono, I. N. (2008). Automatic seasonal auto regressive moving average models and unit root test detection. *International Journal of Management Science and Engineering Management*, *3*(4), 266–274. doi:10.1080/17509653.2008.10671053

Hall, D. L., & Llinas, J. (1997). An introduction to multisensor data fusion. *Proceedings of the IEEE*, *85*(1), 6–23. doi:10.1109/5.554205

Hallikainen, M. T., Ulaby, F. T., Dobson, M. C., El-Rayes, M. A., & Wu, L. K. (1985). Microwave dielectric behaviour of wet soil, Part 1. Empirical models and experimental observations. *IEEE Transactions on Geoscience and Remote Sensing*, *23*(1), 25–34.

Hamill, T. M., & Whitaker, J. S. (2006). Probabilistic quantitative precipitation forecasts based on reforecast analogs: Theory and application. *Monthly Weather Review*, *134*(11), 3209–3229.

Haralick, R. M., & Shanmugam, K. (1973). Textural features for image classification. *IEEE Transactions on Systems, Man, and Cybernetics, 3*(6), 610–621.

Hasan, N., Nath, N. C., & Rasel, R. I. (2015). A support vector regression model for forecasting rainfall. *Proc. 2nd International Conference on Electrical Information and Communication Technology (EICT)*, 554-559. 10.1109/EICT.2015.7392014

He, K., Zhang, X., Ren, S., & Sun, J. (2016). Deep residual learning for image recognition. *Proceedings of the IEEE conference on computer vision and pattern recognition*, 770–778.

He, K., Zhang, X., Ren, S., & Sun, J. (2016). Identity mappings in deep residual networks. In *European conference on computer vision* (pp. 630-645). Springer.

Hewitson, B. C., & Crane, R. G. (1994). *Neural nets: applications in geography: applications for geography* (Vol. 29). Springer Science & Business Media. doi:10.1007/978-94-011-1122-5

Higa, M., Tanahara, S., Adachi, Y., Ishiki, N., Nakama, S., Yamada, H., ... Miyata, R. (2021). Domain knowledge integration into deep learning for typhoon intensity classification. *Scientific Reports, 11*(1), 1–10.

Hignett, C., & Evett, S. R. (2002). Neutron thermalization. In J. H. Dane & G. C. Topp (Eds.), *Methods of soil analysis. Part 4. Physical methods* (pp. 501–521). Amer. Soc. Agron.

Holmes, J. W., Taylor, S. A., & Richards, S. J. (1967). In R. M. Hagan, H. R. Haise, & T. W. Edminster (Eds.), *Measurement of soil water in Irrigation of agricultural lands* (pp. 275–303). American Society of Agronomy.

Hoseinzade, E., & Haratizadeh, S. (2019). CNNpred: CNN-based stock market prediction using a diverse set of variables. *Expert Systems with Applications, 129*, 273–285. doi:10.1016/j.eswa.2019.03.029

Hou, B., Luo, X., Wang, S., Jiao, L., & Zhang, X. (2015). Polarimetric SAR images classification using deep belief networks with learning features. *International Geoscience and Remote Sensing Symposium (IGARSS),* (2), 2366–2369. 10.1109/IGARSS.2015.7326284

Hu, P., Tong, J., Wang, J., Yang, Y., & de Oliveira Turci, L. (2019). A hybrid model based on CNN and Bi-LSTM for urban water demand prediction. *2019 IEEE Congress on Evolutionary Computation (CEC)*, 1088–1094. 10.1109/CEC.2019.8790060

Husain, A. (2017). Flood Modelling by using HEC-RAS. *International Journal of Engineering Trends and Technology, 50*(1), 1–7. doi:10.14445/22315381/IJETT-V50P201

Ibrahimi, A. E., & Baali, A. (2018). Application of Several Artificial Intelligence Models for Forecasting Meteorological Drought Using the Standardized Precipitation Index in the Saïss Plain (Northern Morocco). *International Journal of Intelligent Engineering and Systems, 11*, 267–275.

Ip, W. F., Vong, C. M., Yang, J. Y., & Wong, P. K. (2010). Forecasting daily ambient air pollution based on least squares support vector machines. *2010 IEEE International Conference on Information and Automation, ICIA 2010*, 571–575. 10.1109/ICINFA.2010.5512401

Iqbal, K., Ahmad, S., & Dutta, V. (2019). Pollution mapping in the urban segment of a tropical river: Is water quality index (WQI) enough for a nutrient-polluted river? *Applied Water Science, 9*(8), 197–213. doi:10.100713201-019-1083-9

Jain, S., & Ramesh, D. (2020). Machine learning convergence for weather based crop selection. *IEEE International Students Conference on Electrical, Electronics and Computer Science (SCEECS-2020)*, 1-6, 10.1109/SCEECS48394.2020.75

Jalal, D., & Ezzedine, T. (2019). Performance analysis of machine learning algorithms for water quality monitoring system. *Proc. IEEE Int. Conf. on Internet of Things, Embedded Systems and Communications (IINTEC)*, 86–89. 10.1109/IINTEC48298.2019.9112096

Jensen, M. E., Burman, R. D., & Allen, R. G. (1990). Evapotranspiration and Irrigation Water Requirements. *ASCE Manuals and Reports on Engineering Practice,* 70.

Jichang, T. U., Xueqin, Y., Chaobo, C., Song, G. A. O., Jingcheng, W., & Cheng, S. U. N. (2019). Water Quality Prediction Model Based on GRU hybrid network. *2019 Chinese Automation Congress (CAC)*, 1893–1898.

Ji, L., Zou, Y., He, K., & Zhu, B. (2019). Carbon futures price forecasting based with ARIMA-CNN-LSTM model. *Procedia Computer Science*, *162*, 33–38. doi:10.1016/j.procs.2019.11.254

Johnson, D. L., Ambrose, S. H., Bassett, T. J., Bowen, M. L., Crummey, D. E., Isaacson, J. S., Johnson, D. N., Lamb, P., Saul, M., & Winter-Nelson, A. E. (1997). Meanings of environmental terms. *Journal of Environmental Quality*, *26*(3), 581–589. doi:10.2134/jeq1997.00472425002600030002x

Joslyn, K. (2018). *Water quality factor prediction using supervised machine learning*. Academic Press.

K B. P., & S.M, Y. (2019). One Dimensional Unsteady Flow Analysis Using HEC-RAS Modelling Appoach for Flood in Navsari City. *Proceedings of Recent Advances in Interdisciplinary Trends in Engineering & Applications (RAITEA)*. doi:10.2139/ssrn.3351780

Kachroud, M., Trolard, F., Kefi, M., Jebari, S., & Bourrié, G. (2019). Water quality indices: Challenges and application limits in the literature. *Water (Basel)*, *11*(2), 361–387. doi:10.3390/w11020361

Karl, T. R., & Knight, R. W. (1985). Atlas of Monthly Palmer Moisture Anomaly Indices (1931–1983) for the Contiguous United States. Historical Climatology Series, National Climatic Data Center, No. 3–9, 319.

Kavzoglu, T., & Mather, P. M. (1999). Pruning artificial neural networks: An example using land cover classification of multisensor images. *International Journal of Remote Sensing*, *20*(14), 2787–2803. doi:10.1080/014311699211796

Keiner, L. E., & Yan, X.-H. (1998). A neural network model for estimating sea surface chlorophyll and sediments from thematic mapper imagery. *Remote Sensing of Environment*, *66*(2), 153–165. doi:10.1016/S0034-4257(98)00054-6

Keng, C. Y., Shan, F. P., Shimizu, K., Imoto, T., Lateh, H., & Peng, K. S. (2017). Application of vector autoregressive model for rainfall and groundwater level analysis. *AIP Conference Proceedings*, *1870*(1), 60013. doi:10.1063/1.4995940

Keshavkumarsingh, S. (2013). *Design of Wireless Weather Monitoring System.* Department of Electronics & Communication Engineering, National Institute of Technology Rourkela.

Khan, Z. U., & Hayat, M. (2014). Hourly based climate prediction using data mining techniques by comprising entity demean algorithm. *Middle East Journal of Scientific Research*, *21*(8), 1295–1300.

Kiani, F., Amiri, E., Zamani, M., Khodadadi, T., & Abdul Manaf, A. (2015). Efficient intelligent energy routing protocol in wireless sensor networks. *International Journal of Distributed Sensor Networks*, *11*(3), 618072.

Kikon, A., & Deka, P. C. (2021). Artificial intelligence application in drought assessment, monitoring and forecasting: a review. *Stoch Environ Res Risk Assess*. doi:10.1007/s00477-021-02129-3

Kitagawa, G. (2020). *Introduction to Time Series Modeling with Applications in R*. CRC Press. doi:10.1201/9780429197963

Kleinert, F., Leufen, L. H., & Schultz, M. G. (2020). IntelliO3-ts v1.0: A neural network approach to predict near-surface ozone concentrations in Germany. *Geoscientific Model Development Discussions*, *2020*, 1–69. doi:10.5194/gmd-2020-169

Kogekar, A. P., Nayak, R., & Pati, U. C. (2021a). A CNN-BiLSTM-SVR based Deep Hybrid Model for Water Quality Forecasting of the River Ganga. *2021 IEEE 18th India Council International Conference (INDICON)*, 1–6. 10.1109/INDICON52576.2021.9691532

Kogekar, A. P., Nayak, R., & Pati, U. C. (2021c). Forecasting of Water Quality for the River Ganga using Univariate Time-series Models. *2021 8th International Conference on Smart Computing and Communications (ICSCC)*, 52–57. 10.1109/ICSCC51209.2021.9528216

Kogekar, A. P., Nayak, R., & Pati, U. C. (2021b). A CNN-GRU-SVR based Deep Hybrid Model for Water Quality Forecasting of the River Ganga. *2021 International Conference on Artificial Intelligence and Machine Vision (AIMV)*, 1–6. 10.1109/AIMV53313.2021.9670916

Kohl, M. (2015). *Introduction to Statistical Analysis with R*. Bookboon.

Kowalczuk, Z., Swiergal, M., & Wróblewski, M. (2017, September). River Flow Simulation Based on the HEC-RAS System. In *International Conference on Diagnostics of Processes and Systems* (pp. 253-266). Springer. 10.1007/978-3-319-64474-5

Krasnopolsky, V. M., Fox-Rabinovitz, M. S., & Chalikov, D. V. (2005). New approach to calculation of atmospheric model physics: Accurate and fast neural network emulation of longwave radiation in a climate model. *Monthly Weather Review*, *133*(5), 1370–1383. doi:10.1175/MWR2923.1

Krizhevsky, A., Sutskever, I., & Hinton, G. E. (2012). Imagenet classification with deep convolutional neural networks. Advances in neural information processing systems, 1097–1105.

Krizhevsky, A., Sutskever, I., & Hinton, G. E. (2017, June). ImageNet classification with deep convolutional neural networks. *Communications of the ACM*, *60*(6), 84–90. doi:10.1145/3065386

Kumar Jha, B., & Pande, S. (2021). Time Series Forecasting Model for Supermarket Sales using FB-Prophet. *2021 5th International Conference on Computing Methodologies and Communication (ICCMC)*, 547–554. 10.1109/ICCMC51019.2021.9418033

Kumar, D., & Alappat, B. J. (2009). NSF-water quality index: Does it represent the experts' opinion? *Practice Periodical of Hazardous, Toxic, and Radioactive Waste Management*, *13*(1), 75–79. doi:10.1061/(ASCE)1090-025X(2009)13:1(75)

Kumar, D., Aseri, T. C., & Patel, R. B. (2010). EECHDA: Energy Efficient Clustering Hierarchy and Data Accumulation For Sensor Networks. *BIJIT*, *2*(1), 150–157.

Kussul, N., Shelestov, A., Lavreniuk, M., Butko, I., & Skakun, S. (2016). Deep learning approach for large scale land cover mapping based on remote sensing data fusion. In *Geoscience and Remote Sensing Symposium (IGARSS), 2016 IEEE International* (pp. 198–201). IEEE. 10.1109/IGARSS.2016.7729043

Lachure, S., Bhagat, A., & Lachure, J. (2015). Review on precision agriculture using wireless sensor network. *International Journal of Applied Engineering Research*, *10*(20), 16560–16565.

Lagerquist, R., McGovern, A., & Smith, T. (2017). Machine learning for real-time prediction of damaging straight-line convective wind. *Weather and Forecasting*, *32*(6), 2175–2193.

Lardeux, C., Frison, P. L., Tison, C., Souyris, J. C., Stoll, B., Fruneau, B., & Rudant, J. P. (2009). Support vector machine for multifrequency SAR polarimetric data classification. *IEEE Transactions on Geoscience and Remote Sensing*, *47*(12), 4143–4152. doi:10.1109/TGRS.2009.2023908

Lazo, J. K., Morss, R. E., & Demuth, J. L. (2009). 300 billion served: Sources, perceptions, uses, and values of weather forecasts. *Bulletin of the American Meteorological Society*, *90*(6), 785–798.

LeCun, Y., Bengio, Y., & Hinton, G. (2015). Deep learning. *Nature, 521*(7553), 436-444.

LeCun, Y., Bottou, L., Bengio, Y., & Haffner, P. (1998). Gradient-based learning applied to document recognition. *Proceedings of the IEEE*, *86*(11), 2278–2324.

Lek, S., & Gu'egan, J.-F. (1999). Artificial neural networks as a tool in ecological modelling, an introduction. *Ecological Modelling*, *120*(2-3), 65–73. doi:10.1016/S0304-3800(99)00092-7

Lewis, B., DeGuchy, O., Sebastian, J., & Kaminski, J. (2019). Realistic SAR data augmentation using machine learning techniques. *Proceedings of the Society for Photo-Instrumentation Engineers*, •••, 10987.

Lewis, N. (2016). *Deep Learning Made Easy with R: A Gentle Introduction for Data Science*. CreateSpace Independent Publishing Platform.

Lewis, N. (2017). *Neural Networks for Time Series Forecasting with R: Intuitive Step by Step for Beginners*. CreateSpace Independent Publishing Platform.

Li, B., Yang, G., Wan, R., Dai, X., & Zhang, Y. (2016). Comparison of random forests and other statistical methods for the prediction of lake water level: A case study of the Poyang Lake in China. *Hydrology Research*, *47*(S1), 69–83. doi:10.2166/nh.2016.264

Li, J., Huang, X., & Gong, J. (2019). Deep neural network for remote-sensing image interpretation: Status and perspectives. *National Science Review*, *6*(6), 1082–1086.

Lin, W.-C., Liao, D.-Y., Liu, C.-Y., & Lee, Y.-Y. (2005). Daily imaging scheduling of an Earth observation satellite. *IEEE Transactions on Systems, Man, and Cybernetics. Part A, Systems and Humans*, *35*(2), 213–223. doi:10.1109/TSMCA.2005.843380

Lipper, L., Thornton, P., Campbell, B. M., Baedeker, T., Braimoh, A., Bwalya, M., Caron, P., Cattaneo, A., Garrity, D., Henry, K., Hottle, R., Jackson, L., Jarvis, A., Kossam, F., Mann, W., McCarthy, N., Meybeck, A., Neufeldt, H., Remington, T., ... Torquebiau, E. F. (2014). Climate-smart agriculture for food security. *Nature Climate Change*, *4*(12), 1068–1072. doi:10.1038/nclimate2437

Liu, Hu, You, & Chan. (2014). Deep neural network-based feature representation for weather forecasting. Proc. on the Int'l. Conf. on Artificial Intelligence (ICAI), 1-7. doi:10.1109/ICACSIS.2015.7415154

Liu, J., Yu, C., Hu, Z., Zhao, Y., Xia, X., Tu, Z., & Li, R. (2018). Automatic and accurate prediction of key water quality parameters based on SRU deep learning in Mariculture. *2018 IEEE International Conference on Advanced Manufacturing (ICAM)*, 437–440. 10.1109/AMCON.2018.8615048

Liu, Q., Wu, S., Wang, L., & Tan, T. (2016). *Predicting the Next Location: A Recurrent Model with Spatial and Temporal Contexts*. AAAI.

Lloyd-Hughes, B., & Saunders, M. A. (2002). A drought climatology for Europe. *International Journal of Climatology*. Advance online publication. doi:10.1002/joc.846

Long, J., Shelhamer, E., & Darrell, T. (2015). Fully convolutional net-works for semantic segmentation. *Proceedings of the IEEE Conference on Computer Vision and Pattern Recognition*, 3431–3440. doi:10.1117/12.2518452

Lu, W., Wang, W., Leung, A. Y. T., Lo, S-Y., Yuen, R. K. K., Xu, Z., & Fan, H. (2002). Air pollutant parameter forecasting using support vector machines. *IEEE Proceedings of the 2002 International Joint Conference on Neural Networks, 1*, 630–635. 10.1109/IJCNN.2002.1005545

Lusia, D. A., & Ambarwati, A. (2018). Multivariate Forecasting Using Hybrid VARIMA Neural Network in JCI Case. *2018 International Symposium on Advanced Intelligent Informatics (SAIN)*, 11–14. 10.1109/SAIN.2018.8673351

Lu, W., Li, J., Li, Y., Sun, A., & Wang, J. (2020). A CNN-LSTM-Based Model to Forecast Stock Prices. *Complexity, 2020*, 2020. doi:10.1155/2020/6622927

Lv, C., & ... (2020). The Time Series Prediction Algorithm Based on Improved GSA-ELM. *2020 IEEE International Conference on Artificial Intelligence and Computer Applications (ICAICA)*, 91–94. 10.1109/ICAICA50127.2020.9181944

Lv, Q., Dou, Y., Niu, X., Xu, J., Xu, J., & Xia, F. (2015). Urban land use and land cover classification using remotely sensed SAR data through deep belief networks. *Journal of Sensors, 2015*. doi:10.1155/2015/538063

Majumdar, P., Mitra, S., & Bhattacharya, D. (2021). IoT for Promoting Agriculture 4.0: A Review from the Perspective of Weather Monitoring, Yield Prediction, Security of WSN Protocols, and Hardware Cost Analysis. *Journal of Biosystems Engineering*, 1–22.

Malicki, M. (1983). A capacity meter for the investigation of soil moisture dynamics. *Zesty Problemowe Postepow Nauk Rolniczych*, 201-214.

Manikiam, B., & Kamsali, N. (2015). Climate Change Analysis Using Satellite Data. *Mapana Journal of Sciences, 14*(1), 25–39. Advance online publication. doi:10.12723/mjs.32.4

Martínez-España, R., Bueno-Crespo, A., Timón, I., Soto, J., Muñoz, A., & Cecilia, J. M. (2018). Air-pollution prediction in smart cities through machine learning methods: A case of study in Murcia, Spain. *Journal of Universal Computer Science, 24*(3), 261–276.

Mas, J. (2004). Mapping land use/cover in a tropical coastal area using satellite sensor data, gis and artificial neural networks. *Estuarine, Coastal and Shelf Science, 59*(2), 219–230. doi:10.1016/j.ecss.2003.08.011

Mat, I., Kassim, M. R. M., Harun, A. N., & Yusoff, I. M. (2016, October). IoT in precision agriculture applications using wireless moisture sensor network. In *2016 IEEE Conference on Open Systems (ICOS)* (pp. 24-29). IEEE.

McGovern, A., Elmore, K. L., Gagne, D. J., Haupt, S. E., Karstens, C. D., Lagerquist, R., ... Williams, J. K. (2017). Using artificial intelligence to improve real-time decision-making for high-impact weather. *Bulletin of the American Meteorological Society, 98*(10), 2073–2090.

McKee, T. B., Doesken, N. J., & Kleist, J. (1995). Drought monitoring with multiple time scales. *Ninth Conference on Applied Climatology*, 233-236.

McKee, T. B., Doesken, N. J., & Klesit, J. (1993). The relationship of drought frequency and duration to time scales. *Eighth Conference on Applied Climatology Proceedings*, 6(January), 17–22.

Mehrkanoon, S. (2019). Deep shared representation learning for weather elements forecasting. *Knowledge-Based Systems, 179*, 120–128. doi:10.1016/j.knosys.2019.05.009

Melgani, F., & Bruzzone, L. (2004). Classification of hyperspectral remote sensing images with support vector machines. *IEEE Transactions on Geoscience and Remote Sensing, 42*(8), 1778–1790.

Mingyuan, G., & Shiying, Z. (2009). Study on VaR Forecasts Based on Realized Range-Based Volatility. *2009 International Conference on Business Intelligence and Financial Engineering*, 860–862. 10.1109/BIFE.2009.197

Montavon, G., Orr, G., & Mu¨ller, K.-R. (2012). *Neural networks: Tricks of the trade* (vol. 7700). Springer.

Monte, B. E. O., Costa, D. D., Chaves, M. B., Magalhães, L. D. O., & Uvo, C. B. (2016). Hydrological and hydraulic modelling applied to the mapping of flood-prone areas. *RBRH*, *21*(1), 152–167. doi:10.21168/rbrh.v21n1.p152-167

Moskolaï, W. R., Abdou, W., Dipanda, A., & Kolyang. (2021). Application of Deep Learning Architectures for Satellite Image Time Series Prediction: A Review. *Remote Sensing*, *13*(23), 4822. doi:10.3390/rs13234822

Mutavhatsindi, T., Sigauke, C., & Mbuvha, R. (2020). Forecasting Hourly Global Horizontal Solar Irradiance in South Africa Using Machine Learning Models. *IEEE Access: Practical Innovations, Open Solutions*, *8*, 198872–198885. doi:10.1109/ACCESS.2020.3034690

Nana, H., Lei, D., Lijie, W., Ying, H., Zhongjian, D., & Bo, W. (2019). Short-term Wind Speed Prediction Based on CNN_GRU Model. *2019 Chinese Control And Decision Conference (CCDC)*, 2243–2247. 10.1109/CCDC.2019.8833472

Nielsen, M. (2015). *Neural Networks and Deep Learning*. Determination Press. https://static.latexstudio.net/article/2018/0912/neuralnetworksanddeeplearning.pdf

Nikesh, G., & Kawitkar, R. S. (2016). Smart agriculture using IoT and WSN based modern technologies. *International Journal of Innovative Research in Computer and Communication Engineering*, *4*(6), 12070–12076.

Noh, H., Hong, S., & Han, B. (2015). Learning deconvolution network for semantic segmentation. *Proceedings of the IEEE International Conference on Computer Vision*, 1520–1528.

Ojha, T., Misra, S., & Raghuwanshi, N. S. (2015). Wireless sensor networks for agriculture: The state-of-the-art in practice and future challenges. *Computers and Electronics in Agriculture*, *118*, 66–84.

Ongdas, N., Akiyanova, F., Karakulov, Y., Muratbayeva, A., & Zinabdin, N. (2020). Application of HEC-RAS (2D) for Flood Hazard Maps generation for Yesil(Ishim) river in Kazakhstan. *Water (Basel)*, *12*(10), 2672. doi:10.3390/w12102672

Oord, A. V. D., Dieleman, S., Zen, H., Simonyan, K., Vinyals, O., Graves, A., . . . Kavukcuoglu, K. (2016). *Wavenet: A generative model for raw audio.* arXiv preprint arXiv:1609.03499.

Oprea, M., Dragomir, E. G., Popescu, M., & Mihalache, S. F. (2016). Particulate matter air pollutants forecasting using inductive learning approach. *Revista de Chimie*, *67*(10), 2075–2081.

Paetzold, R. F., Matzkanin, G. A., & Santos, A. D. L. (1985). Surface soil water content measurement using pulsed nuclear magnetic resonance techniques. *Soil Science Society of America Journal*, *49*, 537–540.

Palmer, W. C. (1965). *Meteorological drought.* United States Department of Commerce, Weather Bureau, Research Paper No. 45.

Pandey, A. K., & Mukherjee, A. (2022). A Review on Advances in IoT-Based Technologies for Smart Agricultural System. *Internet of Things and Analytics for Agriculture*, *3*, 29–44.

Pang, S., Xie, P., Xu, D., Meng, F., Tao, X., & Li, B. … Song, T. (2021). NDFTC: A New Detection Framework of Tropical Cyclones from Meteorological Satellite Images with Deep Transfer Learning. *Remote Sensing*, *13*(9). Advance online publication. doi:10.3390/rs13091860

Pan, S. J., & Yang, Q. (2010). A survey on transfer learning. *IEEE Transactions on Knowledge and Data Engineering*, *22*(10), 1345–1359. doi:10.1109/TKDE.2009.191

Pan, X., & Zhao, J. (2017). A central-point-enhanced convolutional neural network for high-resolution remote-sensing image classification. *International Journal of Remote Sensing, 38*(23), 6554–6581. doi:10.1080/01431161.2017.1362131

Paola, J. D., & Schowengerdt, R. A. (1995). A detailed comparison of backpropagation neural network and maximum-likelihood classifiers for urban land use classification. *IEEE Transactions on Geoscience and Remote Sensing, 33*(4), 981–996. doi:10.1109/36.406684

Paras, A., & Mathur, S. (2016). Simple Weather Forecasting Model Using Mathematical Regression. Indian Res. *J. Extension Educ, 1*(Special Issue), 161–168.

Pathan, A. I., & Agnihotri, P. G. (2021). Application of new HEC-RAS version 5 for 1D hydrodynamic flood modeling with special reference through geospatial techniques: A case of River Purna at Navsari, Gujarat, India. *Modeling Earth Systems and Environment, 7*(2), 1133–1144. doi:10.100740808-020-00961-0

Peixoto Xavier, L. C., Oliveira da Silva, S. M., Carvalho, T. M. N., Pontes Filho, J. D., & Souza Filho, F. A. d. (2020). Use of Machine Learning in Evaluation of Drought Perception in Irrigated Agriculture: The Case of an Irrigated Perimeter in Brazil. *Water (Basel), 12*(1546), 1–20. doi:10.3390/w12061546

Pereira, L. S., Cordery, I., & Iaconides, I. (2002). *Coping with water scarcity.* UNESCO, International Hydrological Programme, IHP-VI, Technical Documents in Hydrology No. 58.

Podest, E. (2018). *SAR for Mapping Land Cover.* Academic Press.

Pohl, C., & Van Genderen, J. L. (1998). Review article multisensor image fusion in remote sensing: Concepts, methods and applications. *International Journal of Remote Sensing, 19*(5), 823–854. doi:10.1080/014311698215748

Pontes Filho, J. D., Souza Filho, F. A., Martins, E. S. P. R., & Studart, T. M. C. (2020). Copula-Based Multivariate Frequency Analysis of the 2012–2018 Drought in Northeast Brazil. *Water (Basel), 12*, 834.

Pottier, E., & Ferro-Famil, L. (2008). Advances in SAR Polarimetry applications exploiting polarimetric spaceborne sensors. *2008 IEEE Radar Conference*, 1–6. 10.1109/RADAR.2008.4720872

Pradhan, R., Aygun, R. S., Maskey, M., Ramachandran, R., & Cecil, D. J. (2018). *Tropical Cyclone Intensity Estimation Using a Deep Convolutional Neural Network. IEEE Transactions on Image Processing, 27*(2), 692–702.

Prasad, Dash, & Mohanty. (2009). A logistic regression approach for monthly rainfall forecasts in meteorological subdivisions of India based on DEMETER retrospective forecasts. *Int. J. Climatology, 30*, 1577-1588. . doi:10.1002/joc.2019

Praveen, K. B., Pradyumna, K., Prateek, J., Pragathi, G., & Madhuri, J. (2022). Inventory Management using Machine Learning. *International Journal of Engineering Research & Technology (Ahmedabad), 9*(6), 866–868.

Priyank Hirani, V. D. (2019). *Water parameter information.* Academic Press.

Pu, T., & Bai, J. (2014). An auto regression compression method for industrial real time data. *The 26th Chinese Control and Decision Conference (2014 CCDC)*, 5129–5132.

Racah, Beckham, Maharaj, Kahou, & Pal. (2016). *Extreme Weather: A large-scale climate dataset for semi-supervised detection, localization, and understanding of extreme weather events.* arXiv preprint arXiv:1612.02095.

Radhika, Y., & Shashi, M. (2009). Atmospheric temperature prediction using support vector machines. *International Journal of Computer Theory and Engineering, 1*(1), 55.

Ragi, N. M., Holla, R., & Manju, G. (2019). Predicting Water Quality Parameters Using Machine Learning. *2019 4th International Conference on Recent Trends on Electronics, Information, Communication \& Technology (RTEICT)*, 1109–1112.

Rajesh, K., Ramaswamy, V., Kannan, K., & Arunkumar, N. (2019). Satellite cloud image classification for cyclone prediction using Dichotomous Logistic Regression Based Fuzzy Hypergraph model. *Future Generation Computer Systems, 98*, 688–696. https://doi.org/10.1016/j.future.2018.12.042

Rao, B. S., Rao, K. S., & Ome, N. (2016). Internet of Things (IoT) based weather monitoring system. *International Journal of Advanced Research in Computer and Communication Engineering, 5*(9), 312-319.

Rasmussen, T. C., & Rhodes, S. C. (1995). Energy- related methods: psychrometers. In L. G. Wilson, L. G. Everett, & S. J. Cullen (Eds.), *Handbook of vadose zone characterization and monitoring* (pp. 329–341). CRC Press.

Rasp, S., Dueben, P. D., Scher, S., Weyn, J. A., Mouatadid, S., & Thuerey, N. (2020). WeatherBench: a benchmark data set for data-driven weather forecasting. *Journal of Advances in Modeling Earth Systems, 12*(11).

Rawlins, S. L., & Campbell, G. S. (1986). Water potential: thermocouple psychrometry. In A. Klute (Ed.), Methods of soil analysis. Part 1- Physical and mineralogical methods. Academic Press.

Redmon, J., Divvala, S., Girshick, R., & Farhadi, A. (2016). You only look once: Unified, real-time object detection. *Proceedings of the IEEE conference on computer vision and pattern recognition*, 779–788. 10.1109/CVPR.2016.91

Remote Sensing. (2021). https://earthdata.nasa.gov/learn/backgrounders/remote-sensing

Reynolds, S. G. (1970a). The gravimetric method of soil moisture determination, Part 1. A study of equipment, and methodological problems. *Journal of Hydrology (Amsterdam), 11*(3), 258–273.

Reynolds, S. G. (1970b). The gravimetric method of soil moisture determination, Part 2. Typical required sample sizes and methods of reducing variability. *Journal of Hydrology (Amsterdam), 11*, 274–287.

Richards, S. K., & Marsh, A. W. (1961). Irrigation based on soil suction measurements. *Soil Science Society Proceedings*, 65-69.

Riesenhuber, M., & Poggio, T. (1999). Hierarchical models of object recognition in cortex. *Nature Neuroscience, 2*(11), 1019–1025. doi:10.1038/14819 PMID:10526343

Riquelme, J. L., Soto, F., Suardíaz, J., Sánchez, P., Iborra, A., & Vera, J. A. (2009). Wireless sensor networks for precision horticulture in Southern Spain. *Computers and Electronics in Agriculture, 68*(1), 25–35.

Ritchards, L. A. (1928). The usefulness of capillary potential to soil moisture and plant investigators. *Journal of Agricultural Research, 37*, 719–742.

Ronneberger, O., Fischer, P., & Brox, T. (2015, October). U-net: Convolutional networks for biomedical image segmentation. In *International Conference on Medical image computing and computer-assisted intervention* (pp. 234-241). Springer.

Roth, K., Schulin, R., Fluhler, H., & Attinger, W. (1990). Calibration of time domain reflectometry for water content measurement using a composite dielectric approach. *Water Resources Research, 26*(10), 2267–2273.

Russakovsky, O., Deng, J., Su, H., Krause, J., Satheesh, S., Ma, S., ... Fei-Fei, L. (2015). Imagenet large scale visual recognition challenge. *International Journal of Computer Vision, 115*(3), 211–252.

Russell, S. J., & Norvig, P. (2016). *Artificial Intelligence: A Modern Approach.* Pearson Education Limited. https://zoo.cs.yale.edu/classes/cs470/materials/aima2010.pdf

Rusyana, A., Tatsara, N., Balqis, R., & Rahmi, S. (2020). Application of Clustering and VARIMA for Rainfall Prediction. *IOP Conference Series. Materials Science and Engineering*, *796*(1), 12063. doi:10.1088/1757-899X/796/1/012063

Samal, K. K. R., Babu, K. S., Acharya, A., & Das, S. K. (2020). Long term forecasting of ambient air quality using deep learning approach. *2020 IEEE 17th India Council International Conference (INDICON)*, 1–6.

Sampathkumar, A., Murugan, S., Elngar, A. A., Garg, L., Kanmani, R., & Malar, A. (2020). A novel scheme for an IoT-based weather monitoring system using a wireless sensor network. In *Integration of WSN and IoT for smart cities* (pp. 181–191). Springer.

Satellite Data. (2021). https://www.iceye.com/satellite-data

Satyanarayana, G. V., & Mazaruddin, S. D. (2013, April). Wireless sensor based remote monitoring system for agriculture using ZigBee and GPS. In *Conference on advances in communication and control systems* (*Vol. 3*, pp. 237-241). Academic Press.

Saunders, R. (2021). The use of satellite data in numerical weather prediction. *Weather*, *76*(3), 95–97. doi:10.1002/wea.3913

Sawaitul, Wagh, & Chatur. (2012). Classification and prediction of future weather by using back propagation algorithm-an approach. *International Journal of Emerging Technology and Advanced Engineering*, *2*(1), 110–113.

Sayegh, A. S., Munir, S., & Habeebullah, T. M. (2014). Comparing the Performance of Statistical Models for Predicting PM10 Concentrations. *Aerosol and Air Quality Research*, *14*(3), 653–665. doi:10.4209/aaqr.2013.07.0259

Schalkoff, R. J. (1992). *Pattern recognition*. Wiley Online Library.

Schmitt, M., Hughes, L. H., & Zhu, X. X. (2018). The sen1-2 dataset for deep learning in SAR-optical data fusion. *ISPRS Annals of the Photogrammetry. Remote Sensing and Spatial Information Sciences*, *4*(1), 141–146.

Schofield, R. K., & Taylor, A. W. (1961). A method for the measurement of the calcium deficit in saline soils. *Journal of Soil Science*, *12*(2), 269–275.

Selvaraju, R. R., Cogswell, M., & Das, A. (2020). Grad-CAM: Visual Explanations from Deep Networks via Gradient-Based Localization. *International Journal of Computer Vision, 128,* 336–359. doi:10.1007/s11263-019-01228-7

Senekane, Mafu, & Taele. (2020). Weather Nowcasting Using Deep Learning Techniques. In *Data Mining - Methods, Applications and Systems*. . doi:10.5772/intechopen.84552

Sen, Z. (2008). *Wadi hydrology*. CRC Press.

ShahiriParsa, A., Noori, M., Heydari, M., & Rashidi, M. (2016). Floodplain zoning simulation by using HEC-RAS and CCHE2D models in the Sungai Maka river. *Air, Soil and Water Research*, *9*, 55–62. doi:10.4137/ASWR.S36089

Shah, K. A., & Joshi, G. S. (2017). Evaluation of water quality index for River Sabarmati, Gujarat, India. *Applied Water Science*, *7*(3), 1349–1358. doi:10.100713201-015-0318-7

Shakhari, S., Verma, A. K., & Banerjee, I. (2019). Remote Location Water Quality Prediction of the Indian River Ganga: Regression and Error Analysis. *Proc. IEEE 17th International Conference on ICT and Knowledge Engineering (ICT\&KE)*, 1–5. 10.1109/ICTKE47035.2019.8966796

Shakya, S., Kumar, S., & Goswami, M. (2020). Deep learning algorithm for satellite imaging based cyclone detection. *IEEE Journal of Selected Topics in Applied Earth Observations and Remote Sensing*, *13*, 827–839.

Shanmugapriya, S., Haldar, D., & Danodia, A. (2019). Opti-mal datasets suitability for pearl millet (Bajra) discrimination using multiparametric SAR data. *Geocarto International*, 6049.

Sharvane, M., Shukla, V. K., & Chaubey, A. (2021). Application of Machine Learning in Precision Agriculture using IoT. *2nd International Conference on Intelligent Engineering and Management (ICIEM)*, 34-39. 10.1109/ICIEM51511.2021.9445312

Shivang, J., & Sridhar, S. S. (2018). Weather prediction for indian location using Machine learning. *International Journal of Pure and Applied Mathematics*, *118*(22), 1945–1949.

Shi, X., Chen, Z., Wang, H., Yeung, D. Y., Wong, W. K., & Woo, W. C. (2015). Convolutional LSTM network: A machine learning approach for precipitation nowcasting. *Advances in Neural Information Processing Systems*, 28.

Simonyan, K., & Zisserman, A. (2014). *Very deep convolutional networks for large-scale image recognition.* arXiv preprint arXiv:1409.1556.

Simonyan, K., & Zisserman, A. (2014). *Very deep convolutional networks for large-scale image recognition.* arXiv preprint arXiv:1409.1556v6.

Singh, Chaturvedi, & Akhter. (2019). Weather Forecasting Using Machine Learning Algorithms. 2019 Int'l. Conf. on Signal Processing and Communication (ICSC), 171-174.

Singh, V. P., Khedikar, S., & Verma, I. J. (2021). Improved yield estimation technique for rice and wheat in Uttar Pradesh, Madhya Pradesh and Maharashtra States in India. *Mausam (New Delhi)*, *70*(3), 541–550. doi:10.54302/mausam.v70i3.257

Stowell, S. (2014). *Using R for Statistics.* Apress.

Sugawara, E., & Nikaido, H. (2014). Properties of adeabc and adeijk efflux systems of Acinetobacter baumannii compared with those of the acrab-tolc system of escherichia coli. *Antimicrobial Agents and Chemotherapy*, *58*(12), 7250–7257. doi:10.1128/AAC.03728-14 PMID:25246403

Sun, P., Wang, S., Gan, H., Liu, B., & Jia, L. (2017, April). Application of HEC-RAS for flood forecasting in perched river-A case study of hilly region, China. *IOP Conference Series: Earth and Environmental Science* (Vol. 61, No. 1, p. 012067). IOP Publishing. doi:10.1088/1755-1315/61/1/012067

Sun, Q., Wan, J., & Liu, S. (2020). Estimation of Sea Level Variability in the China Sea and Its Vicinity Using the SARIMA and LSTM Models. *IEEE Journal of Selected Topics in Applied Earth Observations and Remote Sensing*, *13*, 3317–3326. doi:10.1109/JSTARS.2020.2997817

Sun, Z.-L., Choi, T.-M., Au, K.-F., & Yu, Y. (2008). Sales forecasting using extreme learning machine with applications in fashion retailing. *Decision Support Systems*, *46*(1), 411–419. doi:10.1016/j.dss.2008.07.009

Swenson, H. A. (1965). A primer on water quality. US Department of the Interior, Geological Survey. doi:10.3133/7000057

Szegedy, C., Liu, W., Jia, Y., Sermanet, P., Reed, S., Anguelov, D., ... Rabinovich, A. (2015). Going deeper with convolutions. In *Proceedings of the IEEE conference on computer vision and pattern recognition* (pp. 1-9). IEEE.

Taillardat, M., Mestre, O., Zamo, M., & Naveau, P. (2016). Calibrated ensemble forecasts using quantile regression forests and ensemble model output statistics. *Monthly Weather Review*, *144*(6), 2375–2393.

Talegaon, N. S., Deshpande, G. R., Naveen, B., Channavar, M., & Santhosh, T. C. (2022). Performance Comparison of Weather Monitoring System by Using IoT Techniques and Tools. In *Intelligent Data Communication Technologies and Internet of Things* (pp. 837–853). Springer.

Thies, B., & Bendix, J. (2011). Satellite based remote sensing of weather and climate: Recent achievements and future perspectives. *Meteorological Applications*, *18*(3), 262–295. doi:10.1002/met.288

Thomas, A. M. (1966). In situ measurement of moisture in soil and similar substances by 'fringe' capacitance. *Journal of Scientific Instruments*, *43*, 21–27.

Thompson, R. D. (1975). The climatology of arid world. University of Reading, Department of Geography, Paper No. 35, 39.

Topp, G. C., Annan, J. L., & Davis, A. P. (1980). Electromagnetic determination of soil water content: Measurements in co-axial transmission lines. *Water Resources Research*, *16*, 574–582.

Traore, V. B., Bop, M., Faye, M., Malomar, G., Gueye, E. H. O., Sambou, H., ... Beye, A. C. (2015). Using of Hec-ras model for hydraulic analysis of a river with agricultural vocation: A case study of the Kayanga river basin, Senegal. *American Journal of Water Resources*, *3*(5), 147–154. doi:10.12691/ajwr-3-5-2

Tyagi, S., Sharma, B., Singh, P., & Dobhal, R. (2013). Water quality assessment in terms of water quality index. *American Journal of Water Resources*, *1*(3), 34–38. doi:10.12691/ajwr-1-3-3

Tyralis, H., Papacharalampous, G., & Langousis, A. (2019). A brief review of random forests for water scientists and practitioners and their recent history in water resources. *Water (Basel)*, *11*, 910.

Ulaby, F. T., Moore, R. K., & Fung, A. K. (1986). Microwave Remote Sensing Active and Passive. *From Theory to Applications,* 2136.

Ullah, S., Farooq, M., Sarwar, T., Tareen, M. J., & Wahid, M. A. (2016). Flood modeling and simulations using hydrodynamic model and ASTER DEM—A case study of Kalpani River. *Arabian Journal of Geosciences*, *9*(6), 439. doi:10.100712517-016-2457-z

UN SDGs. (2015). Retrieved April 6, 2022, from https://sdgs.un.org/goals

Vagropoulos, S. I., Chouliaras, G. I., Kardakos, E. G., Simoglou, C. K., & Bakirtzis, A. G. (2016). Comparison of SARIMAX, SARIMA, modified SARIMA and ANN-based models for short-term PV generation forecasting. *2016 IEEE International Energy Conference (ENERGYCON)*, 1–6. 10.1109/ENERGYCON.2016.7514029

Vanclooster, M., Mallants, D., Diels, J., & Feyen, J. (1993). Determining local-scale solute transport parameters using time-domain reflectometry (TDR). *Journal of Hydrology (Amsterdam)*, *148*, 93–107.

Verma, G., Gautam, A., Singh, A., Kaur, R., Garg, A., & Mehta, M. (2017). IOT Application of a Remote Weather Monitoring & Surveillance Station. *International Journal of Smart Home*, *11*(1), 131–140.

Verzani, J. (2005). *Using R for Introductory Statistics*. Chapman & Hall.

Villarin, M. C., & Rodriguez-Galiano, V. F. (2019). Machine Learning for Modeling Water Demand. *Journal of Water Resources Planning and Management*, *145*, 1–15.

Vogel, P., Knippertz, P., Fink, A. H., Schlueter, A., & Gneiting, T. (2018). Skill of global raw and postprocessed ensemble predictions of rainfall over northern tropical Africa. *Weather and Forecasting*, *33*(2), 369–388. doi:10.1175/WAF-D-17-0127.1

Walker, J., Houser, P., & Willgoose, G. (2004). Active microwave remote sensing for soil moisture measurement: A field evaluation using ERS-2. *Hydrological Processes*, *18*, 1975–1997.

Wang, Y., Zhou, J., Chen, K., Wang, Y., & Liu, L. (2017). Water quality prediction method based on LSTM neural network. *2017 12th International Conference on Intelligent Systems and Knowledge Engineering (ISKE)*, 1–5.

Wang, H., Chen, S., Xu, F., & Jin, Y. Q. (2015). Application of deep- learning algorithms to MSTAR data. *International Geoscience and Remote Sensing Symposium (IGARSS)*. 10.1109/IGARSS.2015.7326637

Wang, J., & Song, G. (2018). A Deep Spatial-Temporal Ensemble Model for Air Quality Prediction. *Neurocomputing*, *314*, 198–206. doi:10.1016/j.neucom.2018.06.049

Wang, W., Men, C., & Lu, W. (2008). Online prediction model based on support vector machine. *Neurocomputing*, *71*(4–6), 550–558. doi:10.1016/j.neucom.2007.07.020

Wankhede, P., Sharma, R., & Pote, C. (2014). A review on weather forecasting systems using different techniques and web alerts. *International Journal of Advanced Research in Computer Science and Software Engineering*, *4*(2), 357–359.

Ward, A. L., Kachanoski, R. G., Bertoldi, A. P., & Elrick, D. E. (1995). Field and undisturbed column measurements for predicting transport in unsaturated layered soil. *Soil Science Society of America Journal*, *59*, 52–59.

Waske, B., & Benediktsson, J. A. (2007). Fusion of support vector machines for classification of multisensor data. *IEEE Transactions on Geoscience and Remote Sensing*, *45*(12), 3858–3866. doi:10.1109/TGRS.2007.898446

Wasserman, P. D. (1993). *Advanced Methods in Neural Computing*. John Wiley & Sons, Inc.

Watts, G. (1997). *Hydrological Modelling in Practice*. John Wiley and Sons Ltd.

Whalley, W. R., & Stafford, J. V. (1992). Real-time sensing of soil water content from mobile machinery: Options for sensor design. *Computers and Electronics in Agriculture*, *7*, 269–284.

White, I., & Zegelin, S. J. (1995). Electric and dielectric methods for monitoring soil-water content. In L. G. Wilson, L. G. Everett, & S. J. Cullen (Eds.), *Handbook of Vadose Zone Characterization and Monitoring* (pp. 343–385). CRC Press.

WHO. (2019). https://www.who.int/emergencies/ten-threats-to-global-health-in-2019

Wimmers, A. J., Velden, C. S., & Cossuth, J. (2019). Using Deep Learning to Estimate Tropical Cyclone Intensity from Satellite Passive Microwave Imagery. *Monthly Weather Review*, *147*, 2261–2282. https://doi.org/10.1175/mwr-d-18-0391.1

Wirth, P. (2021, November 19). *Predicting rain from satellite images*. Medium. Retrieved from https://towardsdatascience.com/predicting-rain-from-satellite-images-c9fec24c3dd1

Wraith, J. M., & Or, D. (1999). Temperature effects on soil bulk dielectric permittiv- ity measured by time domain reflectometry: Experimental evidence and hypothesis development. *Water Resources Research*, *35*, 361–369.

Wu, J., Yao, F., Li, W., & Si, M. (2016). VIIRS-based remote sensing estimation of ground-level PM 2.5 concentrations in Beijing–Tianjin–Hebei: A spatiotemporal statistical model. *Remote Sensing of Environment*, *184*, 316–328. doi:10.1016/j.rse.2016.07.015

Wu, K., Wu, J., Feng, L., Yang, B., Liang, R., Yang, S., & Zhao, R. (2021). An attention-based CNN-LSTM-BiLSTM model for short-term electric load forecasting in integrated energy system. *International Transactions on Electrical Energy Systems*, *31*(1), e12637. doi:10.1002/2050-7038.12637

Wyman, J. (1930). Measurements of the Dielectric Constants of Conducting Media. *Physical Review*, *35*, 623–634.

Xiong, J., & Wu, P. (2008). An analysis of forecasting model of crude oil demand based on cointegration and vector error correction model (VEC). *2008 International Seminar on Business and Information Management*, *1*, 485–488.

Xu, R., Xiong, Q., Yi, H., Wu, C., & Ye, J. (2019). Research on Water Quality Prediction Based on SARIMA-LSTM: A Case Study of Beilun Estuary. *2019 IEEE 21st International Conference on High Performance Computing and Communications; IEEE 17th International Conference on Smart City; IEEE 5th International Conference on Data Science and Systems (HPCC/SmartCity/DSS)*, 2183–2188. 10.1109/HPCC/SmartCity/DSS.2019.00302

Xu, F., Jin, Y. Q., & Moreira, A. (2016). A Preliminary Study on SAR Advanced Information Retrieval and Scene Reconstruction. *IEEE Geoscience and Remote Sensing Letters*, *13*(10), 1443–1447. doi:10.1109/LGRS.2016.2590878

Yadav, A. K., & Chandel, S. (2013). Solar radiation prediction using artificial neural network techniques: A review. *Renewable & Sustainable Energy Reviews*, *33*, 772–781. doi:10.1016/j.rser.2013.08.055

Yang, J., Gong, P., Fu, R., Zhang, M., Chen, J., Liang, S., Xu, B., Shi, J., & Dickinson, R. (2013). The role of satellite remote sensing in climate change studies. *Nature Climate Change*, *3*(10), 875–883. doi:10.1038/nclimate1908

Yang, W., John, V. O., Zhao, X., Lu, H., & Knapp, K. R. (2016). Satellite climate data records: Development, applications, and societal benefits. *Remote Sensing*, *8*(4), 331. doi:10.3390/rs8040331

Yan, J., Liu, J., Yu, Y., & Xu, H. (2021). Water Quality Prediction in the Luan River Based on 1-DRCNN and BiGRU Hybrid Neural Network Model. *Water (Basel)*, *13*(9), 1273. doi:10.3390/w13091273

Yasuda, H., Berndtsson, R., Bahri, A., & Jinno, K. (1994). Plot-scale solute transport in a semiarid agricultural soil. *Soil Science Society of America Journal*, *58*, 1052–1060.

Yerpude, S., & Singhal, T. K. (2017). Impact of internet of things (IoT) data on demand forecasting. *Indian Journal of Science and Technology*, *10*(15), 1–5.

Ysi. (2019). *Water Quality Parameters*. Academic Press.

Yu, T., Yang, S., Bai, Y., Gao, X., & Li, C. (2018). Inlet water quality forecasting of wastewater treatment based on kernel principal component analysis and an extreme learning machine. *Water (Basel)*, *10*(7), 873. doi:10.3390/w10070873

Zhang, H., & Jin, K. (2020). Research on water quality prediction method based on AE-LSTM. *2020 5th International Conference on Automation, Control and Robotics Engineering (CACRE)*, 602–606.

Zhang, R., Liu, Q., & Hang, R. (2020). Tropical Cyclone Intensity Estimation Using Two-Branch Convolutional Neural Network From Infrared and Water Vapor Images. *IEEE Transactions on Geoscience and Remote Sensing*, *58*(1), 586-597. doi:10.1109/TGRS.2019.2938204

Zhang, C.-J., Wang, X.-J., Ma, L.-M., & Lu, X.-Q. (2021). Tropical Cyclone Intensity Classification and Estimation Using Infrared Satellite Images With Deep Learning. *IEEE Journal of Selected Topics in Applied Earth Observations and Remote Sensing*, *14*, 2070–2086. doi:10.1109/JSTARS.2021.3050767

Zhang, C., Yan, J., Li, Y., Sun, F., Yan, J., Zhang, D., ... Bie, R. (2017). Early Air Pollution Forecasting as a Service: An Ensemble Learning Approach. *IEEE International Conference on Web Services*, 636–643. 10.1109/ICWS.2017.76

Zhang, L., Alharbe, N. R., Luo, G., Yao, Z., & Li, Y. (2018). A hybrid forecasting framework based on support vector regression with a modified genetic algorithm and a random forest for traffic flow prediction. *Tsinghua Science and Technology*, *23*(4), 479–492. doi:10.26599/TST.2018.9010045

Zhang, L., Xia, G.-S., Wu, T., Lin, L., & Tai, X. C. (2016). Deep learning for remote sensing image understanding. *Journal of Sensors*, *2016*, 2016. doi:10.1155/2016/7954154

Zhang, L., Zou, B., Zhang, J., & Zhang, Y. (2009). Classification of polarimetric SAR image based on support vector machine using multiple-component scattering model and texture features. *EURASIP Journal on Advances in Signal Processing*, *2010*(1), 960831. doi:10.1155/2010/960831

Zhang, Y., Wang, Y., Gao, M., Ma, Q., Zhao, J., Zhang, R., Wang, Q., & Huang, L. (2019). A Predictive Data Feature Exploration-Based Air Quality Prediction Approach. *IEEE Access: Practical Innovations, Open Solutions*, *7*, 30732–30743. doi:10.1109/ACCESS.2019.2897754

Zheng C., Abd-Elrahman, A., & Whitaker, V. (2021). Remote Sensing and Machine Learning in Crop Phenotyping and Management, with an Emphasis on Applications in Strawberry Farming. *Remote Sensing*, *13*(3), 531, 1-28. doi:10.3390/rs13030531

Zheng, H., Li, H., Lu, X., & Ruan, T. (2018). A Multiple Kernel Learning Approach for Air Quality Prediction. *Advances in Meteorology*, *2018*, 1–15. doi:10.1155/2018/3506394

Zhou, Y., Zhou, Q., Kong, Q., & Cai, W. (2012, April). Wireless temperature & humidity monitor and control system. In *2012 2nd International Conference on Consumer Electronics, Communications and Networks (CECNet)* (pp. 2246-2250). IEEE.

Zhou, Y., Wang, H., Xu, F., & Jin, Y. Q. (2016). Polarimetric SAR Image Classification Using Deep Convolutional Neural Networks. *IEEE Geoscience and Remote Sensing Letters*, *13*(12), 1935–1939. doi:10.1109/LGRS.2016.2618840

Zhuge, X., Guan, J., Yu, F., & Wang, Y. (2015, October). A New Satellite-Based Indicator for Estimation of the Western North Pacific Tropical Cyclone Current Intensity. *IEEE Transactions on Geoscience and Remote Sensing*, *53*(10), 5661–5676. https:// doi.org/ 10.1109/TGRS.2015.2427035

About the Contributors

Rajeev Kumar Gupta is working as an Assistant Professor at Pandit Deendayal Energy University (An Autonomous Institution) Gandhinagar, Gujarat. He has completed his M.Tech and PhD from MANIT, Bhopal and is a recipient of the Best Young Researcher Award by RSRI in 2019. He has published more than 30 referred articles in various book chapters, conferences and international repute peer-reviewed journals of Elsevier, Springer, IEEE etc. He has a total of more than ten years of teaching experience. He is a life member of some of the reputed societies like CSI India, IAENG (Hongkong) etc. He has organized several STTP/FDP and taken several expert lectures at various institutes. He has supervised 20 M.Tech thesis and around 40 B.Tech projects in various domains. He is a TPC member and reviewer of several International Conferences. His area of interest includes Machine Learning, Deep Learning and Cloud computing.

Arti Jain (SMIEEE) is working as Assistant Professor (Sr. Grade), CSE, Jaypee Institute of Information Technology, Noida, India. She has academic experience of 20+ years. She is an editorial board member of the Journal of Technological Advancements, International Journal of Information Technology and Web Engineering, IGI Global; American Journal of Neural Networks and Applications, SciencePG; International Journal of Innovations in Engineering & Technology. She has been an International advisory committee member of the International Conference on Recent Trends in Multi-Disciplinary Research during 2020-2021, Maldives; International Conference on Applied Engineering and Natural Sciences, Turkey; International Conference on Applied Sciences, Engineering, Technology and Management, UAE. She is a reviewer of reputed International Journals- World Scientific, Taylor & Francis, Springer, IGI Global, Wiley, TISA, Inderscience, Tech Science Press, GRD, Premier, and guest-edited Special Issues. She is a PC member of several International Conferences and has conducted Special Sessions. She has Chaired Sessions at International Conferences- 2022 ICICC, 2021- AIMV, FTSE, and KDIR. She has been the invited speaker for various workshops, seminars, and conferences such as 2021 Workshop on Computer Methods in Medicine & Health Care, Transdisciplinary Information Sciences Conferences, China; 2nd International Webinar on Big Data Analytics and Data Science, Coalesce Research Group, USA; National Workshop on Soft Computing and Language Processing. She has also been the key resource person for prestigious FDPs, namely- Deep Learning for Natural Language Processing, Applications of Machine Learning, and International Knowledge Development Program by Special Minds. She has published the book, "Handbook of Research on Lifestyle Sustainability and Management Solutions Using AI, Big Data Analytics, and Visualization", IGI Global, 2022. She has more than 25 research papers in International Journals, Book Chapters, and International Conferences. Her research interests include Natural Language Processing, Machine Learning, Data Science, Deep Learning, Social Media Analysis, Soft Computing, Big Data, and Data Mining.

John Wang is a professor in the Department of Information & Operations Management at Montclair State University, USA. Having received a scholarship award, he came to the USA and completed his PhD in operations research from Temple University. Due to his extraordinary contributions beyond a tenured full professor, Dr. Wang has been honored with a special range adjustment in 2006. He has published over 100 refereed papers and seven books. He has also developed several computer software programs based on his research findings. He is the Editor-in-Chief of International Journal of Applied Management Science, International Journal of Operations Research and Information Systems, and International Journal of Information Systems and Supply Chain Management. He is the Editor of Data Warehousing and Mining: Concepts, Methodologies, Tools, and Applications (six-volume) and the Editor of the Encyclopedia of Data Warehousing and Mining, 1st (two-volume) and 2nd (four-volume). His long-term research goal is on the synergy of operations research, data mining and cybernetics.

Ved Prakash Singh is working as Scientist at Bhopal center of India Meteorological Department, Ministry of Earth Sciences, Government of India. He is heading the Radar, Instrumentation, Seismology and Hydro-meteorological operations of Madhya Pradesh State. He pursued B.Tech. in computer science from Indian Institute of Technology (BHU), Varanasi, India in 2012 and is about to complete Ph.D. in Advanced Machine Learning from Indian Institute of Technology, Patna, India. Earlier, he served as senior software engineer at Samsung R&D Labs, Bangalore and Seoul, South Korea during 2012-2015 and also, as senior member of the engineering team at Hike Ltd. New Delhi in 2015. Initially, He cleared the UPSC (Indian Meteorological Services) exam in 2015 to join IMD. He has been working in the domain of Artificial Intelligence, Advance Machine Learning, Radical Learning, Climate Sc., GIS, Radar & Satellite Meteorology and Agro-meteorology etc. He has vast professional experience of more than 10 years in R&D and service operations. He is life member of several national and international societies such as AMS, IMS and ACM etc. He got awarded several reputed titles at several national and international platforms. He received the title of 'Best employee' by Samsung R&D labs, India in 2014, 'Hikeathon' for best innovation by Hike, Delhi in 2015, Young Scientist Award' in iRad-2019 at IITM, Pune, further by MPCST, Bhopal in 2019 and Best paper award in IISF-2018, Lucknow. Recently, he received 'Best Group-A Officer' award across India on the occasion of 142 Foundation Day of IMD, New Delhi. He has published many emerging art research works, book chapters, social articles etc. in reputed conferences & journals of Govt. of Mausam, Springer, IEEE etc. He is member of AI/ ML, GIS and Radar application groups of IMD and leading many projects in interdisciplinary domains such as Agro-GS, Lightning prediction etc. He is also IMD representative of ML/ DL working group at MoES, Govt. of India. He has been supervising graduate & post-graduate students including the internship, thesis and professional training of students from meteorology, engineering and health domain across various tier-1 organizations including NITs, IITs and AIIMS. He has also filed several patents in the weather and IT security domain and developed many cutting edge designs and architectures in multimedia and ML streams such as Context awareness, Auto-Wind cut, Time-lapse, swift event recording, sticker recommendation, IM chat search engine, 3D wind recorder etc. He participated in training programs of several national and international organizations such as training on Radar and Satellite in IMD, IIT-Kharagpur, IIT-Chennai and on severe weather and disaster management in CMA, China and JICA, Hokkado, Japan.

He was part of the young leaders team of India in 2017 for the Japan-India Cooperation program. He is also an active contributor to various world open-source communities.

Santosh Kumar Bharti received B. E degree in CSE from VisvesvarayaTechnological University Belgaum, Karnataka, India, and M. Tech.in CSE from Graphic Era University, Dehradun, India. He has completed his Ph.D.(CSE) from the National Institute of Technology Rourkela India in April 2019. He has worked as Assistant Profesor in SCET, Palwal, Haryana. Currently, he is working as an Assistant Professor in the Department of Computer Science and Engineering at the School of Technology of Pandit Deendayal Petroleum University, Gandhinagar Gujarat. He has published more than 25 research papers in repute journals and conferences.

* * *

Prasit Agnihotri has total experience of about 28 years. His areas of interest are water resources engineering, geospatial technologies including GIS, GPS, and Remote Sensing. His Research area is Flood Mitigation and Management under Geospatial Platform.

Vaibhavi Bandi is a final year student currently pursuing B.Tech. with majors in Computer Science & Engineering with a specialization in Data Science from Medi-caps University, Indore.

Divyang Dave is currently pursuing in M.Tech in the field of Data Science from Pandit Deendayal Energy University (An Autonomous Institute), Gandhinagar, Gujarat. He has completed his Bachelor's degree from Government Engineering College, Modasa. His area of interest includes Machine Learning, Deep Learning and NLP.

Tanvi Garg is a final year student currently pursuing B.Tech. with majors in Computer Science & Engineering with a specialization in Data Science from Medi-caps University, Indore.

Sunil Gautam is currently working as an Assistant Professor in the Department of Computer Science and Engineering, Nirma University, Ahmedabad, Gujarat, India. He has completed a Ph.D. degree from Indian Institute of Technology (ISM) Dhanbad, Jharkhand, India. He received his Bachelor in Mathematics from CSJMU Kanpur, and Master in Computer Application from Uttar Pradesh Technical University, Lucknow. His research interest includes Intrusion Detection System & Network Security.

Smita Girish is a skillful Assistant Professor with 15 years of experience in teaching and worked as Freelance Trainer too. She's highly efficient at her job. She has strong technical skills. Because of her skills, she is dynamic and gives productive results in assigned tasks. She caters to major contribution in teaching. she works blend with her all colleagues. In particular, her focus is on topics like Data Mining, Data Analytics, Data Science and Artificial Intelligence. Here, she also contributed to the development of University content curation and examination boards. She wants to engage more with influential decision-makers. She's open to challenges, conversations, and an exchange of ideas from the peers. Smita Girish has a lot of significant professional experiences as Assistant Professor. She created several ideas to help her colleagues. Smita Girish had successfully nourished her insatiable passion in upgrading with new technologies and exceled herself by conducting research by assisting students' community.

Shilpa Gite received a Ph.D. using Deep Learning for Assistive driving in Semi-autonomous vehicles from Symbiosis International (Deemed University), Pune, India, in 2019. Currently, she is working as an Associate Professor in the Computer Science Department of Symbiosis Institute of Technology, Pune. She is also working as an Associate Faculty at Symbiosis Centre of Applied AI (SCAAI). She has around 13 years of teaching experience. Her research areas include Deep learning, Machine learning Medical Imaging, and Computer Vision. She is currently guiding Ph.D. students in biomedical imaging, self-driving cars, and natural language processing areas. She has published more than 50 research papers in International journals and 20 Scopus indexed International Conferences. She is also the recipient of the Best Paper award at the 11th IEMERA Conference held virtually at Imperial College, London in Oct 2020.

Ramesh Chandra Goswami is currently working as an Assistant Professor in the Department of Computer Science, IIICT, Indus University, Ahmedabad, Gujarat, India. His research interest includes Internet of Things, Machine Learning & Artificial Intelligence.

Shilpa Hudnurkar is BE Instrumentation and has completed an M. Tech in Electronics & Telecommunication. She is currently working as an Assistant Professor in the Department of Electronics & Telecommunication Engineering at Symbiosis Institute of Technology affiliated to Symbiosis International (Deemed University) and is a research scholar at Symbiosis International (Deemed University). Her research interests include Artificial Intelligence, Machine Learning, Deep Learning, Signal Processing. She is working on predicting Summer Monsoon Rainfall over a small region. Her teaching experience is over 7 years.

Nidhi Jani is studying at Pandit Deendayal Energy University, Gandhinagar, Gujarat. Currently, she is pursuing her masters in Cyber Security. She has completed her BTECH from Vishwakarma Government Engineering College with distinction. She is also working as a teaching assistant at Pandit Deendayal Energy University. Her area of interest includes Cyber Security, Machine learning.

Hiren Joshi is working as a professor of computer science at Dept. of Computer Science, Gujarat University. He has more than 2 decade (20 + years) experience in academic. He was also worked as Director (I/C), School of Computer Science and Controller of Examination (I/C) in Dr. Babasaheb Ambedkar Open University (BAOU), Ahmedabad between 2013-2017. He has published and presented 30+ research papers in reputed national and international journals. He has written 3 books. He has written 3 chapters in editorial book of national and international publishers. He has served as Ph.D. research supervisor in many universities. He has provided his services as resource person in HRDC (earlier known as ASC) Gujarat University and S.P. University, Vallabh Vidya Nagar. He is serving as member of Board of Studies in various universities in Gujarat and other states of India. He has served as NAAC peer team co-ordinator and NAAC peer team member. He has worked as resource person for various programs recorded and broadcasted on DD Girnar and SANDHAN (Live Television Lecture Series from BISAG – Gandhinagar) and in C2C (College to Career program organized by Govt. of Gujarat and Microsoft, telecast by BISAG). He has developed an e-Content on Web Application Development paper for e-PG Pathshala which is an Ministry of Education (MoE) project under NME-ICT initiative. He has served as reviewer for e-Content on Open Source Software for e-PG pathshala also. He has also worked as a co-ordinator for various academic and academic administrative activities like BAOU CCC book content creation, AISHE nodal officer, MHRD refresher course coordinator in ICT at Gujarat University, coordinator for Short Term course on ICT etc. He is professional member ACM Gandhinagar chapter, CSI Life member and other professional bodies. He has served as an interviewer as subject expert for recruitment of teaching and non-teaching post at various universities.

Shirish Khedikar is currently working as Scientist-C, having specialization in Agrometeorology and climate services & operations. He is actively involved in allied research and training activities of IMD.

Jayashree M. Kudari is an accomplished Associate Professor with over 23 years of experience in teaching. She's so efficient at her job that she considers herself a "Good orator." Jayashree has strong technical skills Because of her skills, she can produce productive and long-lasting results in her tasks. She caters to major contribution in teaching and administration, She works blends very well with the all the colleagues. In particular, she focused on topics like Machine Learning, deep-learning. After she graduated, Jayashree worked hard to attain her role to reach this level. Here, she also contributed to the development of University content curation, examination boards and board of study, Internal Quality Assurance Cell, Student Welfare and many more. Jayashree had successfully nourished her insatiable passion in upgrading with new technologies and exceled herself by conducting research by assisting student's community. She wants to engage more with influential decision-makers and thought leaders in the department. She's open to challenges, conversations, and an exchange of ideas from the peers. Jayashree had a lot of significant professional experiences as Associate Professor. She created several ideas to help her organization to evaluate operational requirements more accurately. She also monitors and assesses the performance of student groups.

Lucky Kulshrestha is a final year student currently pursuing B.Tech. with majors in Computer Science & Engineering with a specialization in Data Science from Medi-caps University, Indore.

Bhavana Gowda M. is an accomplished Assistant Professor with over 10.6 years of experience in teaching and currently working as department coordinator. She is highly efficient at her job. She is a "Good leader" for her team. Bhavana has strong technical and management skills. She is dynamic and provides productive results in her tasks. She caters to major contribution in teaching and departmental administration, she blends with her colleagues to assist in all aspects. She adopts new technologies very quickly, Her research area is Data Mining, Machine Learning, Deep Learning, Internet of things. Here, she is also contributed to the development of department and university. She took major role in content curation, examination boards and board of study and many more. Bhavana had successfully nourished her passion in upgrading with new technologies and exceled herself by conducting research by assisting student's community. She wants to engage more with influential decision-makers and thought leaders in the department. She's open to challenges, conversations, and an exchange of ideas from the peers. Bhavana had a lot of significant professional experiences as Assistant Professor. She created several ideas to help her organization to evaluate operational requirements more accurately. She also monitors and assesses the performance of student groups.

Jimson Mathew, Professor in CSE Dept. at Indian Institute of Technology, Patna, pursued PhD from University of Bristol, UK. His research areas are Fault Tolerant Computing, VLSI Design and Methodologies, Reliability Aware Designs, Hardware Security, etc.

Manobhav Mehta is currently pursuing a bachelor's degree in the field of Electronics and Telecommunications Engineering from Symbiosis Institute of Technology, Pune, India. He has co-authored and published a bibliometric survey paper on the use of LSTM networks for Multivariate Time Series Forecasting. His research interests include computer networking, deep learning, Matlab, and Python programming.

Nimra Memon currently works as a Researcher at Computer Engineering Dept., Pandit Deendayal Petroleum University, Gandhinagar Nimra does research in Geoinformatics (GIS), Remote Sensing and Cartography, Microwave Remote sensing, Python, Machine learning, Deep learning.

Vedansh Mishra is currently pursuing a bachelor's degree in the field of Electronics and Telecommunications Engineering from Symbiosis Institute of Technology, Pune, India. He has co-authored and published a bibliometric survey paper on the use of LSTM networks for Multivariate Time Series Forecasting. His research interests include computer networking, machine learning, and object-oriented programming.

Prajnyajit Mohanty is currently working as a Ph.D. research scholar at NIT Rourkela. He has completed his M.Tech in Control and Instrumentation from Veer Surendra Sai University of Technology, Burla. His working area is energy harvesting system design for IoT devices. He also put interest in sensor design, IoT, and low-power embedded systems.

M. N. Nachappa is an accomplished Professor with over 29 years of experience in teaching. Served department Head since 15 Years. Serving Director since 4 years in School of CS & IT from the current institution. He's very efficient at his job and very "Good orator". Dr M N Nachappa is a good leader having administration and management skills. He has strong technical skills because of hir skills, he can produce productive and longlasting results in his tasks. He caters to major contribution in Management and Administration and teaching, He makes his all colleagues very comfortable in working culture. He provides opportunity to all his colleagues in research, adopting new skills, trainings etc. His research areas are Machine Learning, Deep-Learning. He exceled himself in the field of research with 50+ paper published in all reputed journals. Dr M N Nachappa contributed his knowledge in writing Various Books and Book Chapter. He Is guiding Research Scholars with challenging Projects. He had successfully nourished his insatiable passion in upgrading with new technologies and exceled himself by conducting research by assisting student's community. Dr M N Nachappa wants to engage more with influential decision-makers and thought leaders in the department. He's open to challenges, conversations, and an exchange of ideas from the peers. Dr M N Nachappa had a lot of significant professional experiences as Professor. He created several ideas to help her organization to evaluate operational requirements more accurately. He also monitors and assesses the performance of student groups.

Rashmiranjan Nayak received the BTech degree in Electronics and Telecommunication Engineering from the Biju Patnaik University of Technology, Rourkela, Odisha, India, in 2010, and the MTech degree in Electronics and Communication Engineering (specialization: communication and networks) from the NIT Rourkela, Odisha, India, in 2016. Currently, he is pursuing his Ph.D. in the department of Electronics & Communication Engineering, NIT Rourkela, India, in the area of video anomaly detection. His current research interests include Video Anomaly Detection, Computer Vision, Artificial Intelligence, and the Internet of Things. He is a graduate student member of IEEE.

Dhruvesh Patel received his Ph D degree from Gujarat University, L.D. College of Engineering, India in field of Civil Engineering (Water Resources Engineering). After that, research Fellowship has been achieved by him from University of Bristol, U.K. Presently he is working an Assistant professor of civil Engineering at School of Technology (SoT), Pandit Deendayal Petroleum University (PDPU), Gujarat, India. He has published more than 21 papers in leading international Journals and conferences in his research fields. He is author of two books viz., Application of RS and GIS in Water Resources Management published by LAMBERT academic publishing, Germany and Hydrology and Water Resources Engineeing Atul publication, India; and author of book chapter titled Application of Geo-Spatial Technique for Flood Inundation Mapping of Lower Lying Area which is published in Springer International publishing, Switzerland. Dr Dhruvesh Patel has received International Travel Support (ITS) under the young scientist scheme from the Science and Engineering Research Board (SERB), Department of Science and Technology (DST), India. He is reviewer in many journals, such as Advance in Space Research (Elsevier), Water Resources Management (Springer), Arabian Journal of Geosciences (Springer), Environmental Earth Sciences (Springer), Applied Water Science (Springer), Earth Science Information (Springer), Flood Risk Management (Wiley Online Library) and Geocarto International (Taylor & Francis online). He is life meneber of the ISTE, IWRS, ISH, ISG and Indian Society of Remote Sensing (ISRS).

Samir Patel is currently an Associate Professor and Head CSE Department at School of Technology, Pandit Deendayal Energy University. He obtained his Ph.D. Degree from Nirma University in Computer Engineering in the year, 2012. He has published more than 32 papers of National and International Repute. He has authored two books. Before joining PDEU, he had worked as Principal, GMFE, Sr. Associate Professor at Nirma University, Assistant Prof., Senior Lecturer at AESICS and before that as lecturer and programmer cum lecturer at CPICA and PDPICA. He has total 23 years of teaching and administrative experience. He is reviewer of various Journals of repute and is a professional member of bodies like Computer Society of India, ISTE, IEEE, ACM. His area of interest include Big Data Analytics, Parallel Computing, Multimedia Data Processing, Data Mining, etc.

Azazkhan Ibrahimkhan Pathan currently works at the Department Of Civil Engineering, Sardar Vallabhbhai National Institute of Technology. Azazkhan Ibrahimkhan does research in water resources engineering.

Umesh C. Pati is a full Professor at the Department of Electronics and Communication Engineering, National Institute of Technology (NIT), Rourkela. He has obtained his B.Tech. degree in Electrical Engineering from National Institute of Technology (NIT), Rourkela, Odisha. He received both M.Tech. and Ph.D. degrees in Electrical Engineering with specialization in Instrumentation and Image Processing respectively from Indian Institute of Technology (IIT), Kharagpur. His current areas of interest are Image/Video Processing, Computer Vision, Artificial Intelligence, Internet of Things (IoT), Industrial Automation, and Instrumentation Systems. He has authored/edited two books and published over 100 articles in the peer-reviewed international journals as well as conference proceedings. He has served as a reviewer in a wide range of reputed international journals and conferences. He has also guest-edited special issues of Cognitive Neurodynamics and the International Journal of Signal and Imaging System Engineering. Dr. Pati has filed 2 Indian patents. He has visited countries like USA, Italy, Austria, Singapore, Mauritius, etc. in connection with research collaboration and paper presentation. He was also an academic visitor to the Department of Electrical and Computer Engineering, San Diego State University, USA, and Institute for Automation, University of Leoben, Austria. He is a Senior member of IEEE, Fellow of The Institution of Engineers (India), Fellow of The Institution of Electronics and Telecommunication Engineers (IETE), and Life member of various professional bodies like MIR Labs (USA), The Indian Society for Technical Education, Computer Society of India and Instrument Society of India. His biography has been included in the 32nd edition of MARQUIS Who's Who in the World 2015.

Neela Rayavarapu received her BE degree in Electrical Engineering from Bangalore University, Bangalore, India in 1984, MS Degree in Electrical and Computer Engineering from Rutgers, The State University of New Jersey, USA, in 1987, and a Ph.D. degree in Electronics and Communication Engineering in 2012 from Panjab University, Chandigarh. She has been involved with teaching and research in Electrical, Electronics, and Communication Engineering since 1987. Her areas of interest are digital signal processing and its applications and control systems.

Adlin Jebakumari S. is skillful Assistant Professor with over 14 years of experience in teaching. She's highly efficient at her job. she blends easily with everybody. Adlin Jebakumari S has strong technical, skills Because of her skills, she is dynamic and gives productive results in her tasks. She caters to major contribution in teaching. She works blend with her all colleagues. In particular, she focused on topics like Data Mining, ML, DL. Here, she also contributed to the growth of university, content curation, examination boards. Adlin Jebakumari had successfully nourished her insatiable passion in upgrading with new technologies and exceled herself by conducting research by assisting student's community. She wants to engage more with influential decision-makers. She's open to challenges, conversations, and an exchange of ideas from the peers. Adlin Jebakumari had a lot of significant professional experiences as Assistant Professor. She created several ideas to help her colleagues.

Sushma S. is skillful Assistant Professor with over 9 years of experience in teaching. She's skillful at her job that she considers herself as a easy adoptable new technologies." Sushma B S has strong technical skills. Because of her quick upgrading skills, she is focused and provide productive results in her tasks. She caters to major contribution in teaching, She works on research topics like Artificial Intelligence, ML, DL and IoT. Here, she also contributed to the development of University content curation, Sushma B S had successfully nourished her technical skills by constant upgradation with new technologies and excelled herself by conducting research by assisting student's community. She wants to engage more with influential decision-makers. She's open to challenges, conversations, and an exchange of ideas from the peers. Sushma B S had a lot of significant professional experiences as Assistant Professor. She created several ideas to help her colleagues.

Vidur Sood is currently pursuing a bachelor's degree in the field of Electronics and Telecommunications Engineering from Symbiosis Institute of Technology, Pune, India. He has co-authored and published a bibliometric survey paper on the use of LSTM networks for Multivariate Time Series Forecasting. His research interests include computer networking, robotics, machine learning, and analytics.

Mogarala Tejoyadav received the BTech degree in Electronics and Instrumentation Engineering from the Sree Vidyanikethan Engineering College, Tirupati, Andhrapradesh, India, in 2018. Currently, he is pursuing his MTech in Electronics and Communication Engineering (specialization: Electronics and Instrumentation) from the NIT Rourkela, Odisha, India. His current research interest is time series forecasting of water quality.

Akash Upadhyay is currently pursuing a bachelor's degree in the field of Electronics and Telecommunications Engineering from Symbiosis Institute of Technology, Pune, India. He has co-authored and published a bibliometric survey paper on the use of LSTM networks for Multivariate Time Series Forecasting. His research interests include computer networking, machine learning, and deep learning.

Mink Virparia is an M.Tech research scholar at Pandit Deendayal Energy University, Gandhinagar. His area of interest is Machine Learning and Deep Learning.

Index